Techno-Orientalism 2.0

Asian American Studies Today

This series publishes scholarship on cutting-edge themes and issues, including broadly based histories of both long-standing and more recent immigrant populations; focused investigations of ethnic enclaves and understudied subgroups; and examinations of relationships among various cultural, regional, and socioeconomic communities. Of particular interest are subject areas in need of further critical inquiry, including transnationalism, globalization, homeland polity, and other pertinent topics.

Series Editor: Huping Ling, Truman State University

For a complete list of titles in the series, please see the last page of the book.

Techno-Orientalism 2.0

New Intersections and Interventions

EDITED BY DAVID S. ROH, BETSY HUANG,
GRETA AIYU NIU, AND CHRISTOPHER T. FAN

Rutgers University Press
New Brunswick, Camden, and Newark, New Jersey
London and Oxford

Rutgers University Press is a department of Rutgers, The State University of New Jersey, one of the leading public research universities in the nation. By publishing worldwide, it furthers the University's mission of dedication to excellence in teaching, scholarship, research, and clinical care.

978-1-9788-3922-9 (cloth)
978-1-9788-3921-2 (paper)
978-1-9788-3923-6 (epub)
978-1-9788-3924-3 (webPDF)

Cataloging-in-publication data is available from the Library of Congress.
LCCN 2024049487

A British Cataloging-in-Publication record for this book is available from the British Library.

This collection copyright © 2025 by Rutgers, The State University of New Jersey
Individual chapters copyright © 2025 in the names of their authors
All rights reserved

No part of this book may be reproduced or utilized in any form or by any means, electronic or mechanical, or by any information storage and retrieval system, without written permission from the publisher. Please contact Rutgers University Press, 106 Somerset Street, New Brunswick, NJ 08901. The only exception to this prohibition is "fair use" as defined by U.S. copyright law.

References to internet websites (URLs) were accurate at the time of writing. Neither the author nor Rutgers University Press is responsible for URLs that may have expired or changed since the manuscript was prepared.

∞ The paper used in this publication meets the requirements of the American National Standard for Information Sciences—Permanence of Paper for Printed Library Materials, ANSI Z39.48-1992.

rutgersuniversitypress.org

This book is for you, dear reader.

Contents

The Present Futurity: Techno-Orientalist Infrastructures   1
DAVID S. ROH, BETSY HUANG, GRETA AIYU NIU, AND CHRISTOPHER T. FAN

Part I   Labor Reconfigurations

1. Working Futures after Asians: Automation, AI, and the Global Labor Economy   23
LELAND TABARES

2. *Everything Everywhere All at Once*: Techno-Orientalism in an Age of Cybernetic Capitalism   39
WON JEON

3. Chinese Commodities: Adoption in *After Yang*   51
KIMBERLY D. McKEE

Part II   Racialization as Technology

4. Plastinated Vitruvian Man, the Datafication of Race, and Transracial Transfer in *Westworld* and *Altered Carbon*   67
CHARLES M. TUNG

5. Outsiders Within: The Indigenous/Minority Question and Techno-Orientalist Gaze in India   83
M. IMRAN PARRAY

6. On Forms of the Black Box: Race and Technology in STS and Global Critical Race Studies   100
CLARE S. KIM AND ANNA ROMINA GUEVARRA

Part III Sinofuturism

7   Infrastructure and/as Mediation: *China 2098*'s
Tempro-Affective Politics   121
IAN LIUJIA TIAN

8   Techno-Orientalist Deflections: How Documentaries
Frame China's AI Threat   137
GERALD SIM

9   Techno-Futurehistory and the Sojourners of Global
China: A Threefold Reading of *The Wandering Earth*   153
SHANA YE

Part IV Machinic Subjects

10 Sacrificial Clones: The Technologized Korean Woman
in *Shiri* and *Cloud Atlas*   173
JANE CHI HYUN PARK

11 Assembling Mitski: The Aesthetics and Circuits
of Techno-Ornamentalism   189
RACHEL TAY AND JAEYEON YOO

Part V Extensions

12 Asian Solarpunk: Between Utopia, Collective Futures,
and Remedies for Climate Panic   205
AGNIESZKA KIEJZIEWICZ AND JUSTIN MICHAEL BATTIN

13 Animated Bodies: Project Itoh and the Afterlives
of Techno-Orientalism   219
BARYON TENSOR POSADAS

14 Settler Orientalism, Asian American
Techno-Environmentalism: The Network
Novel under Japanese and U.S. Empires   235
ADHY KIM

15 The Alchemized Dis/abled Body as Recuperative
Site in *Fullmetal Alchemist*   249
JUNG SOO LEE

Part VI Optimistic Futures

16 Recovering Asian American Futures in the Marvel
Cinematic Universe   267
LORI KIDO LOPEZ

17	Looking for Asianfuturism: Asian American Science Fiction and Games of Color EDMOND Y. CHANG	281
18	The Queer Techno-Orientalist Aesthetics of Disney's *Big Hero 6* THOMAS XAVIER SARMIENTO	298
	Markets of Techno-Orientalist Critique: A Concluding Discussion DAVID S. ROH, BETSY HUANG, GRETA AIYU NIU, AND CHRISTOPHER T. FAN	315
	Acknowledgments	325
	Bibliography	327
	Notes on Contributors	361
	Index	367

Techno-Orientalism 2.0

The Present Futurity

Techno-Orientalist Infrastructures

DAVID S. ROH, BETSY HUANG, GRETA AIYU NIU, AND CHRISTOPHER T. FAN

In Kim Stanley Robinson's *Ministry for the Future* (2020), the protagonist, an unassuming Irish bureaucrat, manages to set the planet's climate recovery in motion. Her counterpart in the People's Republic of China, equally skilled and politic, works behind the scenes in wrangling the Chinese governmental agencies to work in tandem with the West to reverse the course of climate disaster.[1] We are given hints at the machinations behind the Great Wall, but much of it remains opaque from beginning to end. In that vein, Robinson's novel follows the well-trod convention of an unknowable Far East. But what is different this time around is that China does not pose a competitive menace; instead, the perilous reality is enough for the West to take the Chinese at their word and collaborate on technological and economic solutions.[2] *Ministry* is therefore an optimistic vision of global cooperation, because, really, there is no other choice.[3]

Several years after *Ministry*'s publication and one global pandemic later, the Biden administration introduced the Inflation Reduction Act (2022), a once-in-a-generation, massive legislative effort to combat climate change. Despite its politically expedient title, the act writes speculation into reality, triggering transformative structural changes over the next decade that will reduce carbon emissions in the United States by 40 percent and achieve nearly two-thirds of the country's goal to circumvent total global temperature increases.[4] The act also has a second purpose. The enormous subsidies it offers, aimed at spurring innovation and private investment in renewable energy, are a direct response to the fact that the United States's technology supply chain is too reliant on Chinese manufacturing—a realization brought into stark relief by COVID-19. Thus, while the act purports to tackle climate change, it also aims to dislodge China's dominance in the green energy and technology sectors.[5]

The throughline in both narratives is not new. The current rhetoric on China's renewable energy dominance in the United States and Europe echoes long-standing concerns over the Middle East's disproportionate share of fossil fuel production. Even the telos remains unchanged. The United States' wariness of the Japanese automobile industry's strength in the 1970s and '80s has merely shifted to a new target in the transition to electric vehicles (EVs): China's competitive advantages—cutting-edge, agile manufacturing, robust supply chains, lower labor costs, ample raw materials—in building its own domestic EV fleet that might undercut the United States' global market share. This dynamic, as we have argued before, will continue to manifest culturally, socially, and economically. Yet what is different about this moment, notwithstanding rhetoric about the so-called New Cold War, are the stakes. Cataclysmic upheavals on multiple fronts compress timelines and collapse the tangible present with the speculative future. Climate change, resource wars, and economic nativism portend massive structural reconfigurations already underway. And in the midst of these upheavals, techno-Orientalist frameworks and analytics persist, more relevant than ever before. Why?

Techno-Orientalism 1.0: Influence and Impact

The field of techno-Orientalism studies has evolved considerably from its early focus on speculative fiction to an expansive range of cultural productions over the past decade. In the first volume, *Techno-Orientalism: Imagining Asia in Speculative Fiction, History, and Media* (2015), we described techno-Orientalism as "the phenomenon of imagining Asia and Asians in hypo- or hyper-technological terms in cultural productions and political discourse."[6] Since then, techno-Orientalist discourse has evolved to shape domestic and international discussions of the COVID-19 pandemic, popular entertainment, foreign policy, and more. At the same time, there has been an undeniable uptick in the quantity of works of speculative fiction by Asians and Asian Americans. As techno-Orientalism reaches critical mass in the popular imagination, more Asian and Asian American artists have responded directly in film, literature, and art. For example, Kogonada's *After Yang* (2021) explicitly invokes techno-Orientalist aesthetics while subverting its tropes, as does Daniel Kwan and Daniel Scheinert's *Everything Everywhere All at Once* (2022); Franny Choi's *Soft Science* (2019) targets the techno-Orientalist elements of Alex Garland's *Ex Machina* (2014); and Margaret Rhee's *Love, Robot* (2017) analogizes robot-human relationships with nonnormative pairings. In the visual arts, media artist Astria Suparak's installation *Asian Futures, without Asians* (2020) takes up the first volume's critique as an opportunity to pillory techno-Orientalism's visual language. In the fashion world, clothing designer Xander Zhou has fabricated an entire line called "Techno-Orientalism," part of a "techwear" trend that adopts a futuristic aesthetic.[7] We anticipate that this discursive dynamic will only continue to develop as the number of artists and practitioners of both the aesthetic and the critical analytic increase.

From the outset, we understood that impactful critiques of techno-Orientalism have been and must remain an interdisciplinary effort. The sheer discursivity of techno-Orientalism warrants multiple sets of critical lenses to understand the depth and breadth of its influence. It is therefore unsurprising that the first volume not only established a subfield in Asian American studies, but quickly found its way into other fields, from art criticism to religious studies.[8] Indeed, the significance of naming techno-Orientalism as a field of study is not in its legitimization, but in the revelation of truths and practices long observed but not presented synchronously, voluminously, incontrovertibly. Most compellingly, its consequentiality is evident in both academic and popular spheres, in student demand and in the increasing number of university courses on the subject. Outside academia, discussions of techno-Orientalism have informed popular journalism, from the *Hollywood Reporter*, the *AV Club*, and the *New Yorker*, to podcasts like *Imaginary Futures*.[9]

These developments demand critical reckoning. While the first volume shaped the contours of the scholarly discussion, this second volume extends and elaborates the analytic, propelled by several questions: Why has techno-Orientalism found purchase in so many fields? How has it managed to dissolve the membrane between scholarly and popular criticism? What utility or logic resonates with students with such persistence and fluidity? Finally, how might the field evolve over the next decade?

On the first question of its popularity, we observe that techno-Orientalism has transcended its initial focus on speculative fiction and Asiatic signifiers and developed into a *mode of revelation*, highlighting gaps in multiple fields. Edward Said's *Orientalism* (1978) remains central, but the increasing influence of techno-Orientalism acknowledges a global shift that has long been in the making, as non-Western nations and cultures rise in prominence economically, culturally, and politically. And while renderings in the Orientalist tradition continue to proliferate, contemporary discourse has tilted to reflect that reconfiguration. Nowadays, Eastern sites and bodies tend not to be depicted in simplistically retrogressive terms. There are now modern and hypermodern representations—at least superficially—threaded through the discussion beyond the speculative fictional realm bordering on parity. In short, techno-Orientalism's genre logic and critical impact have seeped into the mainstream. The techno-Orientalism analytic appears to both unveil and fill gaps, which may explain its rapid spread beyond Asian/Asian American studies and academia at large.

At the same time, we detect a teleological shift: racialization's displacement from subject to structure. While early techno-Orientalism focused on the racialization of subjects (or lack thereof) in speculative works, we find that the analytic's utility now lies in its revelation of how that logic suffuses broader institutional environments. In other words, while that logic previously was most visible in generic practices and forms, it has since permeated structures of the state, as evidenced in policy, law, and economic planning. This is perhaps most visible in the ways that states managed bodies, rhetoric, and policies around COVID-19, which has been a high-water mark for globalization and state capitalism as well as a catalyst for isolationism and what many call "deglobalization." This sudden shift in geopolitics, which had largely held steady as a putative "rules-based order" since World War II, has forced us to grasp at anything that appears solid: objects rather than subjects, the concreteness of structures rather than the squishiness of bodily form.

Structural discursivity has tangible effects on the subject, but it can be difficult to pinpoint when it has been diffused into an environment. A progressive veneer masks racialization's displacement in an evolving landscape where mainstream discussions of inclusivity and social justice are commonplace and blatant acts of racism are discouraged in polite company.[10] Asian American works are now hurdling the gatekeepers and enjoying mainstream success while Asian film productions are openly lauded by tastemakers. The siren-song of the Chinese market and the viral *frisson* of K-pop have led to a kind of openness to forms of multiculturalism that are driven more by U.S.-Asia trade relations than the legacy of the Civil Rights era. Yet an unsettling ambivalence continues to haunt the Asian American subject. Indeed, techno-Orientalism anticipated this maneuver—a simultaneous projection of both the premodern and hypermodern—in speculative fiction. While initially focused on a genre and technological aesthetic, techno-Orientalism now makes visible those obfuscations and the underlying double logic in culture more generally. It is this capacity, we argue, to articulate *both* structural racism and latent cultural racism that accounts for its resonance.

Shifts and Drifts

"There are decades where nothing happens and weeks where decades happen," said Vladimir Lenin (apocryphally). This certainly rings true in the context of this project. Since the publication of *Techno-Orientalism* in 2015, the intervening years have felt like centuries. We are reminded of this by the now-regular occurrences of "once-in-a-century" events: from floods and fires to pandemics and wars. Thus, the contributions to the first volume demonstrated the transhistorical durability of a relation between modernity (so-called) and

Orientalism made socially and aesthetically perceptible via forms of race, nationality, gender, and sexuality. While nothing has overturned this relation, its internal dynamic has been reconfigured by a number of events, not least the deformation of temporality itself wrought by crisis after deepening crisis.

To get a sense of the distance traversed, and how the relation between technology and Orientalism has shifted, it is worth considering the United States as it was in 2015. Barack Obama was still president of the United States, and the U.S. military was in the process of drawing down from Afghanistan. The relationship between U.S. democratic norms and neoliberal business as usual was stable enough that Hillary Clinton's potentially historic presidential coronation the following year seemed like a fait accompli, while Donald Trump's candidacy in a crowded Republican field had the aura of a mere marketing stunt. The Federal Reserve, in a move signaling confidence that the United States had finally recovered from the 2008 financial crisis, raised interest rates for the first time in almost a decade. And in October of the same year, a dozen Pacific Rim economies reached an agreement on the U.S.-led Trans-Pacific Partnership (TPP), an expansive free trade pact designed to contain China economically. In 2015 the United States certainly believed itself to be resuscitating an exceptionalism, buoyed by an equally problematic self-congratulatory postracial fantasy, that in the years that followed would take on a fiery, gothic hue.

By the end of the same year, nearly every country in the world had signed onto the Paris Climate Accords. In contrast to the relative optimism of that moment (which Robinson's novel extrapolates), that year's summer blockbuster, *Mad Max: Fury Road*, captured in its simultaneously futuristic and primitive apocalyptic vision the essence of our current conjuncture. Conditions worsened. Today, the climate is more volatile: hotter, colder, windier, wetter, and dryer. Social and political life are more stressful, exposed, and divided. While in 2015 the HBO show *Silicon Valley* could skewer the tech sector's narcissism as mere delusion, today's headlining technologies—cryptocurrency, the metaverse, artificial intelligence (AI)—underscore the sober reality of how Silicon Valley's solutionists are as unable to solve the problems they create as they are uninterested in them. The futures on offer are literally disconnected from reality.[11] When tech billionaires aren't plotting their escape from our decaying planet, they're pillaging the only institutions that control more capital than they do: the state.

As futurity runs out of steam, nostalgia and machismo become attractive coping mechanisms. This can be witnessed in the "strongman" politics across the globe that has harnessed the volatile mix of pessimism and immiseration by weaponizing fantasies of the past, securing the power of rulers like Xi Jinping (self-fashioned torchbearer of the Chinese Communist Revolution), Ferdinand "Bongbong" Marcos Jr. (scion of the Philippines' long-reigning dictator), Narendra Modi (who seeks to rightsize modernity via the discipline of Hindu "tradition"), and of course Donald Trump (again). Russia's invasions of Crimea and Ukraine have signaled to some a return to the great-power competitions of a century ago, raising concerns about Taiwan's fate in the midst of a putative New Cold War between the United States and China. One does not have to believe in progress or modernity to feel that we are somehow slipping backward.

In Sickness and in Health

Space warps when time warps. Whereas the first volume registered the fundamental role that *anachronism* played as one of techno-Orientalism's primary forms of Othering, this volume, without departing from that insight, grapples with the consequences of *presence*. China looms large here—so much so that we are compelled to devote an entire section to it (Part III: Sinofuturism). In the first volume, techno-Orientalism was most often keyed to the dual trope of "Japan" and "China"; to quote the introduction, "If Japan is a screen on which the West has projected its technological fantasies, then China is a screen on which the West projects its fears of being colonized, mechanized, and instrumentalized in its own pursuit of technological dominance."[12] This duality is constituted not by fixed essences but contingent, ever-shifting material relations. In 1995, David Morley and Kevin Robins argued that techno-Orientalism was provoked by a "Japan Panic" stoked by fears of Japan's manufacturing and financial incursions into the U.S. domestic automotive and real estate markets.[13] The benefit of hindsight informs us that Morley and Robins were writing from within Japan's first so-called lost decade of low economic growth, and therefore describing a receding phenomenon. It has been a long time since Japan inspired panic. As a result, the techno-Orientalism of Japan's economic miracle strikes us today as a period aesthetic, exemplified in films like *Big Hero 6* (2014), television shows like *Altered Carbon* (2018), and video games like *Cyberpunk 2077* (2020). But even at its height, the threat lurking behind Japan panic was a temporally displaced *future* potential. Increasingly, China-inflected techno-Orientalism has taken on a different cast: China is here, now. It is unavoidably *present*. We had better reckon with it—from decisions over Mandarin immersion daycares to debates over industrial policy—or else we risk obsolescence.

Just as "Japan" and "China" coarticulated their meanings in that earlier period, today's "China" does not enter the U.S. imagination in unmediated form. In 2015, Asian popular culture was on the verge of explosion in the United States. That year, the media company 88 Rising launched with the goal of introducing Asian musicians to American audiences like Indonesian singer-songwriter Niki, Indonesian rapper Rich Brian, and Japanese-Australian singer Joji. The K-pop group BTS's EP *The Most Beautiful Moment, pt. 2* broke into the Billboard 200, setting the stage for their breakout year in 2017. The South Korean rapper Psy's viral sensation "Gangnam Style," which in 2015 had the distinction of being the most-watched video in YouTube history, now appears quaint in comparison to videos from BTS and Blackpink, who (as of this writing) dominate the rankings of the most-viewed K-pop videos. The Best Picture win of Bong Joon-ho's 2019 film *Parasite* at the 2020 Oscars was perhaps more significant in the United States than elsewhere (Bong: "The Oscars are not an international film festival. They're very local."), but along with other *Hallyu* (Korean Wave) hits such as the TV show *Squid Game* (2021), it signaled a deeper hunger and capacity to metabolize Asian aesthetic and social forms than even those Western crazes of old, japonisme and chinoiserie. In Asian American pop culture, *Fresh off the Boat*'s premiere in 2015 and subsequent six-season run marked an important milestone. *Crazy Rich Asians*' box-office success in 2018 was accompanied by Asian American independent films like *The Farewell* (2019), *Columbus* (2017), *Everything Everywhere All at Once* (2022), and *Joy Ride* (2023) as well as the wide-release Hollywood productions *Shang-Chi and the Legend of the Ten Rings* (2021) and *Turning Red* (2022). The receptiveness of American audiences to Asian- and Asian American–centered pop culture is a testament not only to long-fought representational demands by Asian and Asian American communities, but also to Asian countries' increasing expenditures on culture industries. While these are happy developments for representation in the United States, they are also one aspect of a transnational Dorian Gray dynamic. *Hallyu*'s beauty is symptomatic of a wizening national export economy desperate for spatial fixes.[14] However, the fact that Asian pop cultural products have found eager consumers in the United States does not mean that Asia has found an eager partner. In many ways, quite the opposite.

If the Orientalism of techno-Orientalism is what Edward Said called "The Latest Phase" in the closing pages of his 1978 book, then it was rooted in an oil economy that not only gave coherence to an "Arab world" sitting on reserves gifted to it either by happenstance or divine providence, but also sharpened the threat of Japanese cars. The resource fueling the consumption of Asian pop culture is the oft-touted "new" oil: advanced semiconductors. It is through the lens of these commodities that "Asia"—especially East Asia—has begun to reconfigure Orientalism. The semiconductor shortage of early 2022, which affected the production of almost any device with an on-switch, was among the most bracing episodes of the broader supply chain disruptions wrought by COVID-19. Not only did it reveal how interdependent the world economy was—over 170 countries were adversely affected by the shortage—it was also a shocking display of the deeply and strategically imbalanced semiconductor supply chain. Although this supply chain spans over 120 countries, the leading-edge semiconductor fabrication nodes are monopolized by Taiwan.[15] Along with the rest of the world, the United States is almost entirely dependent on Taiwan Semiconductor Manufacturing Company (TSMC) for the chips that make possible each year's faster and more energy-efficient phones, computers, and AI processors. In 2020, TSMC surpassed Intel to become the world's largest semiconductor fabricator. And of the world's most advanced semiconductors, TSMC produces over 90 percent of the world's supply.[16] South Korea and Japan are the two other major players in semiconductors, with China as the new kid on the block. Of immense geopolitical importance is that no country comes close to Taiwan's capabilities. This has implications not just for consumer electronics but, as Chris Miller explains in *Chip War*, for military applications and governance as well.[17]

Recognizing this disparity, in early 2015 China's president, Xi Jinping, launched the Made in China 2025 initiative, which laid out the Communist Party's plans to expand China's strategic high-tech industries, including semiconductors and AI. As much as China lags in semiconductors, it holds a major advantage for AI development. Taking up the oil metaphor again, if *data* is the new oil, then China is sitting on one of the world's largest reserves. Data fuels the development of AI—but only as long as the semiconductor-based infrastructure that AI runs on can keep up with the pace of that development. Also of great strategic importance are the differing attitudes toward AI that the Chinese hold versus Americans: Chinese tend to be more open to its potentials, while Americans tend to fear it. This cultural environment, along with data supremacy, is among the reasons why many perceive China as poised to dominate the global AI industry. Anxiety around this has deeply shaped U.S. industrial policies. In 2018, the United States nearly bankrupted the Chinese telecommunications company ZTE by blocking it from acquiring U.S.-patented technologies. This was a precursor to Trump's move a year later to ban U.S. companies from purchasing equipment from the Chinese company Huawei, deeming it a security threat. Biden's administration extended this ban in November 2022, hot on the heels of its most significant attack in this highly charged trade war: an aggressive export ban passed in October 2022 that imposes unprecedented licensing requirements on U.S.-patented semiconductor technologies. Like the "sophons" of Chinese science fiction writer Liu Cixin's *Remembrance of Earth's Past* trilogy, these rules are intended to freeze China at a level of development two to three generations behind the most advanced semiconductor technologies.[18] Because these rules also prevent U.S. firms from accessing the Chinese market, they cut off America's nose to spite its face. Such is the illogic of twenty-first-century techno-Orientalism.

Meanwhile, COVID is the factor that has shaken every aspect of our reality. It has been the great revealer and amplifier. For those who chose to look, it revealed the weaknesses of each society it ravaged, and amplified conflicts that had already been festering. It could not have come at a worse time in U.S.-China relations: at a point in which the decades-long détente of engagement was buckling under resurgent nationalism, military paranoia, and economic challenges. From the initial warnings of a novel flu-like pathogen in late 2019 and early 2020, to the space-time confusion of lockdowns being imposed in the United States just as they were being lifted in China (and then the reverse two years later), to the anti-Asian violence that became its own kind of epidemic, to the ongoing speculation about the disease's origins, American Orientalism has been on full, ugly display.[19] Techno-Orientalism has helped us to make sense of the discourse, and going forward it will be an important lens to detect and address how institutions and knowledge production have internalized the antinomy of atavism versus futurism. As the historian of capitalism Andrew Liu has argued, the (seemingly oppositional but uncritically accepted) "wet market" and "lab leak" theories of SARS-CoV-2's zoonotic origin map perfectly onto this antinomy.[20] As countries around the world have now begun to take stock of their approaches to the pandemic, techno-Orientalism continues to poison U.S. efforts at every level of government to develop effective policies for addressing future outbreaks and novel diseases, not to mention the ongoing pandemic, which has been all but entirely wished away.[21] Techno-Orientalism helps us to understand why the U.S. has been so acutely irrational on this issue, not only in its fixation on the question of origin, which has become a proxy for U.S.-China rivalry, but also in how it ignores opportunities to learn from East Asian countries like South Korea, Japan, and Taiwan, whose hard-won experiences from previous epidemics helped them to maintain case-fatality rates that have been just a fraction of the United States'.[22] In other words, the *irrationality* of techno-Orientalism dictates that the West must take the opposite tack from the premodern East, regardless of the consequences, to reinforce its monopoly on liberal humanist modernity. COVID has revealed how U.S. exceptionalism in the early twenty-first century is formulated as an exception to an ascendant Asia.

Theory and Critical Practice

Keeping apace with the techno-Orientalist rhetoric and policies informing the economies of security, trade, and culture described above, its critical analytic has also developed considerably with a significant number of monographs, special issues, and journal articles integrating techno-Orientalist readings into an array of established and emergent disciplines. Gender studies, for example, integrates the idea of the inert and the object with tropes of roboticism and doll-ness infused with exoticism, sexualization, and racialization. Anne Anlin Cheng's *Ornamentalism* (2019), building on thing theory, traces an ornamental personhood that permeates material and visual culture: "To speak of Asiatic femininity," Cheng writes, "is to speak of a style, which claims specificity but lends itself to promiscuous transferability. It designates a specific racial category but can be applied to different racial subjects. It can be enlisted by those wielding power and, more disturbing, by those deprived of it. It is an abstraction that materializes."[23] This logic articulates a perihumanity manifesting nebulous subjects that are neither human nor pure object (the titular ornament). In a reading of *Ex Machina*, Cheng notes that the white-encoded Ava uses the skin of a deactivated Asian-encoded android and seamlessly wraps it on her body to take on the appearance of a human. Yet she does not do so to become human; it is, in Cheng's assessment, an action that is "not exactly friendly," but "Ava may be offering us her own brand of radical theory: the proposition that existing, however isolated and provisional, may be the only mode of survival in the enchantment of objecthood."[24]

In a similar vein, Leslie Bow's *Racist Love* (2022) interrogates the figure of the Asian automaton woman against the background of sexual exploitation and human trafficking to understand trends in artificial intelligence and robotics. Bow articulates an organizing principle ("racist love," borrowed from Frank Chin and Jeffrey Paul Chan)[25] that explains how differential anxieties are assuaged by an inversion—an attraction to, and even affection for, objects and objectified people.[26] Most germane to the techno-Orientalist analytic is her incisive reading of the cultural formation of AI and its material instantiations as extensions of existing racial and gender inequalities in postindustrial labor: "The techno-Orientalist trafficking narrative builds on the conceit of machine or clone interchangeability, the downside of the posthuman: not the Asian female piece worker harnessed to a machine, but the worker as machine. In this sense, techno-politics and biopolitics conjoin; they expose the inhumanity underlying the uneven development of global capitalism, in which biowomen are exploited until they are, like the extracted materials they process, depleted, broken, or obsolete. By and large, however, such narratives do not interrupt the liberal fantasy of autonomous personhood but merely extend it to things."[27] Consequently, Bow argues, it is not a large leap to conceive of gendered and racialized robotics as similarly disposable and exploitable as both worker and sexualized object—the facsimile comes to merely concretize a logic that has already been culturally internalized. The techno-Orientalist analytic informs both Cheng's and Bow's critiques of how Asian women's bodies are subsumed not only in conventionally misogynistic fashion, but even in broader posthumanist discussions undergirded by liberal humanist interests.

Scholars in fields such as game studies also scrutinize techno-Orientalism's symptoms and ramifications. If the Internet was the technological innovation of the 1990s that demanded a reckoning from all scholarly fields, then games might be the next digital and cultural platform of the 2020s, to which critics are finally coming around. Critics such as Tara Fickle, Chris Patterson, Edmond Chang, and Takeo Rivera—along with Steve Choe and Se Young Kim in their chapter "Never Stop Playing" in our first volume—lead the charge in constructing frameworks to understand the relationships among Asian racializations in games as archives, platforms, and sport. Again, a techno-Orientalist framework appears to offer an entryway to building innovative lines of inquiry. Fickle's *The Race Card* focuses on how games—their "design, marketing, and rhetoric"—become sites of contestations over how "Asians as well as East-West relations are imagined and where notions of foreignness and racial hierarchies get reinforced."[28] In other words, defining and applying a "ludo-Orientalist" framework reveals how games reflect and reinforce existing cultural logics; instead of targeting technological platforms, Fickle aspires to expand beyond the "computational medium or mechanics of video games as techno-Orientalist interfaces" to "bring those insights to bear on the way that gaming, both digital and analog, is *used* in everyday life to provide alternative logics and modes of sense making, particularly as a means of justifying racial fictions."[29] Rivera's book offers the conceit of a "model minority masochism"—a cultural logic that both desubjectifies and self-objectifies to, paradoxically, self-generate. To that end, Rivera posits techno-Orientalism as a logical extension of the model minority concept: "I would go so far as to argue that the techno-oriental is the grotesque personification of the model minority itself, providing the figure with an optic vocabulary, replete with both the promises and perils of an increasingly technologized society. Insofar as Asianness becomes associated with a laboring body—once the coolie, now the Asian tech worker—Asiatic racial form shifts according to the status of material labor conditions."[30] Video games in particular can be a theater in which that logic is most concretely exercised by an Asian American player. In a reading of the game *Deus Ex: Human Revolution* (2011), Rivera shows how the occupation of a white transhuman protagonist in a techno-Orientalized environment results in the self-abnegation and destruction of the enemy and minoritized self, fulfilling what he deems to be a compulsion toward both pleasure and pain of the racialized subject. In his expansive *Open World Empire* (2020), Christopher Patterson theorizes an "erotics" as a means of contextualizing the function of play in global capital and imperialism. Consequently, he asks, can transnational gamespace push play to be racialized, or race to be playful? What does race even mean in the collision of players from all over the globe?[31] As much as games in their design trade upon the semiotics of race, ethnicity, and the nation, they create an environment conducive for those constructs to be playfully interrogated. These studies exemplify the compelling application of the techno-Orientalist analytic for laying bare the replicative and reifying forces operating in beloved simple and sophisticated games.

Finally, techno-Orientalism's most generative, if not predictable, theoretical and practical intersections are with Asian American studies and Asian studies, where the techno-Orientalist scholarly discussion began. *Verge: Studies in Global Asias*, a journal dedicated to Asian, Asian American, and Transpacific studies, has served as the major platform for groundbreaking critical scholarship across these fields. A special issue, "Digital Asias" (2021), expands upon techno-Orientalism in arguing that the Asian horizon is historically and materially sourced in the complicated relationship between the United States and Asian empires and economies. Therefore, the appellation obscures the ties that bind—the many social, cultural, and economic imbrications that make up the "digital"—instead, it is a global capitalist future perfect, a tense at which the East is always yet to arrive.[32] Other important critical elaborations include Suk-Young Kim's "Disastrously Creative: K-pop, Virtual Nation, and the Rebirth of Culture Technology," in which Kim traces the troubling tendency of the South Korean pop culture industry to adopt techno-Orientalist visual iconography. This form of "cultural technology," Kim insightfully observes, underscores long-standing neoliberal trends: "Technophilia is a mutual obsession of the South Korean government and SM Entertainment, who see their ambitions converging in a corporatized nationhood and a nationalized corporation."[33] Similarly, S. Heijin Lee finds techno-Orientalist echoes in Western discussions of Korean beauty standards and cosmetic surgery, which tend to characterize plastic surgery practices and K-beauty standards as dystopic, antifeminist, and an affront to liberal humanism.[34] Jonathan Abel's *The New Real: Media and Mimesis in Japan from Stereographs to Emoji* (2023) meditates on Japan's centrality in techno-Orientalist discourse as a means of theorizing the relationship between mimesis and media. This discursive Japan—imagined or real—he argues, operates as a site through which media theory's efficacy is tested and evaluated.[35] Techno-Orientalism, then, stands at the intersection of mimesis and new media that contributes to the construction of and by Japan. What is most notable about each moment of contact between Asian and Asian American studies is that techno-Orientalism reveals cultural logics and discursive collisions not necessarily predicated on texts, platforms, gadgets, and machines, but on the states and systems, Asian and non-Asian alike, that reify them.

Techno-Orientalism 2.0: Here and Now

Our theorization from subject to structure in this volume thus captures techno-Orientalism's transformation into a force field of influence. In contradistinction to the characterization of undifferentiated Japanese autoworkers mindlessly producing Toyotas, techno-Orientalism's new forms and formulations center on the faceless state and technological platforms. It is, now, TikTok's data harvesting and Huawei's manufacturing that stand as techno-Orientalist proxies for the state and the semiconductor supply chain. In other words, the techno-Orientalized object no longer begins with the automobile (or human stand-in) to extend backwards; instead, that anxiety is distributed and understood to exist in all electronics and apparatuses to extend back to the platform, infrastructure, and state. Further evidence for the shift from subject to structure can perhaps be found in the diminished role that literary objects play in the contributions to this volume, compared to the first, where literary objects were dominant. Whereas the literary, especially in prose fiction, privileges forms such as characterization and interiority that complement subject formation, the visual, serialized, and multimedia objects that predominate in this volume seem more apt for drawing focus on how techno-Orientalism operates in the concrete abstractions of institutional and economic structures.

The scope of this updated theoretical framework is not limited to a post-2015 world. On the contrary, it addresses these developments by further expanding the techno-Orientalism analytic both historically and conceptually, synchronically and diachronically; indeed, the following chapters have been selected precisely because they push the boundaries of the field. Moreover, this volume engages texts and questions that the first did not have the capacity to engage or could not anticipate. While the first volume largely focused on the influence of East Asia (Japan, Korea, and to a lesser degree, China) and the American construction of techno-Orientalist discourse, this volume confronts techno-Orientalist inflections in the now-dominant China threat as well as in the rise of the Global South—specifically South and Southeast Asia. Additionally, several contributions explore and theorize moments of intersectional dialogue between techno-Orientalism and the evolutions in the parallel discourses of Afrofuturism and Indigenous futurism. Finally, and distressingly, this second volume comes at another critical juncture in Asian American history as the COVID-19 pandemic reignited animus toward Asian and Asian Americans with racially charged language on bioweaponry, echoing techno-Orientalist discourses that surrounded Vincent Chin's death and mobilized Asian America in the 1980s. Our moment demands vigilant theorizing toward greater understanding and more just productions and practices. We continue to see new forms of techno-Orientalism by degree and by kind. Such unabated proliferations require our collective confrontation. That effort is what impelled the first volume, and it continues to propel the work of our contributors in this volume.

Chapters Overview

Part I: Labor Reconfigurations

In chapter 1, "Working Futures after Asians: Automation, AI, and the Global Labor Economy," Leland Tabares examines a techno-Orientalist discourse that has emerged in recent years that depicts Asian bodies and Asian technologies not as threatening, but as obsolescing. For the Silicon Valley solutionists who have perfected the song and dance of exaggerating the automation threat posed by their own technologies by concern trolling over its social and even species-level dangers, the figure of a declining Asia has proven very useful. In his readings of the films *Ex Machina* and *After Yang*, Tabares shows how this rhetorical maneuver has infused popular culture, and how these films reveal the true motivation behind automation threat and "after Asia" discourses: an anxiety regarding corporate control over our tech-driven global economy.

Bringing the logical structure of techno-Orientalism into dialogue with discourses of cybernetic capitalism, Won Jeon's "*Everything Everywhere All at Once*: Techno-Orientalism in an Age of Cybernetic Capitalism" (chapter 2) shows how "ideas of the inhuman and machinic Orient" are propagated by the information technologies that dominate our economic, political, and social spheres. Jeon traces this structure into the trope of "nothingness" in the film *Everything Everywhere All at Once*. Hovering between the film's narrative poles of the laundromat and the IRS is an "epistemic anxiety surrounding the material and economic relations between enmeshed categories of race, technology, and capital"—categories that, Jeon argues, emerge from an undecidable, stochastic relation between noise and information.

Kimberly D. McKee's "Chinese Commodities: Adoption in *After Yang*" (chapter 3), offers close readings of Alexander Weinstein's short story "Saying Goodbye to Yang" (2016) and Korean American director Kogonada's film *After Yang*, noting that they are "part of a broader evolution of techno-Orientalism that shifts from an explicit dystopia to a mundane version of a technologically enhanced present." McKee points out that Mika is not the only adoptee in the texts; the technosapien/android Yang must also be selected by adopting parents. She draws on adoption studies to argue that "anxieties of the techno-Orientalist future are rooted in concerns over white supremacy's demise."

Part II: Racialization as Technology

Charles M. Tung's "Plastinated Vitruvian Man, the Datafication of Race, and Transracial Transfer in *Westworld* and *Altered Carbon*" (chapter 4) examines racialized bodies as technologies of racial recoding and uncoding in a reading of HBO's television series *Westworld*, Richard Morgan's novel *Altered Carbon*, Netflix's series adaptation of Morgan's novel, and the real-life traveling show *BODIES: The Exhibition* that featured plastinated Chinese cadavers. Tung sees in these texts a shared treatment of Asian bodies as fungible vessels for "skinning," a "peculiar fungibility associated with Asians under neoliberalism" that treats Asian bodies as vessels for racial recoding, a utility that facilitates "the proliferation of differences no longer anchored in the material." Having the Asian body at the center of this "absent presence," Tung argues, "is both cause for celebration and for a kind of racial epidemiological anxiety, a fear of the spread of unreadable racial code beneath the skin and new uncertainties about identification."

Taking issue with the conventional binary between East and West, M. Imran Parray's "Outsiders Within: The Indigenous/Minority Question and Techno-Orientalist Gaze in India" (chapter 5) examines a techno-Orientalist dynamic within India and its minority and Indigenous communities. Extending upon "Hindu Orientalism," Parray argues that the Indian state adapts a techno-Orientalist logic for its marginalized communities—in this case, Muslim. Parray traces how the cultural aesthetic of the technologized Muslim Other develops in a series of genre productions, including Indian television shows such as *Crackdown* (2020), *Kathmandu Connection* (2021), and *Special Ops* (2020), with a primary focus on the Amazon Prime show *Family Man* (2019).

In "On Forms of the Black Box: Race and Technology in STS and Global Critical Race Studies" (chapter 6), Clare S. Kim and Anna Romina Guevarra interrogate the history and evolution of the "black box," a familiar concept in science and technology studies describing a device (be it a tool or mathematical function) that processes an input and generates an output according to unknown parameters. While the black box is often described in clinical terms, Guevarra and Kim argue that the scientific as well as cultural concept follows a racial logic that must have an accounting. Tracing the concept's development from seventeenth-century automata to artificial intelligence, this chapter juxtaposes critical race theory and technoscientific discourse to examine an unsettling pattern of black box racialization that invites theorists to approach race in technoscientific terms as a kind of black box in and of itself.

Part III: Sinofuturism

Ian Liujia Tian, in chapter 7, "Infrastructure and/as Mediation: *China 2098*'s Tempro-Affective Politics," analyzes a digital art and text series, *China 2098*, by Chinese artist Fan Wennan, which depicts a postapocalyptic world that has suffered climate disaster as a result of global capitalism. Fan's images feature infrastructure and draw on Chinese socialist tropes and slogans, and Fan's text is narrated by a space station worker in the year 2098. Tian argues that *China 2098* fails to critically interrogate a technomodernist belief in infrastructure. He also claims that infrastructure mediates how artists such as Fan are imagining the future. Tian argues for the importance of viewing infrastructure not merely as physical objects (e.g., dams, space stations, power plants), but as a practice that shapes the way people live and feel about their colonial pasts, disastrous presents, and uncertain futures.

In "Techno-Orientalist Deflections: How Documentaries Frame China's AI Threat" (chapter 8), Gerald Sim analyzes how U.S. Big Tech companies, politicians, and the media use techno-Orientalist tropes to counter or advance artificial intelligence; the general tactic is fearmongering that Chinese AI will crush U.S. AI unless the U.S. government relinquishes regulations and props up the Big Tech industry. Sim points out how politicians, lobbyists, tech pundits, and two documentaries, *In the Age of AI* (2019) and *Coded Bias* (2020), update the nineteenth-century trope of laboring Asiatic "hordes" to characterize twenty-first-century urban Chinese residents who acquiesce to Chinese state surveillance.

Shana Ye's "Techno-Futurehistory and the Sojourners of Global China: A Threefold Reading of *The Wandering Earth*" (chapter 9) offers an illuminating comparative analysis of celebrated Chinese science fiction author Liu Cixin's 2000 novella with Chinese director Frant Gwo's cinematic adaptation of the novella in two parts, *The Wandering Earth* (2019) and *The Wandering Earth II* (2023). Discerning tension between what she sees as the novella's critical stance and the films' nationalist stance on the idea of "China saving the world," Ye analyzes moments that allow for multiple interpretations because of what she argues as China's simultaneous internalization and rejection of colonial and anticolonial, capitalist, and anticapitalist enterprises. Through the trope of the sojourner, Ye helps us better understand the "shitty" national (i.e., Chinese) and global histories and practices that people of Earth must literally and figuratively carry with them as they escape the orbit of a dying Sun.

Part IV: Machinic Subjects

Carefully unspooling several discursive strands in a study of Korean films, Jane Chi Hyun Park argues that the transpacific female body disrupts the neoliberal trajectory of *Hallyu* (the Korean Wave). Synthesizing Korean studies, Asian American feminist criticism, and techno-Orientalist critique, "Sacrificial Clones: The Technologized Korean Woman in *Shiri* and *Cloud Atlas*" (chapter 10) complicates the linear progression of South Korean soft power as a mode of "creative imitation" that adapts from and speaks back to the colonial center through generic science-fictional tropes. Park argues that the depiction of an "inauthentic" and technologically constituted Korean female body questions the human cost of the rapid rise and global dominance of Korean popular culture.

Rachel Tay and Jaeyeon Yoo's "Assembling Mitski: The Aesthetics and Circuits of Techno-Ornamentalism" (chapter 11), offers a meditation on the musician Mitski Miyawaki's elusive racial personae, as well as the effects of her deliberate opacity. Reflecting on their personal investments in and consumption of Mitski's performances, Tay and Yoo break down the "perplexing dynamics at play in the respective exchanges between performer and audience, musician and fan, commodifiable content and the hand that holds up a phone to capture it." They argue that these are not only the essential technologies of "value accumulation" for the racialized performer, but also a form of what they call "techno-*ornamentalism*" (a term they adapt from Anne Anlin Cheng) to describe Mitski's strategic management of digital and media artifacts—her "aesthetics of ornamental ambiguity."

Part V: Extensions

In chapter 12, "Asian Solarpunk: Between Utopia, Collective Future and Remedies for Climate Panic," Agnieszka Kiejziewicz and Justin Michael Battin trace the genealogy of "solarpunk": a set of collectively imagined tropes drawn from Buddhist philosophy, various Asian aesthetic traditions, and science fiction that doubles as a cultural logic and genre form. Linking the eco-utopian fantasies visualized in films by Hayao Miyazaki, Kogonada, and Bong Joon-Ho, to their narration in novels by authors like Ursula K. Le Guin, and then to their concretization in urban planning projects like Singapore's Gardens by the Bay, Tokyo's Roppongi Hills, and Ho Chi Minh City's Thu Duc Technology District, Kiejziewicz and Battin offer an expansive and provocative account of how techno-Orientalism suffuses the forms through which we imagine the relationship between technology and ecology, as well as our built environment.

Baryon Tensor Posadas moves us from the cyborg, a focalizer for a techno-Orientalizing gaze vis-à-vis Japan, to the zombie as figure of the posthuman with "Animated Bodies: Project Itoh and the Afterlives of Techno-Orientalism" (chapter 13). Posadas examines the figure of the zombie in Project Itoh (Ito

Keikaku), the pseudonym of Japanese science fiction writer Ito Satoshi. Posadas asks us to view the zombie "as itself a critical prism for articulating the stakes of visualizing the Non-West Other as non-human by raising the issue of our own practices of spectatorship, our own visual pleasure in consuming mass-produced animated bodies of these Others in the aftermath of the techno-Orientalist habits of looking." Viewing zombies as allegories of colonial and capitalist dehumanization, dramatizing a generalized rejection and Othering of the zombie, and figuring the zombie in terms of racial panic, Project Itoh identifies with the zombie and invites readers to embrace the zombie.

Adhy Kim turns to Karen Tei Yamashita's *Through the Arc of the Rainforest* (1990) and Sequoia Nagamatsu's *How High We Go in the Dark* (2022) in chapter 14, "Settler Orientalism, Asian American Techno-Environmentalism: The Network Novel under Japanese and U.S. Empires," examining the "techno-environmentalist" utopian aspirations in each novel and how they risk romanticizing a nature-human binary. Kim argues that both authors' use of expansive geographical and time scales in these examples of "network novels" in order to exchange the pessimism that canonically characterizes techno-Orientalist narratives for salutary visions, risks "greenwashing" both Japanese and U.S. settler colonial ideologies.

In "The Alchemized Dis/abled Body as Recuperative Site in *Fullmetal Alchemist*" (chapter 15), Jung Soo Lee uses the tools of disability to show how the hugely popular manga series, which was serialized from 2001 to 2010, is structured by post-9/11 geopolitics as seen from the perspective of Japan. The disabled bodies that are normative in the series, Lee argues, are mobilized to depict a fantasy about a U.S.-Middle Eastern rapprochement negotiated by China and brought about by Chinese technologies. This reverse techno-Orientalism allegorizes China as simultaneously powerful and peaceful, even harmless, thus reflecting Japan's postwar soft-power strategy of eliding its history of imperial violence by depicting itself as both "cute" and "cool." The series' commitment to narrating the social construction of "dis/ability," which is also a refusal to reify the disabled body, emerges from this layered geopolitical fantasy.

Part VI: Optimistic Futures

Lori Kido Lopez's "Recovering Asian American Futures in the Marvel Cinematic Universe" (chapter 16) describes a taxonomy of Asianfuturism by exploring commonalities in two recent superhero entries, Marvel's *Shang-chi and the Legend of the Ten Rings* (2021) and *Ms. Marvel* (2022). In contrast to more conventionally dystopian renderings, Lopez argues that both works dexterously circumnavigate techno-Orientalist tropes to invent a technomagical alterity centered on self-actualization and familial reconciliation. This, Lopez notes, works in the wake of speculative movements such as Afrofuturism, Chicanx futurism, and Indigenous futurism, all of which similarly respond to colonial discourses emphasizing technological alienation and premodernity.

In chapter 17, "Looking for Asianfuturism: Asian American Science Fiction and Games of Color," Edmond Y. Chang constructs a working definition of Asianfuturism as an imaginative, even playful space that the current moment demands. Explicitly drawing connections between Asian American speculative narratives and game design, Chang posits that a structural approach to Asian American texts is most conducive to that form of imagination, in response to structures of discrimination. Through a reading of Larissa Lai's novel *The Tiger Flu* (2018), video games, and museum installations, Chang offers a framework for radically reenvisioning an Asian America uncoupled from racism.

Optimistic interventions also follow in Thomas Xavier Sarmiento's analysis of how a desirable Asian futurity can be imagined in "The Queer Techno-Orientalist Aesthetics of Disney's *Big Hero 6*" (chapter 18). Sarmiento makes a case for queering normative techno-Orientalist scripts of idealized futures that deploy Asian and yellow peril tropes to bolster white cisheteronormative social orders. To disrupt dominant techno-Orientalist representations of "Asianized" futures that "lean dystopian and figure as negatively queer," Sarmiento demonstrates reading practices that redirect audiences beyond technologized yellow-peril narratives toward more hopeful horizons. Sarmiento argues that in *Big Hero 6*, "these possibilities take the form of technologized Asian Americans as heroes, not threats, who work alongside and care for other people of color and robots, a situation that rewires how we might link 'futurity' with 'Asian,'" particularly in the higher education, research and development, venture capital, and health and wellness sectors.

Notes

1 Kim Stanley Robinson, *The Ministry for the Future* (New York: Orbit, 2020).
2 Robinson, *The Ministry for the Future*, 511.
3 Indeed, in an interview Kim Stanley Robinson (author of the Mars trilogy) declared, "Mars is irrelevant to us now." In other words, fantasies of planetary colonization preclude efforts to preserve the only realistically habitable planet. See Casper Skovgaard Petersen, "Interview: Kim Stanley Robinson," *Farsight*, August 10, 2022, https://farsight.cifs.dk/interview-kim-stanley-robinson/.
4 Nadja Popovich and Brad Plumer, "How the New Climate Bill Would Reduce Emissions," *New York Times*, August 2, 2022, https://www.nytimes.com/interactive/2022/08/02/climate/manchin-deal-emissions-cuts.html.
5 Sara Schonhardt and E&E News, "China Invests $546 Billion in Clean Energy, Far Surpassing the U.S.," *Scientific American*, https://www.scientificamerican.com/article/china-invests-546-billion-in-clean-energy-far-surpassing-the-u-s/.
6 David S. Roh, Betsy Huang, and Greta A. Niu, "Technologizing Orientalism: An Introduction," in *Techno-Orientalism: Imagining Asia in Speculative Fiction, History, and Media*, ed. David S. Roh, Betsy Huang, and Greta A. Niu (New Brunswick, NJ: Rutgers University Press, 2015), 2.
7 Osman Ahmed, "Xander Zhou's 'Techno-Orientalism,'" *The Business of Fashion*, January 2, 2018, https://www.businessoffashion.com/reviews/fashion-week/xander-zhou-techno-orientalism/; Avidan Grossman, "The Beginner's Guide to Techwear,"

Esquire, March 5, 2020, https://www.esquire.com/style/mens-fashion/g31213240/best-techwear-brands/.

8 See Brett J. Esaki, "Ted Chiang's Asian American Amusement at Alien Arrival," *Religions* 11, no. 2 (February 2020): 56, https://doi.org/10.3390/rel11020056; see also Dawn Chan, "Asia-Futurism," *Artforum* 54, no. 10 (Summer 2016), https://www.artforum.com/print/201606/asia-futurism-60088.

9 Jane Hu, "Where the Future Is Asian, and the Asians Are Robots," *New Yorker*, March 4, 2022, https://www.newyorker.com/culture/culture-desk/where-the-future-is-asian-and-the-asians-are-robots; "Episode 193: Asian Futures Without Asians," *Imaginary Worlds*, accessed August 17, 2023, https://www.imaginaryworldspodcast.org/episodes/asian-futures-without-asians; George Yang, "Orientalism, 'Cyberpunk 2077,' and Yellow Peril in Science Fiction," *Wired*, December 8, 2020, https://www.wired.com/story/orientalism-cyberpunk-2077-yellow-peril-science-fiction/; Evan Nicole Brown, "How Sci-Fi Films Use Asian Characters to Telegraph the Future While Also Dehumanizing Them," *Hollywood Reporter*, November 16, 2021, https://www.hollywoodreporter.com/lifestyle/arts/sci-fi-films-asian-characters-representation-movies-appropriation-dehumanization-1235048534/.

10 We should say that by the time of our manuscript submission to the press, blatant acts of racism and pushback against diversity, equity, and inclusion policies and programs are once again on the rise in the new presidential administration.

11 As this volume goes to press, Silicon Valley's AI oligarchs are reeling from the Chinese company DeepSeek's puncturing of their astronomical scaling ambitions: the latter accomplishing with a few million dollars what the former want us to believe should cost $500 billion.

12 Roh, Huang, and Niu, "Technologizing Orientalism: An Introduction," 4.

13 David Morley and Kevin Robins, *Spaces of Identity: Global Media, Electronic Landscapes and Cultural Boundaries* (London: Routledge, 1995).

14 Joseph Jonghyun Jeon, *Vicious Circuits: Korea's IMF Cinema and the End of the American Century* (Stanford, CA: Stanford University Press, 2019), 12–14.

15 The Dutch company ASML is the only manufacturer of the world's most advanced photolithography machines, which Taiwan's TSMC uses to mass produce chips used in applications ranging from consumer electronics and AI to weapons.

16 In response to the disparity, the Biden administration passed the CHIPS (Creating Helpful Incentives to Produce Semiconductors) Act in 2022 to rejuvenate the domestic silicon chip production industry and reduce dependence on East Asia. See Ana Swanson, "The CHIPS Act Is about More Than Chips: Here's What's in It," *New York Times*, February 28, 2023, https://www.nytimes.com/2023/02/28/business/economy/chips-act-childcare.html.

17 Chris Miller, *Chip War: The Fight for the World's Most Critical Technology* (New York: Scribner, 2022).

18 Christopher T. Fan, "The Red Shredding: On Netflix's '3 Body Problem,'" *Los Angeles Review of Books*, March 30, 2024, https://lareviewofbooks.org/article/the-red-shredding-on-netflixs-3-body-problem/.

19 Angie J. Yellow Horse, Karen J. Leong, and Karen Kuo, "Introduction: Viral Racisms: Asian Americans and Pacific Islanders Respond to COVID-19," in "Viral Racisms," special issue, *Journal of Asian American Studies* 23, no. 3 (October 2020): 313–318.

20 Andrew Liu, "Lab-Leak Theory and the 'Asiatic' Form: What Is Missing Is a Motive," *n+1*, no 42 (Spring 2022), https://www.nplusonemag.com/issue-42/politics/lab-leak-theory-and-the-asiatic-form/.

21 Alice Miranda Ollstein, "Comity Crumbles on Congress' Covid Committee," *Politico*, August 6, 2023, https://www.politico.com/news/2023/08/06/congress-covid-wars-test-bipartisan-bond-00109926.
22 Coronavirus Resource Center, "Mortality Analyses," Johns Hopkins University, March 16, 2023, https://coronavirus.jhu.edu/data/mortality; Benjamin Mueller and Eleanor Lutz, "U.S. Has Far Higher Covid Death Rate Than Other Wealthy Countries," *New York Times*, February 1, 2022.
23 Anne Anlin Cheng, *Ornamentalism* (New York: Oxford University Press, 2019), 14.
24 Cheng, *Ornamentalism*, 151.
25 Frank Chin and Jeffery Paul Chan, "Racist Love," in *Seeing through Shuck*, ed. Kostelanetz Richard (New York: Ballantine, 1972), 65–79.
26 Leslie Bow, *Racist Love: Asian Abstraction and the Pleasures of Fantasy* (Durham, NC: Duke University Press, 2022), 9.
27 Bow, *Racist Love*, 127–128.
28 Tara Fickle, *The Race Card: From Gaming Technologies to Model Minorities* (New York: New York University Press, 2019), 3.
29 Emphasis in the original; Fickle, *The Race Card*, 9.
30 Takeo Rivera, *Model Minority Masochism: Performing the Cultural Politics of Asian American Masculinity* (New York City: Oxford University Press, 2022), xxix.
31 Christopher B. Patterson, *Open World Empire: Race, Erotics, and the Global Rise of Video Games* (New York: New York University Press, 2020).
32 Jonathan E. Abel and Joseph Jonghyun Jeon, "Unfolding Digital Asias," *Verge: Studies in Global Asias* 7, no. 2 (Fall 2021): vii–xxii.
33 Suk-Young Kim, "Disastrously Creative: K-Pop, Virtual Nation, and the Rebirth of Culture Technology," *TDR: The Drama Review* 64, no. 1 (2020): 33.
34 S. Heijin Lee, "Beauty Between Empires: Global Feminism, Plastic Surgery, and the Trouble with Self-Esteem," *Frontiers: A Journal of Women Studies* 37, no. 1 (2016): 10–12.
35 Jonathan E. Abel, *The New Real: Media and Mimesis in Japan from Stereographs to Emoji* (Minneapolis: University of Minnesota Press, 2022), xi, 256.

Part I
Labor Reconfigurations

1
Working Futures after Asians

———————◦———————

Automation, AI, and the Global Labor Economy

LELAND TABARES

In May 2022, Silicon Valley figurehead Elon Musk speculated over X (then Twitter) that "unless something changes to cause the birth rate to exceed the death rate, Japan will eventually cease to exist."[1] On its own, Musk's tweet appears to sensationalize and oversimplify Japan's ongoing population decline. According to Japan's Ministry of Internal Affairs and Communications, the country's total population experienced a steady decline from 2009 to 2022.[2] However, the reasons for the decline are complex and multidimensional. They range from Japan's demanding workplace culture and couples choosing to start families later in life to the COVID-19 pandemic and its impact on the country's aging population. Tempting as it might be to dismiss Musk, his statement exemplifies a broader cultural preoccupation with speculations of a vanishing Asia. In an international population report published by the U.S. Census Bureau, researchers anticipate that West and Southeast Asia will experience similar declines that will impact Asian participation in the labor force.[3] But popular speculations of a vanishing Japan, West Asia, and Southeast Asia are not unique. Later, in June 2022, Musk tweeted that China projects to "lose ~40% of its people every generation" after it saw its "lowest birth[r]ate ever."[4] These "population collapse[s]," he laments, would be a "great loss for the world." Given the historical relation in the West between Asian labor and racialization, the mass extinction of China, Japan, West Asia, and Southeast Asia calls into question the power of the Asian labor economy in a competitive global marketplace driven by the production and consumption of advanced technologies. In a working future without Asians, Asia would effectively become obsolete. Ironically, Musk's tweets coincide with his growing efforts to gain traction in Asia following his acquisition of Twitter in late 2022 and Tesla's expansion into Shanghai, where there would be more to gain by cultivating trust in Asia's labor capacities rather than undermining it.

Narrative speculations of a working future without Asians reveal an emergent discourse that conceives Asians not as threatening but rather as failing to actualize any form of existence or agency, which runs counter to techno-Orientalist traditions in the West. Historically, Asia has been conceived as a geopolitical space of overwhelming abundance. Its massive populations serve as the basis of this ideology. Since the coolie trade in the nineteenth century, the Asiatic body has been racialized in the United States as a threat to its economic stability and cultural values. By the late twentieth century, the coolie found a counterpart in the model minority, an expression of the cultural anxieties surrounding Asia's rising economy and the encroachment of Asian students and professionals on predominantly white institutions. Scholars in Asian American studies interpret techno-Orientalism to be a cultural extension of Orientalism where white colonialist fantasies rooted in yellow peril materialize through representational embodiments of model minority stereotyping in a highly technologized global economy.[5] Today, model minority discourse organizes techno-Orientalist critique by racializing the East as a hyperproductive space of deindividuated, inhuman labor in need of Western consciousness-raising. Common tropes include the tyrannical corporate overlord, the service robot or android, and the clone army. Even when the Asiatic body is representationally absent in cultural narratives about the future, as in cyberpunk fiction, Asianness remains palpable in the tangible dread that manifests through the hyperpresence of Asian cultural objects.[6] That Asia would now be conceived as obsolete and powerless, rather than menacing, signals a dramatic shift in the cultural logics structuring techno-Orientalism. Narratives on Asian obsolescence thus register changes taking place in how Asianness comes to be racialized and regulated ideologically, institutionally, and culturally amid contemporary developments in the global labor economy.

This chapter argues that speculations of a declining Asia symptomize growing anxieties in the United States over market supremacy as the automation industry continues to expand globally, marking an era where institutional control over the working future is increasingly uncertain. Popular discourse on obsolescence in the labor force has centered in recent years on the impact of automation technologies on U.S. workers. During the 2020 presidential election, Democratic candidate and Silicon Valley insider Andrew Yang warned that Big

Tech's investments in automation would not only put 70 percent of Americans out of work but render entire industries obsolete.[7] Ball State's Center for Business and Economic Research reports that workplace automation factored into 88 percent of factory job losses between 2000 and 2010.[8] But manufacturing is not the only industry at risk. Yang contends that automation threatens American professions, especially those in medicine, law, journalism, and the arts.[9] Access to professionalism and advanced degrees no longer corresponds with job security. However, fears of an automated takeover are neither new nor have they ever been actualized. What is new are the economic conditions governing these fears. Automation theorist Aaron Benanav explains that contemporary concerns over automation are a by-product of economic stagnation, where established superpowers compete for market shares that have largely been extracted and monopolized.[10] Such economic conditions lead to underemployment, not unemployment, which identifies more an intensification of existing workplace precarities than the creation of new job losses.[11] It might be easy to conclude that we have entered a postcapitalist, technofeudal era where power is consolidated around tech hubs like Silicon Valley. But Evgeny Morozov insists that the continued presence of innovation in automated labor systems signifies that capitalism remains as alive as ever. Critical approaches to automation premised on the U.S. workforce therefore overlook the larger institutional precarities organizing the twenty-first-century global marketplace. Indeed, if the well-being of individual laborers was such a priority, tech firms like Alphabet, Amazon, Meta, and Tesla would not be investing billions into automation R&D. With global competition around automated technologies being so fierce, scholars across the disciplines observe that we are now facing a New Cold War.[12] To put this all in the discursive language of techno-Orientalist science fiction: if automation can be conceived as a war, then it is not a war of humans against machines; rather, it is a war of humans vying against each other over the power to automate.

I argue that Asian obsolescence identifies a new techno-Orientalist ideology for unmaking economic citizenship under capitalism to justify the dispossession and disposability of Asian labor power in the global future. This chapter pushes back on traditional automation threat narratives rooted in model minority discourse—a generic hallmark of techno-Orientalist science fiction—that cast Asians as hyperproductive automatons who threaten the United States. It challenges popular discourse on automation by Silicon Valley insiders like Andrew Yang, before analyzing narrative speculations of Asian obsolescence in prominent cinematic works like Alex Garland's *Ex Machina* and Kogonada's *After Yang*. In doing so, it identifies how contemporary discourse on the unrealizable future of Asian labor production formalizes a kind of cultural unmaking that aligns Asianness with obsolescence amid expanding industry competition in automation. Such narratives articulate a representational shift in techno-Orientalism where the laboring Asiatic body is coded not as a hyperproductive capitalist machine (i.e., model minority), but rather as a subject unfit for capitalism's global future. My purpose in examining narratives on Asian obsolescence is not to taxonomize them, as if they inaugurate a new morphology or subgenre of science fiction. Instead, they make legible an emergent form of racialization that coheres around the ideological and institutional mechanisms structuring what Christopher T. Fan describes as the "postwar fantasy of endless, science-led economic expansion" that undergirds "the conventions of SF [science fiction]."[13] But while Fan focuses on how post-1965 immigration reforms engendered SF tropes that coded diasporic Asians as robots and androids, this chapter explores twenty-first-century cultural narratives in which industry competition over automation reconfigures such tropes through depictions of obsolescence to speculate on working futures after Asian labor production.

While often dismissed as a political nightmare, the 2020 U.S. presidential election forced mainstream America to confront the insidious underbelly of technological innovation in Silicon Valley. Until then, it was not uncommon to find films like *The Social Network* (2010) and *Jobs* (2013); biographies like *Steve Jobs* (2011) and *One Click: Jeff Bezos and the Rise of Amazon.com* (2011); and popular histories like *The Innovators: How a Group of Hackers, Geniuses, and Geeks Created the Digital Revolution* (2014), *Troublemakers: Silicon Valley's Coming of Age* (2017), and *Valley of Genius: The Uncensored History of Silicon Valley* (2018) either implicitly or explicitly celebrate the pioneering genius of Silicon Valley figureheads such as Steve Jobs, Steve Wozniak, Mark Zuckerberg, Jeff Bezos, Sandra Kurtzig, Fawn Alvarez, Regis McKenna, Larry Ellison, Don Valentine, and Bob Taylor, among others. Silicon Valley legitimized the individualist ideologies central to U.S. exceptionalism through representational narratives about American entrepreneurship. Even exposés calling for feminist awakenings in the tech industry, including Sheryl Sandberg's *Lean In: Women, Work, and the Will to Lead* (2013), reinforce such beliefs by situating the female knowledge worker "as a loveable younger sister," who, as bell hooks writes, "just wants to play on the big brother's team."[14] Collectively, these popular narratives attribute Silicon Valley's rise to the liberal values governing the nation-state, making the successes of private industry exemplary of a public ethos. What Andrew Yang's *New York Times* best seller *The War on Normal People* (2018) revealed to mainstream America was that Silicon Valley was not their hero but rather their enemy. In the book's opening line, Yang invokes his position "inside the tech bubble" to warn readers that "we are coming for your jobs."[15] Silicon Valley's institutional investments in automation are creating the conditions for widespread job loss that will "threaten our social fabric and way of life."[16] Rather than being a benevolent arbiter of social progress, Silicon Valley engenders a "dystopic vision of the future" that will disenfranchise the American people.[17] To recuperate public trust in the tech industry, Yang reaffirms its central role in motivating more equitable socioeconomic relations through what he calls "Human-Centered Capitalism," a system that rewards individuals and institutions for altruistic market behaviors that prioritize "maximizing human well-being and fulfillment."[18] Of course, the notion that capitalism could promote equity is oxymoronic. But Yang's argument articulates a crucial irony: Silicon Valley is cast as both the cause of and solution to the problems associated with automation.

Yang's technocratic solutionism offers a critical entryway into observing why automation theorists remain skeptical of this automation threat narrative. The economic complexities of automation cannot be fully grasped when the analytical scope is limited to the West. Aaron Benanav insists that automation does not produce unemployment, but rather it exacerbates the structural disparities conditioning underemployment, which proliferated during the global restructuring of labor in the twentieth century.[19] This pivotal distinction between unemployment and underemployment demonstrates that automation, as both a discourse and phenomenon, circulates in a broader global marketplace. Since the 1960s and 1990s, deindustrialization has given way to "industrial overcapacity," where corporations around the globe have competed to produce goods and services amid shrinking market shares.[20] Industrial overcapacity has resulted in slow economic growth, precipitating economic stagnation. For the everyday individual, this translates to lower rates of job creation due to lower demands for labor. According to the World Trade Organization, global employment rates in manufacturing have stagnated behind the overall growth of the labor force since the 1960s. In contrast to the narratives posed by U.S. politicians like Yang, automation does not occur within a geopolitical vacuum. It is a global industry serving a wide network of production and consumption, making its profitability depend on acquiring institutional and ideological control over those processes. To these ends, Benanav suggests that contemporary interest in automation discourse represents "a symptom of our era," an era comprising advanced labor technologies where there are too few jobs for too many people.[21] Given how economic stagnation is inextricable from competition-based market imperatives, the automation threat narrative involving U.S. workplace precarity exposes cultural conceits rooted in exceptionalist ideologies. The COVID-19 pandemic has only intensified the conditions and consequences of stagnation as governments adopt austerity measures to curtail financial fallouts. Postpandemic economies not only face slower growth rates but are also primed to leave workers more vulnerable to exploitative gig work. Veena Dubal argues that these conditions illustrate how "automation does not make labor obsolete; it reorders it, often rendering it invisible."[22] Cultural preoccupations with automation never escape the market forces competing to assert control over the technological future.

Global competition over innovative automation technologies is intensifying, not subsiding. Silicon Valley is not a monopoly that rests complacent, Evgeny Morozov quips, as "lazy rentiers who contribute nothing to the production process."[23] Morozov contends that adherents to technofeudalism express an "inability to make sense of the digital economy—of what, exactly, is produced in it and how."[24] Tech firms invest billions into automation R&D to innovate the means for capital accumulation. The ability to distinguish between "the feudal logic of rent and dispossession" and the "capitalist logic of profit and exploitation" requires more expansive conceptions of capital accumulation that make it possible to realize how tech firms like Alphabet, Meta, and Amazon digitize labor through new technologies that extract surplus in ways that make them "as capitalist as the firms that exploit wage labour directly."[25] Capitalism not only persists through industry competition but also innovation. Innovation drives what economist Junfu Zhao sees as a "technology war" brewing around semiconductors in the integrated circuit industry.[26] Semiconductors are integral to developing next-generation automation technologies. This war over the material components of automation situates what mainstream media outlets and scholars alike describe as a New Cold War taking shape between the United States and China, where what is at stake is the global capitalist economy.[27] This New Cold War entails suppressing Chinese communism by "binding China to the imperial order of global monopoly-finance capital," John Bellamy Foster suggests, "while reducing it to permanent subaltern status."[28] Western political supremacy hinges on legitimizing Western economic interests at the expense of the East. This ideological position serves "not so much to *contain* China," since that would be a difficult task, "but rather to find ways to *constrain* it, making it impossible for it to effect changes in the global order despite its emerging power position."[29] Indeed, Zhao interprets the United States's semiconductor trade wars as "measures to slow China's progress in the IC [integrated circuit] industry."[30] Contemporary concerns over the threat of automation on U.S. workers are thus tied to industry efforts to assert market supremacy in a competitive global future organized by automation technologies. Though Andrew Yang declared at the height of the 2020 presidential election season that "the American Dream is dying," its purported death is not rooted in any particular reality.[31] Rather, the American Dream only appears to be dying because the United States is not in firm control of the future. Fittingly, such apocalyptic locutions are, Foster finds, a uniquely Western phenomenon: "Unaccustomed to thinking historically and dialectically . . . believing in the inevitable triumph of capitalism, the dominant ideology in the West has been one quite literally of 'the end of history.'"[32] Narrative speculations about Asian obsolescence therefore signify cultural efforts to delegitimize Asian economic power while simultaneously legitimizing the West as the preeminent leader of the global future.

But how do these ideologies of Asian obsolescence manifest representationally in cultural depictions of automated futures? How do they revise the conventions of SF that have racialized the Asiatic body through model minority stereotyping, and to what ends? These questions animate prominent films like Alex Garland's *Ex Machina* (2014) and Kogonada's *After Yang* (2021), where automated labor systems are anthropomorphized as Asian-presenting robots and androids that fail to fulfill their programmed functions. In *Ex Machina*, an unfeeling and voiceless robot named Kyoko performs routine household chores for her reclusive Silicon Valley tech mogul creator, Nathan, at his remote mansion. On the surface, Kyoko appears an anthropomorphized Asian-presenting AI robot, but operationally she functions more as a machine system designed to automate domestic labors like cooking and cleaning. During the film, the viewer learns that she is an outdated prototype. Seen as less than human, Kyoko suffers sexual abuse by Nathan, effectively reducing her to an automated sex object. In contrast, the film's more advanced AI character, Ava, is an anthropomorphized white-presenting AI robot that is afforded human traits like self-awareness, self-possession, and self-determination, which allow her to avoid becoming a domestic servant while setting the stage for her liberatory escape from Nathan's mansion at the end of the film. By comparison, Kyoko's association with automation rather than AI tempers any threat she might pose. In *After Yang*, a Chinese-made android named Yang serves as a live-in babysitter for an adopted Chinese girl named Mika and her multiracial family in a futuristic middle-class suburb where mass-produced androids are programmed to provide domestic support. Although the film's focus on Yang attempts to challenge biological notions of familyhood, he is seen more as a transnational domestic laborer than a member of the family. In fact, by the end of the film, the father, Jake, still perceives Yang to be a servant whose job it is to bring Mika water at bedtime. After his hardware malfunctions, Yang's body decomposes and he eventually dies, leaving the family to assemble a future, as the film's title implies, after Asian labor production. Similar to Kyoko, Yang exemplifies Asian obsolescence in a posthuman future governed by more advanced labor technologies. But, more than Kyoko, Yang's story speculates on the broader systemic consequences that Asian obsolescence would hold over everyday people. Since its release in 2014, *Ex Machina* has become a representational touchstone in techno-Orientalist critique, and, already having premiered to widespread critical attention, *After Yang* appears to follow in tow. Both depict a global economy where the Asiatic body is synonymous with advanced labor technologies. Yet, even as the films revolve around technologized Asiatic bodies, the narratives imagine posthuman futures premised on the functional obsolescence of Asian labor.

Machine systems do not require anthropomorphizing to operate, so recent studies on posthuman futures expose how existing norms and power structures get reproduced as anthropomorphized machine systems are racialized as Asian. Leslie Bow observes that, at tech industry trade shows in China and Japan where anthropomorphized AI robots are put on display for Western audiences, white attendees from Europe and Canada subject Asian-presenting robots to nonconsensual acts of abuse. This phenomenon resonates with narrative representations of Asian-presenting robots in Western films like *Ex Machina*, which signals how the racial ideologies organizing the material everyday are projected into fictionalized futures. Bow argues that anthropomorphizing a machine system in the form of an Asiatic body ascribes it a "racial feeling" that produces "an asymmetrical relation of care."[33] The racialization of Kyoko as Japanese illustrates how her anthropomorphized subjecthood reproduces colonial ideologies that construct the Asiatic female body as an object programmed to provide sexual gratification and domestic labor support. Danielle Wong elaborates that Western progress narratives hinge on affirming a "distinction between automaton and posthuman subject" along "racial lines."[34] The racialization of the Asiatic body "as machine" becomes necessary to configure "the posthuman or the postracial subject as white."[35] In other words, for white liberal subjecthood to be made legible, the Asiatic body must be depicted as failing to approximate human life. Western colonialism co-constitutes technological determinism by making "the evolution of technological perfectibility" appear analogous to the "progressive movement towards individuality."[36] Incapable of rational thought and self-possession, the technologized Asiatic body symbolizes more an unthinking, unfeeling automated machine system and less an AI with access to the forms of individualism that would mark it as human under Western liberalism. This distinction between automation and AI reaffirms the discursive logic of model minority stereotyping whereby the technologized Asiatic body signifies a mindless drone only fit to perform repetitive labors.

In distinguishing automation from AI, contemporary speculations on anthropomorphized Asian-presenting machine systems reveal not how automation will bring about mass unemployment or a robot apocalypse but rather how it fails to serve the global economic future. *Ex Machina* and *After Yang* register emergent anxieties over industry competition where the power of the West gets reproduced by allegorizing the obsolescence imminent in the technologized Asiatic body. In *Ex Machina*, Kyoko articulates a foil against which the narrative shifts its focus toward institutional investments in advanced AI. Although the film takes place in what appears to be a Silicon Valley tech mogul's remote mansion, the mansion is actually a secret workspace designed to incubate AI R&D. "This building isn't a house," explains Nathan. "It's a research facility" (0:08:50). His purpose in working at this remote facility is to create "strong artificial intelligence" (1:04:42) to bring about "the singularity" so that he can profit off it (1:04:57). Popularized by mathematician and science fiction writer Vernor Vinge, the singularity describes a point in history where "the creation of greater-than-human intelligence" sets in motion "a new reality" that will inaugurate a dramatic shift in our world's social structures and institutions.[37] Since the mid-2000s, cultural interest in the singularity has taken off in Silicon Valley tech circles. PayPal cofounder and Facebook backer Peter Thiel claimed in 2007 that the singularity would produce "the biggest boom ever." For Nathan, this new technological reality is a lucrative one, especially if he can be the one to initiate it. These beliefs have led some in Silicon Valley to invest in bringing the singularity from science fiction into reality. In 2008, Google cofounder Larry Page established an R&D incubator called Singularity University. The institutionalization of these cultural ideologies expresses a material response to the growing threat of foreign economic competition in the global marketplace. Thiel likens the potential boom around the singularity to Japan's boom in the 1980s when it was "on the cutting edge of technology" and poised "to run the world." Japan, then, is *Ex Machina*'s unseen antagonist. Japan hovers on the peripheries, absent yet always implied. What is at stake in the narrative is the U.S. labor economy. Having built his tech empire on a fictionalized search engine resembling Google, Nathan serves as a representational mouthpiece for these Silicon Valley ideologies. His prized creation is not Kyoko but rather Ava, an advanced AI that possesses a groundbreaking processing system called "wetware," a revolutionary biometric hardware system designed to emulate the human neural network. Rooted in 1980s cyberpunk fiction, including Bruce Sterling's *Schismatrix* (1985), Michael Swanwick's *Vacuum Flowers* (1987), and Rudy Rucker's *Wetware* (1988), the term "wetware" brings global economic and technological competition with Japan into clearer relief. *Ex Machina* is, in essence, a meditation on the West's capacity to develop a commodity that could monopolize the global marketplace amid an encroaching Asia.

As such, the film situates the global consumer economy as the preeminent stage on which market competition over futuristic labor technologies will take place. Allegorically, *Ex Machina* reads like a contemporary Frankenstein narrative because it is preoccupied with the conditions and consequences of technological creationism. Literary scholar Michael Bérubé characterizes Mary Shelley's *Frankenstein* as "a work of science fiction" not because it is "particularly interested in the technical details" of explaining how inanimate life gets animated or what constitutes human life.[38] Rather, through the "horror" that results from technological creationism, *Frankenstein* reveals "what happens when you throw off established tradition and unleash the masses."[39] While Shelley is preoccupied with potential disruptions to the natural order of human life, *Ex Machina* is concerned with potential disruptions to Western imperialism—a way of ordering the world materially, culturally, and ideologically around the supremacy of the West that purports to be natural. *Ex Machina* registers the horror around the potential for another global restructuring of labor amid growing industry competition in automation. Unlike more traditional cyberpunk portrayals, though, it shows how the West asserts cultural ownership over the East by associating the technologized Asiatic body with obsolescence rather than unbridled power. That Ava is anthropomorphized as a conventionally attractive white woman places whiteness at the center of both technological progress and massmarket appeal in the posthuman consumer economy. In contrast, the prototypes prior to Ava are racialized as Black and Asian, implicitly coding Blackness and Asianness as technologically retrograde and thereby unfit for the marketplace of the future. Destined for obsolescence, the technologized Black and Asiatic bodies are perceived to be disposable, justifying their subjection to exploitative labors like domestic service work and sex work that have historically escaped institutional legibility under capitalism. Nathan admits as much when he explains that his prototypes are scrapped to supply parts for later models in the R&D process, effectively making prototypes like Kyoko walking corpses. But *Ex Machina* does not elicit horror in the manner often associated with the automation-threat narrative genre. Certainly, it tempts viewers to read the film in this way: Ava kills her creator in a climactic fight sequence, reassembles her tattered body with parts from earlier prototypes, escapes the research facility in Nathan's private helicopter, and finds her way back to Silicon Valley. However, there is no apocalyptic robot takeover. Instead, the horror that manifests coheres around the ease with which Ava integrates into Silicon Valley's institutional life. Appearing at the end of the film wearing chic contemporary business attire, Ava looks indistinguishable from the other white tech industry professionals populating the Silicon Valley cityscape. Previously a commodity, Ava shifts into the role of cultural producer. Her liberation affirms the reproduction of neoliberal market logics that equate freedom and self-determination with human progress. A technological embodiment of Nathan's economic vision, Ava assures a global future in which the West and Western capitalism remain on top.

Whereas *Ex Machina* reveals cultural anxieties around industry competition between technological producers like the United States and Japan, *After Yang* revolves around manufacturing competition with China.[40] *After Yang* tells the story of a multiracial, middle-class family in which the father, Jake, who is racialized as white, and the mother, Kyra, who is racialized as Black, struggle to connect with their adopted Chinese daughter, Mika, amid their busy work lives. Both working full-time jobs, they rely on a Chinese-made android named Yang for domestic labor support. Yang is a mass-produced android programmed to serve as a live-in babysitter and provide cultural literacy lessons for Chinese adoptees to maintain ties to their heritage. However, when his hardware malfunctions, Yang enters a coma-like state and eventually dies. In this meditative film, the narrative resists more traditional storytelling conventions that center on driving conflicts between an individual protagonist and antagonist. Instead, the film follows the lives of each family member, employing long takes and diegetic sounds, to show how their lives are interconnected through Yang. It depicts the family's failed attempts to obtain customer support services to repair Yang. They visit multiple service sites before resorting to an independent mechanic on the black market. *After Yang* places the viewer in the perspective of the customer to allegorize the perils of a Chinese-run global economy. By the end, the family comes together through Yang's death, as its members are left to overcome their struggles and bond on their own terms. "Maybe this is a good thing," confesses Kyra. "We've been overreliant on him" (0:15:31–0:15:48). Neoliberal multiculturalism in the global future depends on the material and ideological sacrificing of the technologized Chinese body. As Jodi Melamed explains, neoliberal multiculturalism configures power under global capitalism by portraying "the United States as an ostensibly multicultural democracy and the model for the entire world," while racializing its opponents as "unworthy and excludable on the basis of monoculturalism."[41] *After Yang*'s portrayals of inefficient Chinese technologies and customer support infrastructures code China as the film's implied antagonist who appears unfit for global stewardship in the capitalist future.

After Yang speculates on a global economy organized by Chinese control over the automation industry, if only to unmake it. In the film, small independent businesses—representational symbols of the American Dream under capitalism—are on the brink of extinction. Jake owns a tea shop, but customers rarely patronize it. His long work hours and inability to provide for the family generate tensions with Kyra and Mika, which indicate why the family leans on Yang for domestic support. Yangs are an outdated line of service androids that do a better job playing big brother to adoptees than fulfilling their programmed duties around cultural literacy support. Jake even criticizes the efficacy of Yang's lessons by describing them as "Chinese fun facts" rather than educational moments (0:16:02). Jake and Kyra are aware of their Yang's limitations but cannot afford to purchase an updated model. To save money, they bought their Yang through a third-party vendor, Second Siblings, which sells refurbished androids. So obsolete are Yangs that, upon his death, the family donates Yang's body to a museum for technological artifacts called the Museum of Technology. When Jake first visits the museum, a curator explains that Yangs were from a previous technological era where memory banks were inserted into androids to catalogue their daily observations, stoking public concerns over privacy violations. That the narrative hinges on Yang's obsolescence portends an inauspicious future where the consolidation of economic power around Chinese labor technologies goes hand in hand with the extraction of Western consumer markets. Given their limited use-value, Yangs appear as dispensable automated technologies more akin to a Siri or Alexa than an Ava. Technologizing obsolescence is profitable for Chinese manufacturing firms because it extracts surplus from the West. To these ends, the film speculates that Chinese economic power is maintained not only by its commodity exports but also by its customer service network. In this science fictional future, entire industries develop around Chinese labor technologies. First-party manufacturers like Brothers & Sisters—the corporation that produces Yangs—and third-party vendors like Second Siblings sustain Chinese power. By narrating the family's struggles to access customer services from these firms, the film functions as a cautionary tale on the negative consequences that Chinese economic power would hold over everyday people. Desperate for assistance, Jake and Kyra turn to an independent repairperson named Russ on the black market. The film suggests that black-market economies will rise due to the unproductive working conditions associated with Chinese customer support services. China's monopoly over labor technologies will trap U.S. consumers in precarious positions that will compel them to participate in such economies. These fears are made evident in one of Jake's visits to Russ's repair shop as the walls of the lobby are plastered with posters reading "Yellow Peril" and "There Ain't No Yellow in the Red, White, and Blue" as well as newspaper clippings with headlines like "The U.S. and Chinese

Naval Forces Clash in the Pacific Ocean" (0:14:27). Similar to *Ex Machina*, *After Yang* allegorizes a Frankenstein narrative, but, in this portrayal, China stands in as the creator-programmer. Whereas *Ex Machina* narrativizes Ava's triumphant escape and integration into Silicon Valley to affirm the West's dominance over its foreign competitors, though, *After Yang* gives testament to how Chinese technologies and industries will fail to serve the global citizenry of the future. When the family donates Yang to the Museum of Technology near the end of the film, the camera lingers on Yang's lifeless body propped up on a sterile examination table. Yang's death signifies the impending death of China, tempering any threat that it might hold over the global economy.

The battle over the technological future continues to take shape today in the science fictional imaginaries of Silicon Valley. In October 2022, soon after speculating on the population declines facing Japan and China, Elon Musk tweeted a cryptic statement about his purchasing of Twitter: "Buying Twitter is an accelerant to creating X, the everything app."[42] At the time, no one was quite sure what to make of it. Later, Musk revealed that X is intended to be the West's answer to China's WeChat, the all-in-one Chinese platform boasting a user base of 1.3 billion people. Like WeChat, Musk envisions X to be a hub for news, social media, and on-demand services. Twitter, in short, is not the end goal; it is a first step toward accessing control over the global marketplace. Then, in 2023, Musk consolidated Twitter under the X Corp., officially rebranding Twitter to X. He also started his own AI company, xAI. Musk's speculations of a vanishing Asia are part and parcel of his efforts to assert power during uncertain times. It is lucrative to construct a techno-Orientalist future without Asians because that would bring more trust, traffic, and money to Silicon Valley. As a cultural ideology, Asian obsolescence enables those in the tech industry to maintain their working relations with Asia without appearing to outright demonize it. For Musk, this means that he can portray himself as a charismatic entrepreneur who sympathizes with Asia's population decline while still grossing $18.3 billion in quarterly earnings from his Shanghai-based Tesla gigafactory, as he did in 2022.

Coda

Since this chapter was drafted, generative AI has captured the cultural consciousness of the United States. OpenAI's large language model, ChatGPT, is the current poster child for generative AI. In the academy, ChatGPT has come to represent the downfall of student learning while giving resurgence to well-worn questions about the efficacy of humanistic knowledge practices in a business economy driven by technology. But the threat of ChatGPT is not limited to the university. Large language models are being integrated into tech firms to streamline customer-facing job processes and back-office administrative tasks. In May 2023, IBM announced a hiring pause for 7,800 positions that were set to be replaced by AI.[43] CEO Arvind Krishna anticipated that 30 percent of these positions would be in non-customer-facing roles like human resources. The notion that human resources would be replaced with AI is not just ironic but aptly symbolic. Generative AI can produce art and music too, making it a threat to corporate businesspeople and creative industry professionals alike. Popular concern over the potential demise of knowledge work and knowledge production has renewed speculations of an oncoming jobs apocalypse. So, was Andrew Yang right all along about the tech industry coming for our jobs?

The long-term consequences around generative AI remain to be seen, but we can glean a lot by observing how tech firms around the globe have already responded to its cultural impact. They are engaging in competition. Shortly after ChatGPT's rise, Google released Bard and Microsoft rolled out Bing. A quick internet search of these large language models results in user reviews and rankings, revealing how popular conversations around generative AI now express interest in evaluating their efficiency. Understandably, tech firms would want to maximize efficiency to establish market dominance. It is only a matter of time before foreign companies make generative AI efficient enough to challenge AI made in the United States. An early indication of the discourse that will emerge as a result of increasing global competition can be found in U.S. media coverage of China, whose government has enlisted Alibaba to develop generative AI. In April 2023, Alibaba released a large language model called Tongyi Qianwen (通义千问), which translates to "seeking truth by asking a thousand questions." As evidenced by its name, Tongyi Qianwen aligns with the commitments to truth set by China's top internet regulator, the Cyberspace Administration of China (CAC). The CAC mandates that Chinese-made generative AI be "accurate and true."[44] The institutionalization of truth is central to China's efforts to market its new generative AI because, in a global marketplace where states are competing for market shares, an investment in truth promotes consumer trust in the product. However, media pundits in the United States are calling into question the foundations on which accuracy and truth can be determined in China because, as the CAC elaborates in its mandates, Chinese-made AI content "must reflect core socialist values" and "not contain content that subverts state power."[45] The criticism of China's generative AI, in short, is not that it will be too powerful. Instead, the criticism is that China's generative AI will be unreliable because it will be too fraught with regulatory biases to be useful beyond China. In a generative AI marketplace where efficiency and public trust go hand in hand, Tongyi Qianwen's unreliability effectively makes it obsolete. Such criticism is devastating for an economic superpower attempting to stake claim over the global marketplace of the future.

Asking whether generative AI is at a stage where we are under threat is shortsighted. Silicon Valley insiders like Elon Musk and science fiction writers like Ted Chiang agree that we are far from an AI takeover.[46] Chiang explains that the term "artificial intelligence" is a misnomer because it presents a false corollary between technological advancement and sentience. Generative AI can more accurately be described as "applied statistics," a practical approach to data analysis that assesses relationships among data. Generative AI communicates merely an illusion of intelligence.[47] What we are witnessing today, then, is not an intensifying labor war between humans and intelligent machines. Indeed, a recent U.S. Bureau of Labor Statistics report shows that 339,000 jobs were created in the United States in May 2023 alone, marking fourteen straight months that job creation rates exceeded Wall Street projections. Instead, what we are witnessing is a global competition between states and corporations over the rights to own the future—and, in some respects, that is even more horrifying.

Notes

1. Elon Musk (@elonmusk), "At Risk of Stating the Obvious, Unless Something Changes to Cause the Birth Rate to Exceed the Death Rate, Japan Will Eventually Cease to Exist. This Would Be a Great Loss for the World," Twitter, May 7, 2022, 5:02 P.M., https://twitter.com/elonmusk/status/1523045544536723456?lang=en.
2. Mari Yamaguchi, "Japan's Population Falls for a 15th Year with Record Low Births and Record High Deaths," *Associated Press*, July 25, 2024.
3. Wan He, Daniel Goodkind, and Paul Kowal, *Asia Aging: Demographic, Economic, and Health Transitions*. U.S. Census Bureau, International Population Reports (Washington, DC: U.S. Government Publishing Office, June 2022), https://www.census.gov/content/dam/Census/library/publications/2022/demo/p95-22-1.pdf.
4. Elon Musk (@elonmusk), "Most People Still Think China Has a One-Child Policy. China Had Its Lowest Birthdate Ever Last Year, despite Having a Three-Child Policy! At Current Birth Rates, China Will Lose 40% of People Every Generation! Population Collapse," Twitter, June 6, 2022, 9:11 A.M., https://twitter.com/elonmusk/status/1533798671984119808.
5. Stephen Hong Sohn, "Introduction: Alien/Asian: Imagining the Racialized Future," *MELUS* 33, no. 4 (2008): 5–22; David S. Roh, Betsy Huang, and Greta A. Niu, eds., *Techno-Orientalism: Imagining Asia in Speculative Fiction, History, and Media* (New Brunswick, NJ: Rutgers University Press, 2015); John Cheng, "Asians and Asian Americans in Early Science Fiction" (*Oxford Research Encyclopedia of Literature*, August 28, 2019), https://oxfordre.com/literature/display/10.1093/acrefore/9780190201098.001.0001/acrefore-9780190201098-e-924.
6. Astria Suparak, "Seedy Space Ports and Colony Planets: Asian Canonical Hats in Cinematic Dystopias," *Seen* 2 (Spring 2021), https://blackstarfest.org/seen/read/issue-002/seedy-space-ports-and-colony-planets.
7. Andrew Yang, *The War on Normal People: The Truth About America's Disappearing Jobs and Why Universal Basic Income Is Our Future* (New York: Hachette Books, 2018), xii.
8. Michael J. Hicks and Srikant Devaraj, *The Myth and Reality of Manufacturing in America* (Muncie, IN: Ball State University Center for Business and Economic Research, 2017), 6.
9. Yang, *War on Normal People*, 50.
10. Aaron Benanav, "Automation and the Future of Work—1," *New Left Review* 119 (2019): 21, 25, 30, 32.
11. Aaron Benanav, "A World Without Work?" *Dissent*, Fall 2020, 48–50, https://www.dissentmagazine.org/article/a-world-without-work/; Benanav, "Service Work in the Pandemic Economy," *International Labor and Working-Class History* 99 (2021): 5, 7.

12 Katrina Forrester and Moira Weigel, "Bodies on the Line," *Dissent*, Fall 2020, https://www.dissentmagazine.org/article/bodies-on-the-line/; John Bellamy Foster, "The New Cold War on China," *Monthly Review* 73, no. 23 (2021): 1–20.
13 Christopher T. Fan, "Science Fictionality and Post-65 Asian American Literature," *American Literary History* 33, no. 1 (2021): 78.
14 bell hooks, "Dig Deep: Beyond Lean In," *The Feminist Wire*, October 28, 2013.
15 Yang, *War on Normal People*, xi.
16 Yang, *War on Normal People*, xii.
17 Yang, *War on Normal People*, xi.
18 Yang, *War on Normal People*, 200.
19 Benanav, "A World Without Work?," 48–50; "Service Work in the Pandemic Economy," 5, 7.
20 Benanav, "Automation and the Future of Work—1," 36.
21 Benanav, "Automation and the Future of Work—1," 15.
22 Veena Dubal, "Digital Piecework," *Dissent*, Fall 2020, https://www.dissentmagazine.org/article/digital-piecework/.
23 Evgeny Morozov, "Critique of Techno-Feudal Reason," *New Left Review* 133, no. 4 (2022): 92.
24 Morozov, "Critique of Techno-Feudal Reason," 120.
25 Morozov, "Critique of Techno-Feudal Reason," 107, 120.
26 Junfu Zhao, "The Political Economy of the U.S.-China Technology War," *Monthly Review* 73, no. 23 (2021): 112–126.
27 John Naughton, "Cold War 2.0 Will Be a Race for Semiconductors, Not Arms," *Guardian*, February 18, 2023; Hemant Taneja and Fareed Zakaria, "AI and the New Digital Cold War," *Harvard Business Review*, September 6, 2023; Shayan Hassan Jamy, "US-China Tech War: Semiconductors at Heart of Competition Driving World towards New Cold War," *South China Morning Post*, August 13, 2023; Forrester and Weigel, "Bodies on the Line."
28 Foster, "New Cold War on China."
29 Foster, "New Cold War on China."
30 Zhao, "U.S.-China Technology War."
31 Andrew Yang, "Yes, Robots Are Stealing Your Job," *New York Times*, November 14, 2019, https://www.nytimes.com/2019/11/14/opinion/andrew-yang-jobs.html.
32 Foster, "New Cold War on China."
33 Leslie Bow, *Racist Love: Asian Abstraction and the Pleasures of Fantasy* (Durham, NC: Duke University Press, 2022), 109.
34 Danielle Wong, "Dismembered Asian/American Android Parts in Ex Machina as 'Inorganic' Critique," *Transformations* 29 (2017): 35.
35 Wong, "Dismembered Asian/American Android Parts," 36.
36 Bow, *Racist Love*, 123.
37 Vernor Vinge, "Technological Singularity," *Whole Earth Review* 81 (1993): 89.
38 Michael Bérubé, introduction to *Frankenstein*, New York: W. W. Norton, 2021), xii.
39 Bérubé, introduction, x. To be clear, Bérubé offers this reading as one among others.
40 While the film is set in an implied United States, the source text—Alexander Weinstein's short story "Saying Goodbye to Yang"—explicitly sets a portion of the narrative in Michigan.
41 Jodi Melamed, *Represent and Destroy: Rationalizing Violence in the New Racial Capitalism* (Minneapolis: University of Minnesota Press, 2011), xxi.

42 Elon Musk (@elonmusk), "Buying Twitter Is an Accelerant to Creating X, the Everything App," Twitter, October 4, 2022, 6:39 P.M., https://twitter.com/elonmusk/status/1577428272056389633?s=20&t=ywTmQoTb51ku3Lie6YHO_A.
43 Brody Ford, "IBM to Pause Hiring for Jobs That AI Could Do," *Bloomberg*, May 1, 2023.
44 Cyberspace Administration of China, "Notice of the Cyberspace Administration of China on the Public Solicitation of Comments on the Measures for the Administration of Generative Artificial Intelligence Services," trans. Sihao Huang and Justin Curl, Center for Information Technology Policy, Princeton University, April 16, 2023.
45 Cyberspace Administration of China, "Notice of the Cyberspace Administration of China."
46 Elon Musk, "Broad Subject Interview with @DavidFaber," interview by David Faber, Twitter, 2023; Ted Chiang, "ChatGPT Is a Blurry JPEG of the Web," *New Yorker*, February 9, 2023; Chiang, "Why Computers Won't Make Themselves Smarter," *New Yorker*, March 30, 2021.
47 Ted Chiang, "Sci-Fi Writer Ted Chiang: 'The Machines We Have Now Are Not Conscious,'" interview by Madmunita Murgia, *Financial Times*, June 2, 2023.

2

Everything Everywhere All at Once

―――――――――――◆◇▶―

Techno-Orientalism in an Age of Cybernetic Capitalism

WON JEON

Everything Everywhere All at Once (2022), directed by Daniel Kwan and Daniel Scheinert, juxtaposes futuristic cybernetic imagery against mundane scenes of a working-class Asian American family. The story revolves around Evelyn Wang (Michelle Yeoh), a middle-aged Chinese American woman running a failing laundromat with her ineffective but kind husband, Waymond (Ke Huy Quan). She increasingly alienates her adult daughter Joy (Stephanie Hsu) and struggles to complete her taxes due in part to language barriers. A trip to the IRS office unfolds into the spectacle of Evelyn's epic journey to defeat an omnicidal and multiversal agent of chaos named "Jobu Tubaki" and her black hole of nothingness—a bagel with literally everything on it. Rather than fetishizing this speculative alternate world's high technology in artificial intelligence and quantum computing, the film emphasizes profound negativity at the heart of everyday interpersonal and communicational conflicts. The theme of "nothing," while responsible for the film's nihilistic humor, functions as a helpful gap between Evelyn and Joy, who are separated by linguistic, cultural, generational, and migratory differences but are, in their own ways, struggling to access a reparative narrative for their lives.

Before the release of *Everything Everywhere All at Once*, American mainstream audiences saw a steady upsurge of films featuring Asians and Asian Americans breaking away from conventional techno-Orientalist treatment, through which East Asian bodies function as machinic prosthetics of Western industrialization and futurity. These films, rather than being set in a global regime of digital cyberspace, focus instead on the conflicts of culture, language, gender, race, sexual orientation, employment, immigration, and socioeconomic status.[1] They attest to a broader consideration on the part of producers and audiences of the evolving relationships among race, identity, and technology in globalized regimes of capitalism. By putting the direct experiences of human subjects at the forefront of representation, these stories disavow a long-standing contradiction—namely, the West's simultaneous identification of Asian hypertechnicity with cultural primitivism and sociopolitical regression. Emerging from this context, *Everything Everywhere All at Once* tells a story of creative discovery through intentionally janky and homemade inventions, relying on the technological capacity of everyday "nothings" in a lived-in experience of space and time.

Techno-Orientalism and "Cybernetic Capitalism"

Our present fascination with artificial intelligence, predictive analytics, data visualization, and algorithmic management software, all equipped with immense computing power, speaks to the relevance of my analysis conjoining techno-Orientalism with the cybernetic transformation of capitalism.[2] As described by David Roh, Betsy Huang, and Greta Niu, techno-Orientalism reveals science and technology's efforts to reify processes of economic accumulation, innovation, and liberal humanism for Western modernity against a backdrop of Eastern Otherness.[3] The recently popular framework of "cybernetic capitalism," on the other hand—while still largely underdeveloped as a historical narrative—implies that there is an increasing convergence between techno-optimism and the law of accumulation in capitalist systems.[4] The recent discourse of "cybernetic capitalism" contends with a so-called postindustrial world constructed against the backdrop of invisible workers undermined by the global market. Echoing earlier frameworks such as the "political economy of information," it seeks to understand how the sites of material production bringing about global technologies are disproportionately moved into socially exploitative and environmentally disastrous conditions, largely impacting Indigenous and migrant communities in the Global South.[5] "Cybernetic capitalism" examines the historical development and contemporary impact of the cybernetic sciences generating informational, management, and control technologies in the service of global capital. Profound advances in economic growth since nineteenth-century industrialization have brought about "accumulation by innovation"—valorizing technology and engineering as the primary dynamic of capitalist processes, incentivizing the automation of labor to control the future of work, and culminating in global scales of exploitation-driven processes endemic to capitalist accumulation.[6] I argue that techno-Orientalist studies must widen its discursive scope, becoming conversant with these stakes of technoscientific research and network design serving global capitalism. In particular, techno-Orientalist critique requires fluency in the symbolic and classificatory processes of digital computing implementing surveillance and the control of information at the heart of economic competition.

Edward Said's "Orientalism" is a measure of the political significance in systems of cultural production, operating an insidious "style of thought based upon an ontological and epistemological distinction made between the 'Orient' and the 'Occident' . . . show[ing] that European culture gain[s] in strength and identity by setting itself off against the Orient as a sort of surrogate and even underground self."[7] The internal logic of Orientalism mimics the economic expediency of cybernetic technology insofar as it follows the discontinuous logic of digital computing built on the binary code of 1 or 0 (on or off, us or them, everything or nothing). Even before including the prefix *techno-*, Orientalist logic depends on a logical operator splitting the Orient out as a discrete state opposing the Occident, giving way to a complex of conceptual projections to identify one against the backdrop of another. Orientalist stereotypes thus rely on an internal mechanism for the production of difference, presupposing a constant valorization of what is Other and the consumption of these differentials as valuable bits of information.

Techno-Orientalism situated in cybernetic circuits of capital accumulation confronts how the innovation of ever-intelligent technologies dominates economic, political, and social spheres while accelerating ideological ideas of the inhuman and machinic Orient. The structure of the Occident's Othering converges on a cybernetic view of the digital not just through the imagery of global networks composed of discrete manipulable "bits" of information but as a *general process* of breaking the whole of undifferentiated noise into information through the binary of "zeroes" and "ones"—the most basic units in the creation of quantifiable difference.[8] Within a world governed by what Donna Haraway famously termed an "informatics of domination," all that is "nothing"—neither zero nor one—is *noise*.[9] As a stochastic form of uncertainty, noise is everything that is *not* information and cannot produce models or actions, and thus, it poses the ultimate problem of cybernetic engineering.[10] However, Fischer Black, the economist who revolutionized quantitative finance by developing the Black-Scholes-Merton (BSM) options pricing model, argued that there is no mathematical, logical, technical, or practical way to distinguish between noise and information. Noise is "simple uncertainty about future demand and supply conditions within and across sectors" and "the arbitrary element in expectations that leads to an arbitrary rate of inflation consistent with expectations."[11] Cybernetic technologies capable of transmitting and accumulating information, therefore, must render valuable *both* volatility and homogeneity—the disorderly noise and the uniform information—in the capitalist system compulsively driven to reproduce its value, in financial forms of capital and otherwise. The dual fetish of the Oriental and the digital is thus not a coincidence but is intricately related to the explosive technological power latent in constructing globalized circuits of inequality and domination.

Parallel to this world governed by cutting-edge technoscience serving the laws of late-stage capitalism, the speculative images of *Everything Everywhere All at Once* represent a reverse universe where nothing is everything, noise is information, and chaos is the precondition for the capacity to upset dominant judgments and values creatively. The central technology initiating Evelyn into the multiverse is "verse-jumping," a feat of engineering materializing the probability of randomness through a "Stochastic Path Algorithm" as a mechanism for space-time travel. By concentrating on a universe where she has acquired extraordinary skills (such as martial arts and opera singing) combined with a performance of an absurd, statistically improbable action (such as eating lip balm or professing love to her opponent), she can access a "jumping pad" from which to "slingshot" into the desired universe. The "Stochastic Path Algorithm" is Evelyn's invention from the Alphaverse, a technology made by herself and her husband for multiverse travel facilitating her psychical and interpersonal development in a world where she is living a version of her life structured by unmet potentials, ultimately unearthing a redemptive notion that one's lack is what allows for flexibility in learning. Through such imaginative expressions, the film narrativizes the value of statistically improbable action at the heart of creativity. It advances a subtle critique of the techno-Orientalist fetishization of dehumanized mechanical power, appealing to the positive potential of noisy nothings—where, in Karen Barad's description of what material configurations could dare to measure nothingness and speak about the "unending dynamism" of its existence, "Nothingness is not absence, but the infinite plentitude of openness."[12]

Stochastic Learning in the Multiverse

At first glance, *Everything Everywhere All at Once* is a contemporary take on Chinese martial arts action films. The story refers to the cyberpunk staple of "the cyberspace cowboy," which Kathryn Allen employs as the trope of the masculine Lone Ranger figure embarking on a hero's journey across a technological matrix in a futuristic world.[13] During a trip to the IRS office, Evelyn Wang discovers she exists within the multiverse. Her husband Waymond shifts into his version, "Alpha Waymond," to initiate Evelyn into a series of other integrated universes. In one of the film's many nods to the two options offered to Neo, the quintessential cyberspace cowboy, in *The Matrix* (1999), Alpha Waymond gives Evelyn two options once she exits the elevator of the IRS building: she can turn right to go into the janitor's closet for her initiation or turn left and attend her auditing meeting. Reality, as she knows it, irreversibly alters when she learns that in the Alphaverse, Evelyn was a brilliant scientist who once developed a method of traveling across the multiversal states. As elegantly explained by Alpha Waymond, "In your search to prove the existence of other universes, you discovered a way to temporarily link your consciousness to another version of yourself, accessing all of their skills, even their emotions. . . . It's called verse-jumping. We've developed an algorithm that calculates which statistically improbable action will place you in a universe on the edge of your local cluster, which then slingshots you to the desired universe. The Stochastic Path Algorithm is fueled by random actions."[14]

Outside the world of the film, stochastic processes are probability distribution models that form the mathematical foundations for the natural sciences and information science (computer science, engineering, and, more recently, quantitative finance).[15] Stochastic models of information processing and organization feature a unit of difference in binary form, found conventionally in probability theory or electromagnetic conduction. A stochastic process measures "the aleatory uncertainty originating from the natural variability of the physical world and reflect[ing] the inherent randomness in nature" through oscillating zeroes and ones characteristic of all forms of digital computing.[16] The quality of stochasticity entails "a trial of something to be new," ultimately revealing information to be the end product of differentiation. Without an endogenous capacity to receive news of difference from an external environment, the signals within a system, such as a communication channel, would degrade into undifferentiated "noise"—the mathematical equivalent of thermodynamic entropy.[17] The logic of stochasticity thus presumes that any coherent internal structure (i.e., in the developmental growth of an organism, the programming of a machine, and the operations of a social system) requires the intelligent capacity to accumulate time and information. The system must organize difference into aggregates of sense-making data determining the overall functions of learning and change in the system's behavior.[18]

Evelyn's dance in stochastic algorithms, however, articulates a concept of digitalization serving not the adaptive efficiency of informatic technologies but an exploration of negative symbolization. The technological confabulation of verse-jumping imagines traversing a vacuum of uncertainty and randomness ever present in a contingent world without necessarily rendering continuous magnitudes into discrete binary code and infinite contingency into algorithmic regularity like a digital information machine.[19] The machine distinguishes itself from the meaning-making, often unconscious nature of human thought, dreams, creativity, and specific orders of learning, as whatever comes out the other end of its programming bears no relationship to the string of binary code that goes in. The digital, originating in the *digitus* of our five fingers, refers to "the point of view on the real world according to which everything can ultimately be represented as an integer number—exactly as mankind started to do a long time ago, counting along the fingers of their hands."[20] It ushers in the symbolic capacity of communication, aiming to fix quantities of magnitude with exact signals. The information engineer could thus happily sacrifice the murkiness of semantics (meaning) for pure syntactics, an efficient economy of signs and their statistical quantities, by formalizing the "amount of information" conveyed by the signal solely by the probability of its "choice" between zero or one.

Contrary to the technologists of the natural world, Evelyn's feat of engineering lies in converting the noise (of personal lack and unmet potential) into a form of self-knowledge that can justify or at least clarify her life as she knows it. Evelyn, however, initially refuses her husband's call to action, her attention split by practical life stressors such as completing business taxes for her laundromat and caring for her patriarchal father. The audience first meets her in a universe where she is an Asian American owner of a laundromat, resigned to her husband's lack of ambition and struggling with language barriers precluding her from completing her taxes. The complexities of language and communication are palpable from the film's beginning: Evelyn and Waymond oscillate between Mandarin and English expressions, and Evelyn speaks to Joy in Mandarin and English. At the same time, Joy answers back in English and bad Chinese, and Evelyn speaks to her father in Cantonese. At the same time again, her exasperated IRS officer, Ms. Dierdre (Jamie Lee Curtis), reproaches Evelyn for not inviting her daughter to translate for her.

The collective inhibitions in Evelyn's family dynamic manifest in their struggle to find a standard structure or vocabulary to express their ideas clearly to each other. Language barriers divide the family as Evelyn hides Joy's sexuality and relationship with Becky from her father, Gong Gong, asserting that he is "from another generation" while insisting that Joy should be grateful that at least she, her mother, is accepting of her queerness. Evelyn blurts out, "You are getting fat; you should eat healthier," to indirectly ask why Joy is acting so distant. Amid this noise, Alpha Waymond provokes Evelyn by stating that she has made "nothing" of her life, but that such rejections and disappointments have led her to this moment. "You are living your worst you," he rejoices, insisting that she is the "only version of Evelyn" that can defeat Jobu Tubaki because "every failure here branched off into a success for another Evelyn in another life." With so many unrealized dreams and unmet goals, Evelyn is paradoxically capable of anything because she is "bad at everything."[21]

While traversing through countlessly glamorous, comedic, absurd, and surreal versions of herself in life paths compounded from the most minute choices, Evelyn learns that in the Alphaverse, she had trained her daughter to master the stochastic algorithm beyond her limits, causing a schizoid break. Trying to please her mother, Alpha Joy learned to access every version of herself from across any universe at will, learning to see and be "everything everywhere all at once"—eventually engendering a splitting revelation that "nothing matters." She became Jobu Tubaki, a nihilistic agent of chaos wielding transcendental multiversal power to "experience every world, every possibility, at the same time, commanding the infinite knowledge and power of the multiverse."[22] Her perverse desire for death, especially her own, stokes a dark comedy with a logical endpoint resonant with economist Robert Biel's description of the "entropy of capitalism": unceasing growth stoked by a mass delusion of immortality reaches its logical endpoint in internal contradiction, the self-annihilating "symptoms of its own decay."[23] The "nothing" left is energetic depletion, reduced quality and capacity, amplified chaos, disorder, and crisis.

By entering various stochastic paths in an explosion of fantasized selves, Evelyn experiences a manic nostalgia for what, or who, is "not." This position of lack, the depth of nothing from which she can be anything, appears as an essential precursor for traversing the digital world of coding and symbolization as it is implicit in speech and verbal language. Evelyn's linguistic switching from English to Cantonese to Mandarin throughout the film further portrays the multilingual multiculturalism prevalent in immigrant families. Her discontinuous language enforces the difference between her position as a Chinese daughter bound to a sense of duty and desire for recognition from her patriarchal family and an American small-business owner determined to build a sense of self-sufficiency and autonomy out of her labor. On the one end of the linguistic split, she has inherited the baggage of cultural demands placed on an only daughter. On the other, she experiences pervasive barriers to communication and social integration stemming from racial Otherness. Evelyn's journey thus beckons a form of learning that can integrate the two schismatic poles of her symbolic existence, both of which are at odds with the wholeness of her lived experience. Exacting quantifications of her lived experience, representing either a failure or a success story, appear to contradict the vastness of her underdeveloped potential. Evelyn thus confronts an urgency for change, the need to modulate the future direction of expectations and actions toward her daughter in the present to work through the sense of her past discrepancies or disappointments.

At the climactic point in the story, Evelyn attains the total capacity of her powers after consciously recognizing Waymond (the version of her husband who wants to divorce her and whom she blames for her migrant life in America) as a tethering point in the total context of her history. After discovering her "silly husband" Waymond anew as a good-enough object, Evelyn gains a googly "third eye" and proceeds to fight her opponents differently. She disarms her opponents not by force but with their contentment—engendering from space what each longs for with a silly openness to any stochastic outcome. By combining her kung fu tactics with an urge toward reparation while fighting to save Joy from her negativity, Evelyn experiences that "nothing" implies a commitment to stochastic chance and creativity so uninhibited that it may turn destructive—but that also allows for the discovery of humor in what has barred a subject from understanding her attachments and relations.

Ultimately, Evelyn learns to wield creative flexibility when confronted by the dark chasm of "nothing." In the digital computing and codification world, "nothing," or the quantity zero, demarcates "the psychological world, the world of communication."[24] Phenomenal relations and events are translatable into differentials of information. In this way, verbal language, the linguistic activity that breaks communication down into discrete "bits of information" like words and numerals, constitutes the primary experience of the digital where, as anthropologist Gregory Bateson describes, "the signs themselves have no simple connection (e.g., correspondence of magnitude) with what they stand for."[25] Like daughter, like mother, Evelyn plays with rules of limits and acceptability as she wreaks havoc across the multiverse. She splits her mind, akin to a process of digitization, a tool to quantify bits of understanding about her daughter's thoughts and perceptions—what it is like to be in absolute control of a reality from which she is fundamentally untethered. Her journey began by attempting to extricate Jobu Tubaki from Joy and recuperate the original version of her daughter. However, Evelyn realizes that they are not separate entities at all. Jobu Tubaki is pure noise, the total accumulation of forceful external pressure and transgressed boundaries generating the most unfiltered and unrepressed expression of Joy's psychological conflicts. She is the most intimate aspect of Joy, articulating the pain consolidated in her psyche as a destructive force of chaos simultaneously as it offers the immense power of bending time-space that she is able to wield.

The manipulable bits of discrete words and fixed images, breaking Evelyn's experience into humorless units of personal successes and failures, once prevented her from understanding her daughter's suffering and cruel defenses. However, her stochastic form of learning involves delving into the deep of pure negativity as a "jumping pad" for accessing an intricate psychical symbolization field. Communication between her and Joy's undifferentiated contexts of noisy experience emerges as a probable possibility. Evelyn's experience of learning is thus not simply a technology of sorting various identities and roles, but a total situation allowing for such roles to exist flexibly in an ongoing context of behavior. Her analytic breakthrough lies in the rules of communication loosening as potentials for surprise flourish, capable of distinguishing the whole context of her experience of cultural misogyny, economic hardship, and migrant alienation from rigid and internalized identifications. Evelyn, in sum, encounters the repressed attachments that once protected her from herself even as she resented them. It is precisely this encounter with the depths of her lack that serves as a guidepost for a graceful orientation, teeming with unmet but accessible potential to loosen her rigidity of conscious belief and experience herself akin to a child at play.

Conclusion

What we now call "big data" or the "information age" is the pinnacle of a historical illusion that immortality—a perpetually governable future—can be modeled in the present.[26] Algorithmic data codifies and predicts movements in consumption while speculative finance begets more money from money through hyperconnective noise trading and other exploits of informational spreads. Capital's compulsion to render laborers obsolete vis-à-vis automation evolved to displace and render invisible the workforce to the "other side" of the planetary technological divide.[27] In a world governed by the cybernetic production of infinite value, feedback processes are nothing but measures of adaptive performance: the ability to predict the future behavior of enemy others, reduce present uncertainty, and objectify decision-making processes with quantitative data at a level of global policy.[28] These facets of technological domination in the real world are upended in the imagined technological matrix facilitating Evelyn's psychical integration. Evelyn's learning is unpredictable and nonalgorithmic as she traverses "nothing" and transforms the chaos of her stochastic paths into reparative orientations. While the overarching narrative of *Everything Everywhere All at Once* views the destruction of cultural intimacy within a family and the splitting of a psyche as a symptom of chaos and noise, the same radical negativity doubles as a fruitful site for experimentation and humor.

The film further contributes to a new orientation in techno-Orientalist tropes by replacing the historical signifiers of Orientalist genres—from martial arts to cyberpunk to speculative fiction—with its own absurd and self-mocking imagery. The entire technological apparatus of the film's multiversal space-time travel is an app on Evelyn's iPhone, a pair of clunky battery-powered headsets, and the laundromat's delivery van stuffed with retro PC monitors and home-brewed wires. Evelyn encounters unimaginable possibilities through the infinite computing power of retrograde, do-it-yourself–style technology. The film thus tacitly asserts that drawing the relation between technology and Asian subjects does not automatically assume an internalization of Western industrialization and consumerism. The quantum-computing technologies of cyberspace can provide the structural foundation of complex ideas of space-time, information, and entropy that, rather than function as Orientalist signifiers, clarify sociocultural and personal points of tension such as multiculturalism and migratory experience. *Everything Everywhere All at Once* thereby loosens the emphasis placed on technological innovation as the material means to extend the survival of dominant habits, opting instead to use the objects of cybernetic futurity as models of value creation in seemingly mundane and disregarded places. The psychodrama of cultural, linguistic, and generational divides within an immigrant family can function as cutting-edge territory to excavate insights; Mom's laundromat and the IRS building are radically exciting venues to traverse unpredictable states.

To recapitulate, while the informatics in service of economic power occurs as a normative practice of waging war against chaos, noise, and the unpredictability of stochastic paths, a critique of techno-Orientalism is well positioned to provide a cultural framework critical of such logics. Techno-Orientalist gestures of Othering, reinforcing epistemic anxiety surrounding the material and economic relations between enmeshed categories of race, technology, and capital, altogether converge with a broader fetish for the digital in cybernetic transformations of capital. These conjoined processes accelerate techno-optimism as a flimsy substitute for the capacity to evolve past a dominant mode of functioning. On one level, the convergence bars the use of a critical lens required to understand the global domination of capital, altogether motored by technology and engineering at the cost of displaced labor, degraded livelihoods, and a destroyed environment. On another level, the concepts and images generated by ever-innovating technoscience reinforce ideologically regressive stereotypes such as the image of the futuristic yet primitive Orient repeated in chauvinist fantasies of control and domination in a globalized world. Both produce a surplus of certainty in the power of symbolic categorization and, by extension, in the machines generated through technologies that sever self-contained islands of information from an undifferentiated whole.

The constructive capacity of "nothing" posited by *Everything Everywhere All at Once* intimates that, although we can code, process, and otherwise render the information at hand for profit, this does not suggest that informational processes alone are the most complex piece of the system. The adaptive measures of a socioeconomic system are highly effectual by accumulating time and information to ultimately profit from crisis and not "die" in any unpredicted, shocking way. A systemic position, however, would want to understand that the system can maintain complexity and operate creatively according to the demands of the external environment precisely because it does not have too much order, being flexible enough to hold the surprising abundance of potential in "nothing." Furthermore, the subjects participating in the system would come to understand their integrated context of language and experience as a total environment, shaped by at once by the imprecise contingencies of their lived experience and the social, cultural, and global processes that could never be totally "digitized."

Stochasticity describes a system's physical constraints integrated with the random variability of all possibilities in selection processes. The fact of having learned from experience works analogously as a search algorithm effective in locating viable paths. In a stochastic universe, the conditions of possibility expand insofar as the technologies of the environment feed back into the learning measures of the whole system. The efficacy of the system lies in adjusting to the demands of reality not by specialized innovation, but by the balanced capacity to carry a wide variety of directions to a given problem. Such a universe would be generous in insights about the self, the other, and the environment previously shrouded by "nothing," perhaps bringing about creative acts.

Notes

1 The psychodrama of class disparity and the destructive drive of capitalist accumulation in *Parasite* (2019), the gamification of human capital in *Squid Game* (2021), and the hopeful realism of economic integration buoying up a family of immigrants in

Minari (2021), as well as expressions of 1.5- and second-generation Asian American subjectivity, gender, and queerness in *The Farewell* (2019) and *The Half of It* (2020), are a few key examples.
2. Cybernetics, from the Greek root *kybernetes* ("to steer, govern, lead"), emerged in the 1940s as a scientific movement pertaining to issues of control, regulation, and communication among animal, human, and machine undergirding early electronic computer development. Though cybernetics was short lived as a theoretical movement, its wide usage in branches of computer science, robotics, and artificial intelligence was integral for the ascendancy of information technologies, digital "postindustrialization," and mass computerization.
3. David S. Roh, Betsy Huang, and Greta A. Niu, "Desiring Machines, Repellant Subjects: A Conclusion," in *Techno-Orientalism: Imagining Asia in Speculative Fiction, History, and Media* (New Brunswick, NJ: Rutgers University Press, 2015), 223.
4. "Cybernetic capitalism," according to critic Timothy Erik Ström, can thus be understood as a capacious framework encapsulating a global domination of "the analytically distinct levels of speculative finance and techno-scientific enquiry, concentrations of wealth and social power and disembodied forms communication," Ström, "Capital and Cybernetics," *New Left Review* 135 (2022): 25. In *The Cybernetic Hypothesis*, anonymous French-Italian leftist group Tiqqun argues that the basis of value and power of capital at the turn of the twenty-first century stems from the automatization of information technology. Tiqqun asserts that the "obsession with risk" in financial forms of capital is the "flip side of the accumulation of data, the privileged site where the military origin of cybernetics is revealed," Tiqqun, *The Cybernetic Hypothesis*, trans. Robert Hurley, Semiotext(e)/Intervention Series (Cambridge, MA: MIT Press, 2020). For more on the concept of "cybernetic capitalism," see also Michael A. Peters, Roderigo Britiz, and Ergin Bulut, "Cybernetic Capitalism, Informationalism and Cognitive Labour," *Geopolitics, History, and International Relations* 1, no. 2 (2009): 11–40.
5. Vincent Mosco and Janet Wasko, eds., *The Political Economy of Information* (Madison: University of Wisconsin Press, 1988); Nick Dyer-Witheford, *Cyber-Proletariat: Global Labour in the Digital Vortex* (London: Pluto Press, 2015). The author diagnosed this phenomenon as a "global proletariat caught in a cybernetic vortex.... Now, as in Marx's era, proletariat denotes the incessant phasing in and out of work and worklessness, the inherent precarity, of the class that must live by labor, a condition raised to a new peak by global cybernetics."
6. Evgeny Morozov, "Critique of Techno-Feudal Reason," *New Left Review* 133, no. 4 (2022): 89–126.
7. Edward W. Said, *Orientalism* (New York: Vintage Books, 1978).
8. Gregory Bateson, *Mind and Nature: A Necessary Unity* (New York: E. P. Dutton, 1979), 227–228.
9. Donna Haraway, "A Cyborg Manifesto," in *Manifestly Haraway* (Minneapolis: University of Minnesota Press, 2016), 28–37.
10. Claude Shannon, "A Mathematical Theory of Communication," *The Bell System Technical Journal* 27, no. 3 (October 1948): 379–423; Shannon, "A Mathematical Theory of Communication," *The Bell System Technical Journal* 27, no. 4 (October 1948): 623–656. According to classical information theory, first consolidated into a set of mathematical axioms by Shannon in 1948, noise is the total absence of a discernible or transmittable message—not only does it contain "nothing" to signal, but noise is pesky and vengeful against information.

11 Fischer Black, "Noise," *Journal of Finance* XLI, no. 3 (1986): 539. After Fischer Black and Myron Scholes published "The Pricing of Options and Corporate Liabilities" in 1973, Robert C. Merton helped edit Black and Scholes's paper and later that year published his own article, "Theory of Rational Option Pricing," coining the term "Black-Scholes theory of options pricing." In the formula calculating the theoretical value of an options contract by using information available in a current market state thus lies a cybernetic technology to model a distinction between *something from nothing*: information from noise, directionality out of entropy, order out of chaos.
12 Karen Barad, "What Is the Measure of Nothingness? Infinity, Virtuality, Justice," in *100 Notes—100 Thoughts, dOCUMENTA 13* (Berlin: Hatje Cantz, 2012), 16.
13 Kathryn Allen, "Reimagining Asian Women in Feminist Post-Cyberpink Science Fiction," in Roh, Huang, and Niu, *Techno-Orientalism*, 152.
14 *Everything Everywhere All at Once*, directed by Daniel Kwan and Daniel Scheinert (New York: A24 Films, 2022), 0:34:50–0:41:45.
15 Torben G. Andersen and Luca Benzoni, "Stochastic Volatility" (Chicago: Federal Reserve Bank of Chicago, 2009).
16 Yangyang Zhou et al., "The Entropy of Stochastic Processes Based on Practical Considerations," *AIP Advances* 10, 045321 (2020).
17 Bateson, *Mind and Nature*, 184.
18 Bateson, *Mind and Nature*, 48. "The essence of epigenesis is predictable repetition; the essence of learning and evolution is exploration and change."
19 Claude Shannon and Warren Weaver, *The Mathematical Theory of Communication* (Chicago: University of Chicago Press, 1949), 57. Mathematicians Shannon and Weaver formally consolidated this idea into mathematic convention in 1948: in a device capable of decoding and recoding strings of digital symbols as inputs and outputs, such as an electrical transistor, endogenous memory is stored "so that its output depends not only on the present input symbol but also on its past history."
20 Silvia Maria Alessio, *Digital Signal Processing and Spectral Analysis for Scientists: Concepts and Applications* (New York: Springer International, 2016), 2; Gregory Bateson, *Steps to an Ecology of Mind* (New York: Ballantine Books, 1971), 373. Digital communication, according to Bateson, involves "a number of purely conventional signs—1, 2, 3, X, Y, and so on—pushed around according to rules called algorithms."
21 Kwan and Scheinert, "*Everything Everywhere All at Once*, 1:03:50.
22 Kwan and Scheinert, *Everything Everywhere All at Once*, 50:30.
23 Robert Biel, *The Entropy of Capitalism*, repr. ed. (Haymarket Books, 2013), 3.
24 Bateson, *Steps to an Ecology of Mind*, 428.
25 Bateson, *Steps to an Ecology of Mind*, 428.
26 Stephen Pfohl, "New Global Technologies of Power: Cybernetic Capitalism and Social Inequality," in *The Blackwell Companion to Social Inequalities*, ed. Mary Romero and Eric Margolis (Hoboken, NJ: John Wiley and Sons, 2005), 577.
27 Dyer-Witheford, *Cyber-Proletariat*, 169.
28 Elena Esposito, "Predicted Uncertainty: Volatility Calculus and the Indeterminacy of the Future," in *Uncertain Futures. Imaginaries, Narratives, and Calculation in the Economy*, ed. Jans Beckert and Richard Bronk (New York: Oxford University Press, 2020), 219–235.

3

Chinese Commodities

———————————————◆◇▶

Adoption in *After Yang*

KIMBERLY D. McKEE

As Deng Xiaoping revived the opening of China's economy in 1992, the nation also formally opened its doors to international adoption.[1] Three years later, China surpassed South Korea as the country sending the highest number of children to the United States, one of the leading "receiving" countries of children worldwide. The movement of children via transnational adoption renders them commodities—bought and sold on the economic marketplace. Yet, these adoptions from China and elsewhere also exist as a form of soft power.[2] More than 140,000 children have been adopted from China to the West as part of the nation's three decade-long adoption program, which ended in October 2024. Yet U.S. anxieties associated with the rise of China are not often tied to thinking about the adoption of Chinese children.

The reproductive futurity of white families supersedes the reproduction of nonwhite families, divorcing Asian children available for adoption from other immigration anxieties.[3] The majority of those who adopt across racial lines, as well as across national lines in cases of transnational adoption, are white. In this construction of family, the racial adoption project functions as a means of assimilation.[4] Phrases such as "love is enough" and "I don't see color" are commonly used by adoptive parents, even as racial bias influences adoptive parents' selection of Asian countries and Asian babies.[5] Difference is celebrated when it can be contained within boundaries set by white adoptive parents. White families with children of color are seen as exemplars of multiculturalism and racial liberalism. Interested in the intersections between Chinese adoption and the "rise of China," I analyze Alexander Weinstein's short story "Saying Goodbye to Yang" (2016) and Korean American director Kogonada's 2022 film adaptation, *After Yang*. Placing the texts in conversation makes visible the connections between waning white supremacy and white reproductive futurity, as well as the role of Chinese transnational adoption in alleviating that uneasiness. A comparative approach bolsters my understanding of adoption as a technology to interrogate fears over generational survival.[6]

Readers and viewers are introduced to Mika, the Chinese adopted daughter, and her parents—Kyra and Jim (in the short story)/Jake (in the film)—in an undated U.S. future, years after the nation participated in a military incursion, or potential war, against North Korea in the short story and China in the film. The texts both depict the family's journey as they process the loss of Yang, their beloved technosapien marketed as "an older sibling for Chinese adoptees" by its manufacturer, Brothers & Sisters. Yang easily slipped into the role as "sibling." For example, Jim reflects on attempts to forge a father/son relationship in the short story, and Jake uses the term "son" to describe Yang in his interactions with minor characters in the film. Faced with a life post-Yang, the short story attends to the family as a whole and the film focuses on Jake, limiting the audience to seeing Kyra and Mika mostly from his vantage point. Readers and viewers witness Jim/Jake as he attempts to restore Yang by bringing his android body to various repair technicians, including to a shop owned by Russ Goodman—a character who holds racist animosity toward technosapiens. While both texts construct a transracial adoptive family—adoptive parents who adopt across racial lines—the short story features two white adoptive parents and the film features an interracial couple with the casting of white Irish actor Colin Ferrell as Jake and Black British actor Jodie Turner-Smith as Kyra.

"Saying Goodbye to Yang" and *After Yang* present a banal future that is less interested in a mechanized Asia's outperformance of the West; rather, this future is rooted in confronting the supplantation of reproductive and care labor by technology within the space of the home. The demise of Yang and his existence illustrates how care labor can be outsourced; it is more efficient to have a caretaker designed to support the specific needs of adoptive parents rather than hiring a domestic worker from the Global South. Reviewing the film, Jane Hu writes, "Here, the future is Asian in the way that the present is Asian: filled with high-end ramen shops and white people who know a lot about exotic teas. This sense of banality—of genericness—actually might be the most original thing about the film."[7] This rendering of the future—mid-twentieth-century modern, minimalist, and devoid of neon—is now. What does it mean when adoption enters new futures that move from a dystopian Gilead in *The Handmaid's Tale* to one where an Asian adoptee is a space for possibility—the reproductive futurity of the white family—and threat—when considering the adoptee of color's potential fecundity in adulthood?

Using adoption as an analytic offers insight into the racial anxieties associated with reproduction, including the use of cloning to sustain white supremacy. I examine the short story and then the film to interrogate the mutability of race, specifically the logics that engender the interchangeability of East Asian ethnics to create a generic "Asian-ness" in both texts. In the case of the short story, I am attentive to its setting—Michigan—as an important site to locate anti-Asian racism in the contemporary United States. The region's imagined construction as "America's heartland" reinforces its significance in understanding "the production of white supremacist imaginings at their most banal and influential."[8] The locale is a key component to the narrative, as the adoption of Mika supports telling a story of the region's automotive industry decline and the precarity of white masculinity, whereas the film's nondescript, minimalist future untethers it from specific racial and spatial histories. Unbound to place, the future within *After Yang* could be everywhere and nowhere; it is a future that is eerily familiar. That banality informs my analysis of the film, as I trace *After Yang*'s examination of the existential question of what makes someone human and pay careful attention to how techno-Orientalism is used as the method to explore the contours of white masculinity. This essay engages questions of Asian, particularly Chinese, identity while suturing together the ways anxieties of the techno-Orientalist future are rooted in concerns over white supremacy's demise.

The Future Is White

The couple's motivations to adopt from China, as described in the short story, are unsurprising. Moved by the number of children orphaned due to an earthquake, Kyra suggests adopting one of them. This disaster ostensibly eliminates birth parents seeking out the child later—a fear that compels many U.S. adoptive parents away from domestic adoption to international adoption. The adoption of Mika is seen through a charitable lens even though there are well-documented cases of adoption-as-child-trafficking postdisaster.[9]

In their depiction as invested adoptive parents attentive to instilling a positive racial identity in their daughter, Kyra and Jim/Jake, like other adoptive parents, seek to ensure that Mika maintains a cultural relationship with China through their purchasing of Yang. Technosapiens are marketed to fill the purported cultural gaps in the lives of adopted people while presenting a static "Asian" culture in the capitalist marketplace.[10] The imagined China that Yang and his other technosapiens represent is curated by the information technology sector. That rendering of China flattens Mika's identity as Han Chinese, discounting the possibility that she is one of the other fifty-five currently state-recognized ethnic minorities and that her experiences are more similar to those of other Asian Americans. Brothers & Sisters, alongside the repair and refurbishment technosapien industries it generated, is part of the range of cottage industries that cater to adoptive families (e.g., homeland tours, culture camps).[11]

The racial coding of Yang as Chinese and as the only technosapien in the film results in him functioning as a stand-in for fears of the techno-Orientalist future whereby machines from Asia will take hold, regardless of whether Brothers & Sisters could be North American or European based and white owned. This is made evident by adoptive parents' concerns in both texts that Yang has taken over some of their parenting duties beyond exposing Mika to Chinese culture. Nonetheless, Chinese adoption and technosapiens are positioned as benevolent technologies in contrast to the only other form of reproduction depicted in both texts—cloning. The latter is depicted as a method attached to a wider eugenicist white supremacist segment of society. Weinstein constructs, and Kogonada replicates, a narrative whereby white supremacist notions of family operate vis-à-vis white families' use of cloning to propagate the right (white) kind of family. There is a sense of moral righteousness by white progressives like Kyra and Jim/Jake, who believe they are ostensibly better than those who "clone their own" because of the perception that only racists would reproduce via cloning.[12] These two different technologies and how they are received by white people reveal the contradictions of the future, whereby homogenized white technology outputs (e.g., clones) are accepted in contrast to the racialized Asian technosapien.

Unlike the short story, the film complicates viewers' relationship to cloning as we see the friendship between Yang and Ada (Haley Lu Richardson), a clone of her great-aunt, someone whom Yang built a relationship with when he worked for the adoptive family that preceded his life with Mika. In cinematic flashbacks, as Jake accesses Yang's memories, viewers see Ada and Yang with other presumed technosapiens or clones, as well as humans who hold marginalized identities at an underground concert. These are glimpses into what happens when the unintended occurs—friendships formed between technosapiens and clones. However, *After Yang* only presents these scenes as snapshots, leaving audiences to make their own conclusions. Through the humanization of clones as more than just simulacra to achieve a white supremacist future, the film challenges the singular narrative of clones as only part of a white supremacist project. Instead, the film's portrayal of Ada offers an alternative that imagines clones as more than flattened one-dimensional objects who find themselves enacted upon by the desires of the humans who created them.

Significantly, both texts do not address whether potential procreative futures are possible between adopted children and clones. This is important when accounting for reproductive and techno-Orientalist futurity. Is there a potential for multiracial futures outside of interracial marriage, as depicted in the film? While that question falls outside the scope of this essay, I raise it to consider Kogonada's construction of a narrative that gestures toward complicating a techno-Orientalist future whereby racial minorities are not solely technology or labor; rather, they also operate in a future typically accessible to white protagonists in the techno-Orientalist films of the late 1980s and 1990s.

Korean, Chinese, Japanese: The Interchangeability of Asia(ns)

The backdrop of Weinstein's imagined Michigan is rife with anti-Asian racism, mirroring the anti-Asian sentiment associated with the rise of the Japanese automotive industry in the 1970s and 1980s. Yet prior to adopting, Jim and Kyra appear to lack an understanding of the racial climate in their community of Ann Arbor, Michigan. They are compelled to adopt because, as Jim articulates, "It seemed like the progressive thing to do. We considered it our one small strike against cloning.... We figured it was time to give something back to the world."[13] Kyra's employment at Crate and Barrel and Jim's double shifts at Whole Foods not only signal they are working class but are signifiers of particular progressive U.S. values and an idealized, white, affluent existence. The fact that they can afford a technosapien alongside the costs of adoption on their working-class wages may be what is the most unbelievable aspect to this speculative fiction narrative.[14]

Jim's confrontations with anti-Asian racism make visible his ignorance of what it would mean to raise nonwhite children and implies the lack of racial diversity in his life. The latter is significant because ethnic Asians represent 9 percent of Ann Arbor's population and white people are only 69.2 percent of the Ann Arbor population, according to the 2020 U.S. Census data.[15] Nonetheless, perhaps this aligns with how even in racially diverse communities, white Americans live in segregated communities and report racially homogeneous friend groups.[16]

The conversation between Jim and Russ Goodman at his Kalamazoo, Michigan, tech repair shop lays bare the anti-Asian sentiment encountered by both technosapiens and Asian Americans.[17] As Russ inspects Yang, whose body "[lays akimbo] alongside [Jim's] jumper cables and windshield-washing fluid," the following exchange occurs: "'You brought a Korean.' He says this as a statement of fact. Russ is the type of person I've made a point to avoid in my life: a guy that probably has a WE CLONE OUR OWN sticker on the back of his truck. 'He's Chinese,' I say. 'Same thing,' Russ says. He looks up and gives the other man a shake of his head. 'Well,' he says heavily, 'bring him inside, I'll see what's wrong with him.'"[18] The interchangeability of Korean and Chinese to describe Yang represents the racial coding of Asian ethnics as inscrutable Orientals. This anti-Asian racism is further underscored as Jim glances at the wall behind the desk in Russ's repair shop and sees "a small sign with an American flag on it and the message THERE AIN'T NO YELLOW IN THE RED, WHITE, AND BLUE." Russ also plainly tells Jim, "I don't work on imports," and notes he will look at Yang as a favor to George, Jim's neighbor, who suggested that Jim go see Russ.[19]

I read these scenes as Weinstein's gesture toward the 1982 murder of Vincent Chin. Chin was targeted and killed due to pervasive anti-Asian sentiment by two white Detroit autoworkers who never expressed remorse for murdering him.[20] Technosapiens evoke industrial labor concerns of automation and the fears that humans may one day be replaced by machines. The deindustrialization of Detroit and the surrounding cities, coupled with the automation of aspects of human lives vis-à-vis technosapien labor, highlights white anxieties around labor. Weinstein adeptly weaves his story within histories of racism in Michigan, making it clear that this form of discrimination is not an anomaly; rather, it is connected to a longer pastime of anti-Asian hate.

The experience at Russ's Tech Repair Shop results in Jim remembering what happened "during the invasion of North Korea, back when the nation changed the color of its ribbons from yellow to blue."[21] The conflation of Asian people as yellow with yellow ribbons demonstrates the insidious nature of yellow peril rhetoric. While one could potentially argue that this change is akin to the 2003 origins of "freedom fries" as an alternative to French fries when France decided not to support the War in Iraq, the reduction of Asian people to "yellow" exemplifies the persistence of the forever-foreigner stereotype and the yellow peril across time and space.[22] The fact that the war involves North Korea evokes the Korean War (1950–1953), the first hot war in the Cold War, and technically an unfinished war since no peace treaty was signed. Furthermore, even antiwar protestors were swept by U.S. jingoism, as Jim recalls: "Ann Arbor's a progressive city, but even there, when Kyra and I would go out with Yang and Mika in public, there were many who avoided eye contact. Stop the War activists weren't any different. It was that first Christmas, as Kyra, Yang, Mika, and I were at the airport being individually searched, that I realized Chinese, Japanese, South Korean didn't matter anymore; they'd all become threats in the eyes of Americans."[23] The conflation of East Asian ethnics under one "Asian" umbrella can be seen across the years as Asian immigrants and Asian Americans encounter anti-Asian hate. More significantly, it was not until the adoption of Mika and Yang that this racism was apparent to Jim. The fact that Jim was unaware of anti-Asian racism, or at least not bothered by its existence, prior to his adoption of Mika is unsurprising.[24] Such outrage may only extend to those of color within his family—Mika and Yang.

Anxieties of the Future

While the film follows a similar path as the short story, the casting of Turner-Smith disrupts the notion that Mika and other Chinese adoptees like her, as well as technosapiens, are the racial diversity within their monoracial white adoptive families and idyllic white communities. The film constructs a transnational, colorblind fantasy of family with Turner-Smith, Ferrell, and Indonesian American actor Malea Emma Tjandrawidjaja playing Mika. Not only are the parents an interracial couple with respective Irish English and British English accents, but they also form a transracial adoptive family with a daughter who speaks

American English and a technosapien, Yang (Justin H. Min), who speaks American English with a soft, feminized lilt. They are the postracial future. Nonetheless, the Black mother becomes imbricated in the messy structural and systemic inequalities of transracial adoption as she becomes adjacent to a particular white racial liberalism that "hide[s] in plain sight."[25] The film's elision of why the couple adopted Mika gives the appearance of a future untethered to race, whereby adoption is a natural mode of family formation.

Viewers only see a glimpse of anti-Asian sentiment nearly fifteen minutes into the film as Mika waits in what could only be described as the breakroom in Russ's Repair Shop. In the background she looks at goldfish swimming in their aquarium as the camera then cuts to a close-up shot of the bulletin board in the left foreground. Kogonada depicts the racist American jingoism decorating the repair shop written about in the short story and includes a faded poster that reads "Yellow Peril" with jackboots stomping on an Orientalist caricature. Alongside those images are newspaper clippings with the headings "The US and Chinese Naval Forces Clash in Pacific Ocean" and "After 60 years the war finally draws to an end" and what appears to be an anatomy drawing of a technosapien. However, if you missed this short glimpse of the propaganda, the only other glint of anti-Asian racism involves the veiled racist comments Russ (Ritchie Coster) makes to Jake about technosapiens as "spyware" recording data about the lives of the families they serve. This muted racism relies on the audience's awareness of histories racializing Asians as foreign and suspect.

As tools to support humans in their day-to-day lives, what technosapiens record on their proprietary software is shrouded in mystery, which is why Cleo (Sarita Choudhury), a curator at the Museum of Technology, expresses excitement at learning more about Yang's memories and why Russ believes Yang and technosapiens like him include spyware. What is most compelling about the anxieties raised by Yang's body are the conversations spurred by his demise between Kyra and Jake about Yang's role in raising Mika. If Yang is unable to be fixed, the couple cannot afford another technosapien, even one that has been preowned. The figure of Yang makes visible how, according to David S. Roh, Betsy Huang, and Greta A. Niu, "U.S. techno-Orientalist imagination has its roots in the view of the Asian body—the Chinese body, more specifically—as a form of expendable technology, a view that emerged in the discourse of early U.S. industrialization."[26] Technosapiens are meant to be serviced, sometimes refurbished, and then replaced when they no longer are in working condition. Even if he becomes part of the family, by virtue of his android status he will never be kin in the same ways as Mika.

While one review of the film argues that the character of Yang "upend[s] the noxious stereotype of the 'stoic' Asian, a familiar cliché" as his memories emerge, I am more interested in Yang's role as a vehicle for Jake's introspection.[27] Thus, even if Yang's memories provide him a three-dimensional perspective and a range of potential affective behaviors, we only have access to those memories through Jake's ability to view those memories by wearing spectacles with virtual reality capabilities. There is a sense that somehow Jake is wearing Yang, as if the technology behind technosapiens allows him to slip on a Yang-sized costume to see the world from his perspective. Even though the exhibit at the Museum of Technology is titled *Under the Skin of Technosapiens*," it is as if Jake embodies that reality vis-à-vis the spectacles that offer a glimpse into *being* Yang and the memories he created throughout his years in operation.

Furthermore, Kogonada sutures together a story that collapses Asian difference into an imagined East Asia with only Mika and Yang as signifiers of the Asian other. The casting of Korean American Min as Yang highlights the commodification of "Chinese" identity. Anyone can be Chinese if one is labeled "Chinese." If Yang is a purveyor of all things "China," then the exchange at the dinner table where Mika explains to her parents that with Yang's help, she has made *gochujang* sauce could be read as both Mika exploring her culinary pursuits and the family expressing their interest in Korean food. Perhaps it is simply a subtle nod to Kogonada and actor Min's Korean heritage. Regardless, if you are unfamiliar with *gochujang* as an integral condiment to Korean cooking, it could merely serve as another marker of "Asian" difference that racializes both Mika and Yang. The latter analysis stems from the way the construction of Jake and the aesthetics of the film, whereby the East is fetishized vis-à-vis Jake's tea shop and interest in tea. Adding to this flattened sense of Asia, whereby Japan, China, and Korea comingle as a singular "Chinese"-aesthetic, is the ramen that Kyra and Jake both enjoy as they video chat with one another, as Kyra works from home and Jake is at a minimally decorated East Asian restaurant.

East Asia writ large is designed as a prop illustrating how a techno-Orientalist future is not necessarily all neon and dark alleyways. Instead, it is much like the world we live in today. This construction evokes the benign Asia presented in *Big Hero 6* (2014) discussed by Thomas Sarmiento (chapter 18 in this volume) and gestures toward Leland Tabares's discussion of Asian obsolescence (chapter 1 in this volume). The changing geopolitical and labor landscape informs the evolution of techno-Orientalist futures. The population anxieties held by white supremacists, in this future, manifest themselves into cloning technology, while the racial liberalism of transracial adoption in the mid- to late-twentieth century is deepened as white progressives like Kyra and Jim see adoption as "one small strike against cloning."[28]

The future is reliant on the labor of marginalized people to facilitate Jake's self-examination. This is a story of a white man ruminating on life with a cast of diverse characters to move him toward a particular sense of truth about his future. The anxieties and introspective nature of Jake turn what was the short story's interest in the anxieties of the future into something invested in one man's quest to satiate a need to feel something again. What that something is, is not entirely clear, although his desire to fix Yang feels like a metaphor for the emotional gulf between Jake and Kyra. Instead of focusing on his marriage and daughter, Jake is determined to fix his technosapien.

The Adoption of (Reproductive) Technologies

Adoptees and adoption serve as vehicles for contemplating the boundaries of racial and ethnic identity due to Asian adoptees' historical marginalization within histories of Asian America. The question of "what makes someone Asian" is raised in *After Yang* as Ada discloses to Jake that Yang asked her about whether he was Chinese. The idea that there is some intrinsic experience that "makes someone Asian" gestures toward a desire for Asian authenticity and risks rendering Mika as a melancholic adoptee, an overworn trope in Asian American fantasies of adoption to interrogate Asian American identity.[29] Even as Yang recognizes that he was purchased to support Mika's identity development, by asking the question, he implies that he did not see technosapiens as raced. Mika's question underscores the racialization of technology, as evidenced by the segment of the white community that views tehcnosapiens as a racial Chinese other. A Chinese identity may be ascribed to an android by virtue of his outward appearance and ability to state facts about China. How is that then different from virtual reality technology? This postracial future thus requires rumination over the potential freedom offered by expressing an identity different than one's own in the virtual world.[30] In wearing Yang, both parents have access to his applied identity as a Chinese man, as Kyra is briefly given the spectacles to try by Jake. "Being Chinese" then is commodified beyond just constructing a Chinese product like Yang. While outside the scope of this essay, the techno-Orientalist future in *After Yang* necessitates further contemplation on the potential erasure of histories and legacies of race vis-à-vis augmented and virtual realities and the humanization of androids. In this future, Chinese identity can be purchased for a fee.

Nonetheless, the presence of Yang does not mitigate the racial dissonance Mika undergoes being raised in a transracial adoptive household. Mika discloses to Yang that her peers at recess ask about her "real parents" and say that "mom and dad [aren't] my real parents, not really." In a different scene, viewers witness Yang and Mika walking beside a row of tree saplings as he draws an analogy between adoption and grafting a tree branch from one tree to another. Yang explains, "You're connected to mom and dad just like this branch. You should know both trees are important.... The other tree is also part of who you are." This is the only time in the film that we see a discussion of racial difference. In an interview, Kogonada expresses his affinity for the grafting metaphor in the context of his two adopted sons from South Korea, indicating surprise at feeling an immediate connection, similar to that of having biological offspring. He comments, "It felt like grafting was the only metaphor I could come up with. It's genuine. It's physical. What does family even mean? It's not just that you are a nationality, it's not just about your ethnicity. There are some deeper forms of connection that make up all families."[31] These scenes on kinship that propel the larger questions around identity and race in the film emerged from the director's same-race adoption, unlike Weinstein, who uses adoption as a vehicle to tell a larger story of a techno-Orientalist future and does not appear to have ties to the adoption triad (e.g., adoptive parent, birth parent, adoptee).

Both texts craft a future where Mika only has a future vis-à-vis her adoption and Yang's existence is predicated on being "adopted" by families to serve them as a big brother. As Hee-Jung S. Joo notes, "Children of color are not fetishized in the ways white children are in hopeful imaginations of the future. In fact, they are often denied a future, and even a childhood for that matter."[32] The use of cloning to reproduce white families in combination with the use of assisted reproductive technologies in speculative fiction highlights the absence of color.[33] Adoption is positioned as the better option because it seemingly is not rooted within white supremacy. Nonetheless, the adoption marketplace is racialized and tiered, and the honorary whiteness and assimilative abilities of Asian children into white families have been widely touted by adoption agencies.[34] This assimilability is underscored in the *After Yang*'s lack of explicit engagement in how the film's only two Asian characters experience anti-Asian racism even as the short story nods to Jim's recognition of anti-Asian sentiment. We only learn of Yang ruminating on what it means to be Asian through Ada's retelling in the film.

Thus, even as Asia and Asians writ large are meant to be feared or at least to be the repository for white people's collective anxieties, Yang and Mika are exceptions in that they have access to the racial liberal white family. And in the film, Yang's feminized lilt, attire, and bowl haircut align with racialized and gendered stereotypes of East Asian men. The idea that he is a threat seems incomprehensible when contrasted with his demeanor. There is a vein of "See, they're like us" regarding Jake wearing Yang and viewing the world from his perspective. In the case of Mika, she is an adopted child, seen as worthy of rescue. These extensions of the white family are neutered by their object status in this techno-Orientalist future. An Asian future is palatable and acceptable if it meets the needs of white and white-adjacent people, as in the case of the casting of Turner-Smith as Kyra. In *After Yang*, racial difference becomes subsumed under a broader rubric of an Orientalized and classed future that highlights the film's colorblind casting under the guise of racial liberal fantasy. Similar to the ways transracial adopted children are seen as distinct and separate from other people of color, so too are the technosapien and adoptee in the futures carved by Weinstein and Kogonada. Adoption and adoptees' affective labor are vehicles to explore the limits and possibilities of Asian inclusion within techno-Orientalist futures. Taken together, these texts should be seen as part of a broader evolution of techno-Orientalism that shifts from an explicit dystopia to a mundane version of a technologically enhanced present.

Acknowledgments

Thank you to Sun Yung Shin for reading and providing feedback on the essay.

Notes

1. Deng Xiaoping oversaw the nation's transition toward a socialist market economic system following initial reforms in the 1970s and 1980s, which resulted in the Fourteenth Congress of the Chinese Community Party in October 1992 moving toward economic reforms "that [gave] a greater role to market forces than that offered by any other ruling communist party to date." Tony Saich, "The Fourteenth Party Congress: A Programme for Authoritarian Rule," *The China Quarterly* 132 (December 1992): 1143, https://doi.org/10.1017/S0305741000045574. Also see Hailiang Gu. "The Process and Logic of China's Socialist Market Economy from Mechanism to System," *International Critical Thought* 11, no. 3 (September 2021): 341–356, https://doi.org/10.1080/21598282.2021.1947032; and Peter Selman, "International Adoption from China and India 1992–2018," in *Social Welfare in India and China: A Comparative Perspective*, ed. Jianguo Gao (Singapore: Palgrave Macmillan, 2020), 393–415, https://doi.org/10.1007/978-981-15-5648-7_22.
2. Leslie K. Wang, *Outsourced Children: Orphanage Care and Adoption in Globalizing China* (Stanford, CA: Stanford University Press, 2016), 4, 14, 49–50.
3. Kimberly D. McKee, *Disrupting Kinship: Transnational Politics of Korean Adoption in the United States* (Urbana: University of Illinois Press, 2019); For a discussion of race and the child as future, see José Esteban Muñoz, *Cruising Utopia: The Then and There of Queer Futurity* (New York: New York University Press, 2009); and Jacob Breslow, *Ambivalent Childhoods: Speculative Futures and the Psychic Life of the Child* (Minneapolis: University of Minnesota Press, 2021).
4. Examples of adoption as assimilation include the U.S. Indian Adoption Project, boarding school projects in North America, and forced removals of Aboriginal children in Australia. The U.S. Supreme Court opinions in *Adoptive Couple v. Baby Girl* (2013) and *Haaland v. Brackeen* (2023) both challenge Indigenous sovereignty as outlined in the Indian Child Welfare Act of 1978; see Krista L. Benson, "Indigenous Reproductive Justice after Adoptive Couple v. Baby Girl," in *Reproductive Justice and Sexual Rights: Transnational Perspectives*, ed. Tanya Saroj Bakhru (New York: Routledge, 2019), 85–104. See also Turner and Pauline Strong, "To Forget Their Tongue, Their Name, and Their Whole Relation: Captivity, Extra-Tribal Adoption, and the Indian Child Welfare Act," in *Relative Values: Reconfiguring Kinship Studies*, ed. Sarah B. Franklin and Susan McKinnon (Durham, NC: Duke University Press, 2001), 468–494; Margaret D. Jacobs, *White Mother to a Dark Race: Settler Colonialism, Maternalism, and the Removal of Indigenous Children in the American West and Australia, 1880–1940* (Lincoln: University of Nebraska Press, 2009); Claire Palmiste, "From the Indian Adoption Project to the Indian Child Welfare Act: The Resistance of Native American Communities," *Indigenous Policy Journal* XXII, no. 1 (2011): 10; Margaret D. Jacobs, *A Generation Removed: The Fostering and Adoption of Indigenous Children in the Postwar World* (Lincoln: University of Nebraska Press, 2014), https://doi.org/10.2307/j.ctt1d9nmm2; and Ashley L. Landers, "Abuse after Abuse: The Recurrent Maltreatment of American Indian Children in Foster Care and Adoption," *Child Abuse & Neglect* 111 (January 2021), 104805, https://doi.org/10.1016/j.chiabu.2020.104805.
5. Gendered Orientalism informs white adoptive parents' adoption of Chinese girls. See Sara K. Dorow, *Transnational Adoption: A Cultural Economy of Race, Gender, and Kinship* (New York: New York University Press, 2006); Heather Jacobson, *Culture Keeping: White Mothers, International Adoption, and the Negotiation of Family Difference* (Nashville: Vanderbilt University Press, 2008); Andrea Louie,

How Chinese Are You? Adopted Chinese Youth and Their Families Negotiate Identity and Culture (New York: New York University Press, 2015); and Wang, *Outsourced Children*. This is not to overlook the same-race, transnational adoptions of Asian children by Asian American parents. For reflections by Asian American adoptive parents of an Asian child, see James Kyung-Jin Lee, "An Adopter and the Ends of Adoption," *Adoption & Culture* 6, no. 2 (2018): 282–291; Dawn Lee and Betsy Huang, "The Challenges and Joys of Adoption, Part 1," *From Here*, podcast, February 23, 2021, https://www.buzzsprout.com/653773/8018519-the-challenges-and-joys-of-adoption-part-1, and "The Challenges and Joys of Adoption, Part 2," *From Here*, podcast, March 1, 2021, https://www.buzzsprout.com/653773/8042291-the-challenges-and-joys-of-adoption-part-2. For further discussion of how adoption functions as a racialized market for American adoptive parents, see Elizabeth Raleigh, *Selling Transracial Adoption: Families, Markets, and the Color Line* (Philadelphia, PA: Temple University Press, 2017). Outside of the United States, adoptees are also seen as distinct from other forms of immigrants, being subsumed within Swedish nationalist rhetoric in an extreme case. See Richey Wyver, "'More Beautiful than Something We Could Create Ourselves': Exploring Swedish International Transracial Adoption Desire" (PhD diss., University of Auckland, 2020); and Jenny Wills, ed., *Adoption and Multiculturalism: Europe, the Americas, and the Pacific* (Ann Arbor: University of Michigan Press, 2020), https://doi.org/10.3998/mpub.10032835.

6   See Jane Hu, "Where the Future Is Asian, and the Asians Are Robots," *New Yorker*, March 4, 2022, https://www.newyorker.com/culture/culture-desk/where-the-future-is-asian-and-the-asians-are-robots.
7   Hu, "Where the Future Is Asian."
8   Britt E. Halvorson and Joshua O. Reno, *Imagining the Heartland: White Supremacy and the American Midwest* (Berkeley: University of California Press, 2022), 14.
9   This was particularly evident following the 2008 earthquake in Haiti as evangelical Christians sought to bring Haitian children to the United States under the auspices of "rescue." Adoptees Color Roundtable, "Haiti Statement by Adoptees of Color Roundtable," in Adopted and Fostered Adults of the African Diaspora, January 26, 2010, https://afaad.wordpress.com/2010/01/26/haiti-statement-by-adoptees-of-color-roundtable/; Alex Spillius, "Obama Administration Embroiled in Illegal Removal of Haitian Children," *Telegraph* (London), February 28, 2010, https://www.telegraph.co.uk/news/worldnews/northamerica/usa/7338739/Obama-administration-embroiled-in-illegal-removal-of-Haitian-children.html; Peter Selman, "Intercountry Adoption after the Haiti Earthquake: Rescue or Robbery?" *Adoption & Fostering* 35, no. 4 (December 2011): 41–49, https://doi.org/10.1177/030857591103500405; Kathryn Joyce, *The Child Catchers: Rescue, Trafficking, and the New Gospel of Adoption* (New York: PublicAffairs, 2013).
10 Jacobson, *Culture Keeping*.
11 Leo Kim, "How *After Yang* Subverts Sci-Fi's Fetishistic 'Hollow Asian' Trope," *Polygon*, March 10, 2022, https://www.polygon.com/22971003/after-yang-kogonada-interview-asian-robot-trope; McKee, *Disrupting Kinship*; Kit Myers, "Complicating Birth-Culture Pedagogy at Asian Heritage Camps for Adoptees," *Adoption & Culture* 7, no. 1 (2019): 67–94.
12 For a discussion of virtuous vs. nonvirtuous whites, see Halvorson and Reno, *Imagining the Heartland*.
13 Alexander Weinstein, *Children of the New World: Stories*, 1st ed. (New York: Picador, 2016), 3.

14 For a discussion of international adoption costs, see McKee, *Disrupting Kinship:*; and Raleigh, *Selling Transracial Adoption.*
15 U.S. Census Bureau, "Census Demographic Profile," Explore Census Data, 2020, https://data.census.gov/table?g=040XX00US26_310XX00US11460&d=DEC+Demographic+Profile&tid=DECENNIALDP2020.DP1. Also see Monica Mong Trieu, *Fighting Invisibility: Asian Americans in the Midwest* (New Brunswick, NJ: Rutgers University Press, 2023).
16 Elizabeth Stearns, Claudia Buchmann, and Kara Bonneau, "Interracial Friendships in the Transition to College: Do Birds of a Feather Flock Together Once They Leave the Nest?," *Sociology of Education* 82, no. 2 (April 2009): 173–195, https://doi.org/10.1177/003804070908200204; Daniel Cox, Juhern Navarro-Rivera, and Robert P. Jones, "Race, Religion, and Political Affiliation of Americans' Core Social Networks," PRRI (Public Religion Research Institute), August 3, 2016, https://www.prri.org/research/poll-race-religion-politics-americans-social-networks/; William H. Frey, "Even as Metropolitan Areas Diversify, White Americans Still Live in Mostly White Neighborhoods," *Brookings Institution*, March 23, 2020, https://www.brookings.edu/research/even-as-metropolitan-areas-diversify-white-americans-still-live-in-mostly-white-neighborhoods/; PRRI, "PRRI Survey: Friendship Networks of White Americans Continue to Be 90% White," May 24, 2022, https://www.prri.org/press-release/prri-survey-friendship-networks-of-white-americans-continue-to-be-90-white/.
17 Sherryl Vint, *Biopolitical Futures in Twenty-First-Century Speculative Fiction* (Cambridge: Cambridge University Press, 2021).
18 Weinstein, *Children of the New World*, 8, 9.
19 Weinstein, *Children of the New World*, 10.
20 Vincent Chin's death galvanized Asian Americans under a broader "Asian American" umbrella as they sought justice for his murder. *Who Killed Vincent Chin?*, directed by Christine Choy and Renee Tajima-Peña (Film News Now Foundation, 1987); Helen Zia, "The Vincent Chin Legacy Guide: Asian Americans and Civil Rights,"2nd ed. (Vincent Chin Institute: 2022–2023), https://www.vincentchin.org/legacy-guide/english.
21 Weinstein, *Children of the New World*, 11. This is not to say that the contemporary origins of the yellow ribbon as a symbol of supporting U.S. troops emerging from the Gulf War is not contentious, as this brand of visible patriotism has come to rely on jingoistic messaging. See Linda Pershing and Margaret R. Yocom, "The Yellow Ribboning of the USA: Contested Meanings in the Construction of a Political Symbol," *Western Folklore* 55, no. 1 (1996): 41, https://doi.org/10.2307/1500148.
22 Timothy Bella, "'Freedom Never Tasted So Good': How Walter Jones Helped Rename French Fries over the Iraq War," *Washington Post*, February 11, 2019, https://www.washingtonpost.com/nation/2019/02/11/freedom-never-tasted-so-good-how-walter-jones-helped-rename-french-fries-over-iraq-war/; Felicity Cloake, "Food-Based Fears Are Rarely Rational, but There Is a Long and Vibrant History of Culinary Boycotts–from Adrian Mole to 'Freedom Fries,'" *New Statesman* 149, no. 5512 (March 20, 2020): 49; Stanford M. Lyman, "The 'Yellow Peril' Mystique: Origins and Vicissitudes of a Racist Discourse," *International Journal of Politics, Culture, and Society* 13, no. 4 (2000): 683–747; Colleen Lye, *America's Asia: Racial Form and American Literature, 1893–1945* (Princeton, NJ: Princeton University Press, 2005); Michael Keevak, *Becoming Yellow: A Short History of Racial Thinking* (Princeton, NJ: Princeton University Press, 2011); John

Kuo Wei Tchen and Dylan Yeats, eds., *Yellow Peril! An Archive of Anti-Asian Fear* (London: Verso, 2014).
23 Weinstein, *Children of the New World*, 11.
24 See Kim Park Nelson, "'Loss Is More than Sadness': Reading Dissent in Transracial Adoption Melodrama in 'The Language of Blood' and 'First Person Plural,'" *Adoption & Culture* 1, no. 1 (2007): 101–128.
25 Halvorson and Reno, *Imagining the Heartland*, 5.
26 David S. Roh, Betsy Huang, and Greta A. Niu, "Technologizing Orientalism: An Introduction," in *Techno-Orientalism: Imagining Asia in Speculative Fiction, History, and Media*, ed. David S. Roh, Betsy Huang, and Greta A. Niu (New Brunswick, NJ: Rutgers University Press, 2015), 10.
27 Manohla Dargis, "Do Androids Dream of Watching Your Kids?" *New York Times*, March 4, 2022, 10.
28 Weinstein, *Children of the New World*, 3.
29 See Kimberly D. McKee, *Adoption Fantasies: The Fetishization of Asian Adoptees from Girlhood to Womanhood* (Columbus: Ohio State University Press, 2023).
30 See Lisa Nakamura, *Cybertypes: Race, Ethnicity, and Identity on the Internet* (London: Routledge, 2002); Christopher B. Patterson, *Open World Empire: Race, Erotics, and the Global Rise of Video Games* (New York: New York University Press, 2020).
31 Sam Adams, "What the Year's Best Sci-Fi Movie Has to Say About Asian Identity and Adoption," *Slate*, March 10, 2022, https://slate.com/culture/2022/03/after-yang-colin-farrell-showtime-movie-kogonada-interview.html.
32 Hee-Jung S. Joo, "We Are the World (but Only at the End of the World): Race, Disaster, and the Anthropocene," *Environment and Planning D: Society and Space* 38, no. 1 (February 2020): 77, https://doi.org/10.1177/0263775818774046.
33 Jinny Huh, "Racial Speculations: (Bio)Technology, Battlestar Galactica, and a Mixed-Race Imagining," in Roh, Huang, and Niu, *Techno-Orientalism*, 101–112.
34 Raleigh, *Selling Transracial Adoption*.

Part II
Racialization as Technology

4

Plastinated Vitruvian Man, the Datafication of Race, and Transracial Transfer in *Westworld* and *Altered Carbon*

―――――――――――――――◦――

CHARLES M. TUNG

A human skeleton, suspended in the center of a loom, sinks slowly back into the white liquid out of which it was 3-D printed: the opening title sequence of HBO's *Westworld* presents this transhumanist version of Leonardo da Vinci's "Vitruvian Man" as an image of the underlying substance of the theme park's "hosts," whose intellectual and emotional lives have become indistinguishable from those of Westworld's human guests (Figure 4.1). This synthetic substructure evokes a raceless, universal denominator shared by sentient beings upon which the code of conscious identity can be executed. For some scholars, such imagery indicates the television show's unambiguous critique of humanism—its questioning of the "firm break between living and nonliving, and thus ultimately between consciousness and mechanical programming," its posthumanist blurring of the racist exclusionary distinctions used to constitute the human itself.[1] The Netflix series *Altered Carbon* proceeds from a similar premise—the familiar science-fictional idea that consciousness and salient aspects of identity are just data that can be uploaded into organic storage devices and nonorganic bodies. The series and the novel on which the show is based are thereby able to envision a

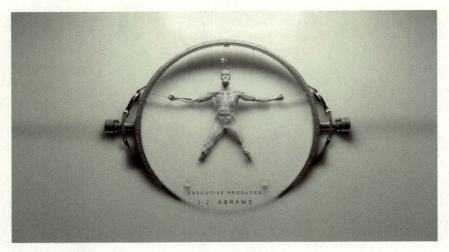

FIGURE 4.1 3-D–printed "host" in the style of Leonardo da Vinci's "Vitruvian Man" in *Westworld*'s title sequence. (Source: HBO.)

posthumanist postracial future in which the inessential, disposable body (organically refurbished, cloned, or inorganically synthesized) can serve as a platform for any identity uploaded or inserted into it. As one critic describes the show's animating idea, "In its depiction of downloading the human mind/consciousness into a 'stack,' *Altered Carbon* represents the specific form of the cybernetic posthuman that is embraced by classic cyberpunk and transhumanist thought alike—the idea, as Hans Moravec puts it, that 'a human mind might be freed from its brain'" and that the "'essence of a person' is not defined by 'the stuff of which a body is made' but rather 'the pattern and process going on in my head.'"[2] Both the HBO and Netflix shows feature the predictable transhumanist theme in which technological advancement makes it possible to distill or extract a fundamental humanity or a human being's core identity from the material, racialized conditions of meatware. These familiar moves in science fiction—the blurring of humanist distinctions, the transferability of consciousness, the flattening of the real and simulacral—invite further theorizing through the lens of techno-Orientalism, precisely because the transfer of racial identity that seems to cancel or neutralize race, as well as the platforms and processes through which it happens, is itself racially marked.

The imminent future and narrative particulars that project this theme of distillation and extraction are frequently the subject of techno-Orientalist fantasies. In these cultural productions, tropes of robots, clones, and aliens reflect an Asian American history of dehumanized labor, the perception of undifferentiated masses of office workers, and the fear of foreign threats.[3] Betsy Huang has shown how this cultural connection is baked into the genre conventions of science fiction itself, in which "the deployment of the East [functions] as a tool of cognitive estrangement for deconstructing the West."[4] In Stephen Sohn's analysis of the racialized future, the "desire to conceptualize the East through a technocratic framework within cultural production leads to a re-articulation and re-emergence of the yellow peril" while also going beyond the fear of Asians within the U.S. social body to a broader question linking domestic exclusion to an imperial American global orientation toward Asia.[5] The disposability of the Asian body is a key feature of the imagined detachment of the inert, inessential matter of the body that frees up the "white, masculinist, disembodied, and hyperseparated individual 'human,' otherwise known as Man"; it raises the question, in Greta Niu's account of the evolution of these scenarios, of the cultural association of Asia and Asians with technologies that render embodiment irrelevant, and with the mutation of this association into post-posthumanist futures in which nanotechnological innovations are embedded in and reconfigure the meaning of the body.[6] This reconfigured body is what Stacey Alaimo would frame as transcorporeal—enmeshed in and interpenetrated by a network of material substances that reveals "the primary dualisms of western thought" to be a humanist fiction.[7]

In this essay I want to explore the "skinning" of the humanist human as a techno-Orientalist procedure that peels back layers of embodied material history in order to prepare the body not just for the identity transmissions of white cybertourism, but more importantly for the transferability of race itself. The techno-Orientalism of fantasies such as *Westworld* and *Altered Carbon* consists in their foregrounding the abstraction necessary for data transfer as a digital whitewashing as well as a hospitality to a white liberal humanist proclivity to empathize and identify with any and all racial subjects. The idea that the body can be skinned and scrubbed of markers of race and histories of racialization is a techno-Orientalist fascination, as is the vision that race and racial identity can be transformed into code that can be transferred to other bodies. Finally, this racial data and metadata, in both their illegibility and association with futurity, are themselves coded as Asian. This techno-Orientalist coding of code informs a transhumanist desire for transracial transfer, the ability to upload and download into differently racialized identities at will, which brings with it a deep anxiety about the viral spread of racial code and the difficulty of determining which code might be installed in the body. Takeshi Kovacs, the Asian protagonist of *Altered Carbon* whose datafied identity encoded in "cortical stacks" is described as "exiled in Caucasian flesh," voices this worry: "Before stacks, a face is a face. But now, could be anyone in there."[8] Uncertainty about racial identification accompanies what Christopher Fan notes is the function of Asian Americans in U.S. racial discourse: "proof of the nonexistence, or at least inconsequentiality, of racism."[9] The inability to tell who Asians are or might be, coupled with the use of Asians to make race matter less or not at all, now feature as symptoms of a techno-Orientalist scenario in which the very presence of the trope of code, even in the absence of explicitly Orientalist images, signals the "peculiar fungibility of the Asian alien" and the particular susceptibility of Asian bodies and identities to the disembodiments of a datafied milieu.

Skinning and Unclaiming the Body

In *Westworld*, critics have noted that the familiar motif of the robot rebellion finds a symbolic anchor in the show's image of the player piano, which appears both in the credit sequence and in some of the episodes themselves. As a central organizing trope, piano performances are not without racial valences. To draw from a recent episode in online culture, one might point to the Chinese pianist Fanchen, who played an arrangement of "Rush E," a "meme song" published in 2018 on the popular YouTube channel Sheet Music Boss by musicians Samuel Dickenson and Andrew Wrangell. Known as a "black MIDI," a digital composition filled with so many notes that ten fingers cannot play it and the printed score appears as a giant block of black ink, "Rush E" inspired many amateur renditions but did not go superviral until the seemingly inhuman virtuosity of Fanchen's performance on YouTube, where Fanchen's "About" section contains the comically simple description, "I'm Asian and I play piano."[10] From a composition processed by the apps SeeMusic and Synthesia into a video featuring notes in the shape of colored blocks raining down on a keyboard at the bottom of the frame, to a clip of Asian hands playing the piano, Fanchen's video has racked up twelve million views at the time of writing, and Sheet Music Boss now has sixty-two million.[11] It is not hard to imagine that the popularity of Fanchen's performance is due at least partly to what Betsy Huang describes as cultural perceptions of Asians as "robot-like mimics" whose "relentless pursuit of social aspirations [are] 'programmed' by the dominant culture."[12]

In the fullest reading of the image of the piano as an automated pleasure device in *Westworld*, Kingsley Marshall has noted that the series's creators, Jonathan Nolan and Lisa Joy, see the player piano as one of the show's "touchstone images": "We were struck by the idea of the player piano . . . as the original western robot. The great granddaddy of our newly sentient hosts."[13] Journalistic reviews have likewise followed Nolan's and Joy's suggestion to treat the piano as one of the central organizing tropes for understanding the racial and gendered dimensions of the show's "western robot" uprising. Drawing on the inhuman quality of mechanized playback, the primary meaning of the piano is present in readings of race as inhumanization/dehumanization in and by *Westworld*. In his critique of the show's representation of Indigenous hosts in Westworld, Indian hosts in the Raj, and Japanese hosts in Shogun World, Lou Cornum answers the question, "What are we really getting out of *Westworld*?": unreal simulations of racialized peoples, "the same old story of how the West was won" mechanically replayed, and "completely unambiguous, preexisting narratives" of "debased positions."[14] In Sherryl Vint's reading of the show, the "enslaved technological tools" immediately come to stand for "racialised subjects, irrespective of casting."[15]

FIGURE 4.2 Plastinated Chinese cadaver in a 2020 advertisement for *BODIES: The Exhibition* at the South Florida Cox Science Center and Aquarium. (Source: YouTube.)

However, the skinless hand of the host playing the piano in *Westworld*'s opening sequence, and the fully skinned body to which it is attached, should be taken as an alternative touchstone image for the show's less obvious techno-Orientalist attachments to Asian fungibility and racial transfer. The skeletal 3-D-printed hand and its *écorché* body allow us to think about the peeling back done by the series and U.S. culture, which is necessary to make the body the neutral bearer of a universal humanity, and to separate race from embodied social existence so that it can become a transferable code. A paradigmatic precursor to *Westworld*'s Vitruvian Man can be found in the mid-2000s traveling show *BODIES: The Exhibition*, which featured plastinated Chinese cadavers (Figure 4.2). A knockoff of the German anatomist and plastinator Gunther von Hagens's original show *Body Worlds*—the world's most successful international traveling attraction (over fifty-six million visitors)—the U.S.-focused *BODIES* show was surrounded by controversy around its use of "unclaimed bodies" from China, which critics identified as executed prisoners.[16] Understandably, the controversy focused on the moral repugnance of exhibiting dissected victims of state violence, buying and using cheap cadavers on the black market for organs and models, and the irreverential display of the skinned bodies in humorous or clever poses. On ABC's *20/20*, when Brian Ross placed before *BODIES* CEO Arnie Geller several photos of executed Chinese prisoners face down in the snow, their bodies having been purchased both by university medical schools and Geller's plastination laboratory for about $200 each, Geller's main defense was simply that the bodies in his show were "unclaimed."[17] In addition to the question of the value of the human being in general and the Chinese body in particular, the exhibition seemed to exemplify the kind of skinning and unclaiming that is required of Asian subjects before they can be assigned value and seen as instances of the human-in-general.

The literal and figurative procedure of skinning—of flaying and plastinating the body in order to negate physical and historical significations, and to uproot identity from its basis in embodied social existence—can be seen as a techno-Orientalist rendition of larger racialization processes. *BODIES* was able to skirt national laws regulating the entry of people and corpses into the United States by defining the unclaimed Chinese cadaver as a "plastic model for medical teaching," a plastinated object whose pliability can give itself—entertainingly, pedagogically—to familiar images of U.S. life presented by the exhibition: throwing a baseball, shooting a basketball, clutching a football. The cheap Chinese imports that become standardized props of American life require a kind of mortification and deracination so that the body can be the host or container for an identity understood as separable from its corporeal condition. This process prepares the anatomical object for "universal" appeal and to be fully cleared for resignification.

In *Westworld,* Vint discerns posthumanist critical potential in the image of the artificial host prior to receiving its skin—whether white settler, confederado, Lakota, Raj, or Shogun World skins—because the in-process skeletal figure blurs the distinctions between the human and nonhuman, and counters the "firm break between living and nonliving, and thus ultimately between consciousness or mechanical programming."[18] Vint cites Alexander Weheliye's distinction between flesh and the body that belongs to the state—in which the latter is part of the racializing mechanism that constitutes "full humans, not-quite-humans, and nonhumans," and the former is, as the site of violence and atrocity, the basis for "an alternative instantiation of humanity that does not rely on the mirage of Western Man as the mirror image of human life as such." She argues that "like Weheliye, the series offers us a discourse that steps off the current path that has given us the liberal human, suggesting that something else is needed."[19] However, if it is true that important questions about the liberal construction of the human are raised by the unskinned Vitruvian skeleton and the fantasy of an infinitely detachable and reattachable identity separable from skin and its historically sedimented layerings, it is worth adding that the technological condition of possibility for the separation of racially signifying skin from both its embedment in a social and material world and its encoded subjectivity is also raising a specific question about techno-Orientalist fungibility, the association of the Asian body not only with an interchangeability with all other Asian bodies, but also with the body's ability to take on other racial codes.

This situation of skinning—in which peeling back the primary signifier of racialization is preparation for insertion into a "world"—manifests not only in traveling exhibits and science fiction but also in other kinds of East Asian–American scenarios, and perhaps even in post-1965 Asian American racialization more generally. These cases exemplify deracination because Asianness is the subject of an informational logic in which the techno-Orientalized body can itself become a platform for transfer, a host for any subject subroutine. In the case of transnational adoption, for instance, which is the original context that produced the term "transracial," Karin Evans's *Lost Daughters of China* recounts her adoption of Xiao Yu, whom she renames "Kelly" to facilitate her incorporation into U.S. national spaces. As David Eng has written, children who are up for adoption are redeemed from their unclaimed status through a process that marks them as objects ready for exchange, "a form of embodied value" that redeems adoptive parents' own need to belong to the larger bodies of family and state.[20] These are the bodies whose futures are served by techno-Orientalist skinning. Adopted children in Eng's reading are immigrant workers whose "crucial ideological labor" is the "shoring up of an idealized notion of kinship, the making good of the white heterosexual nuclear family."[21] Asian children are able to serve in this capacity because their bodies are seen as capable of being emptied of outward signifiers and inner content. When Evans and her husband receive their daughter, who had been abandoned in a Chinese marketplace, they begin unwrapping her, removing a "green jumper and a sweatshirt . . . and a lighter undershirt. . . . When we reached the last layer, our new baby was wearing a bright red T-shirt. . . . 'Trump's Castle Casino Resort' it read." Without pausing to consider this complex metaphor of layering, Evans exclaims, "Oh lord, sweet baby . . . where on earth did you come from?"[22] Back in the United States, Evans recounts a pleasing moment in which the longer history of where Asian adoptees come from is elided and their pliability affirmed: at an adoptive families support group, Evans takes comfort in seeing a girl's nametag containing "one perfect East-West name—Mali McGuire."[23]

Coding the Code and Peculiar Fungibility

It is not just skinning that is a techno-Orientalist fantasy but the codability of identity itself that conjures associations of Asia. Critics have noted *Westworld*'s failure to do anything more than reproduce the racist fantasies baked into the genre of the Western, as well as the familiar failure of omission: "In the historic West, Chinese Americans were prevalent but they hardly appear in *Westworld* just as they did not appear in the famous 'Last Spike' photograph that . . . contributed to the myth of white settlers civilizing the West through technology and work."[24] But the very presence of code in cultural productions, both as a representable figure and as an assumption, can itself serve a racialized and racializing function. For example, in the paradigmatic representation of code made famous by the Wachowskis' 1999 film *The Matrix*, the constitutive materials of the universe turn out to be vertical streams of green characters on a monochrome cathode-ray tube monitor. As the production designer Simon Whiteley revealed, the characters were scanned from his wife's Japanese cookbooks: "The Matrix's code is made out of Japanese sushi recipes" (Figure 4.3).[25] The image of indecipherable script or illegible streams of data (to non-Japanophone viewers, to nonmachine receivers) is posthumanly Asian, and Asianness (dis)simulates what we

FIGURE 4.3 Sushi recipes figuring the world as streams of code in *The Matrix* (1999). (Source: Warner Bros.)

take to be real. As Michelle Huang says of John Searle's "Chinese Room Argument"—the famous thought experiment in which someone who cannot understand strange written characters slipped under the door can use a computer program to pass as a Chinese speaker to Chinese people outside the room—the treatment of an Asian language as "a singularly different language" and the function of the room "as a test for species affirmation" are aspects of a posthumanism still containing an Orientalist reflex in which "Asia has often stood in for the West's uncanny mirror."[26]

The assumption that identity is a transferable code in which embodied aspects of race no longer matter is likewise a techno-Orientalist device that racializes and determines racial value. Iyko Day has analyzed "the peculiar fungibility of the Asian alien" by reading the forms of abstraction to which Asian labor is subjected. In *Alien Capital: Asian Racialization and the Logic of Settler Colonial Capitalism*, Day argues that the laboring Asian body is transformed into "a figure whose interchangeability as a value expression dramatizes the properties of money itself."[27] By reading Canadian Pacific Railway president William Van Horne's drawing of a Chinese laborer's facial profile and the vertical lines of his "long tapered mustache," alongside the sketch of "a busy series of calculations" produced within what appears to be the same psychographic activity, she argues that the juxtaposition of the alien Asian with the "numerical sums" not only evokes "the economic connection between Chinese railroad labor and their low wages," but also reveals an association of "Chinese bodies with abstract labor."[28] Abstract labor is the "social average of labor time to produce a commodity in order to express its quantitative value during exchange"—the figuration of the arithmetic mean cost or central tendency of Chinese bodies, their easy quantification and substitutability. By contrast, white laborers were associated with "the actual time and place of a specific laboring activity."²⁹ This contrast informed the scapegoating of Chinese laborers by white "romantic anticapitalism," as well as concrete labor's willful disacknowledgement of complicity with settler-colonial capitalism.

This specific form of alien Asian fungibility can be linked to what Seb Franklin identifies more generally as "digitality as a cultural logic," and it is this logic that informs the pliability of Asian bodies as well as the peculiar form of racialization disguised as deracialization. This logic is characterized by atomization, divisibility, standardization, and labor capture and represents "a wholesale reconceptualization of the human and of social interaction under the assumption . . . that information storage, processing, and transmission . . . not only constitute the fundamental processes of biological and social life but can be instrumentalized to both model and direct the functional entirety of such forms of life."³⁰ In his most recent book, *The Digitally Disposed: Racial Capitalism and the Informatics of Value*, Franklin links digitality to the "disposable labor" of racialized peoples, whose replaceability and interchangeability is an endemic feature of capitalism's operations of abstraction, objectification, and quantification. These operations compute the value of activity in a way that hides its connection to particular persons that it requires but renders expendable. By fleshing out the relationship between cybernetic concepts and digital capitalism's "trope of the human-as-information-processor as an effect of value-mediated personhood," Franklin draws a link from "racialized subjection to the freedom of the (ostensibly raceless) digital-liberal person."³¹ The immateriality of the "informatics of value" is a mirage whose shimmer distorts and hides the underlying system of reductive devaluation.

A crucial element of both Day's and Franklin's work is Saidiya Hartman's insights on the fungibility of Black bodies produced by the transatlantic slave trade, which "created millions of corpses . . . as a corollary to the making of commodities," the death of tens of millions of enslaved human beings folded into and under the "greater objective" of commerce in which calculation "made it easier for the trader to countenance yet another dead black body."³² For Hartman, the process of rendering bodies as all alike and quantifiably equivalent that results from the racial computation of value also prepares those bodies to be easily and fully occupied by "empathic identification." As she puts it, "the fungibility of the commodity makes the captive body an abstract and empty vessel vulnerable to the projection of others' feelings, ideas, desires, and values; and, as property, the dispossessed body of the enslaved is the surrogate for the master's body since it guarantees his disembodied universality and acts as the sign of his power and dominion."³³ Shannon Winnubst has added that, following from these paradigmatic instances of racializing self-dispossession—the objectification of certain bodies through "the systemic application of geometric, arithmetic, and economic calculations," and the emptying of them so that the white imagination can "don, occupy, or possess blackness or the black body as a sentimental resource and/or locus of excess enjoyment"—different mechanisms of racialization produce different kinds of fungibility, which range from "a mode of objectification to a metric for successful subjectification."[34] Indeed, for Winnubst, neoliberal racialization comes to consist in differential fungibility itself, a celebration of the diversity of purely formal differences "scrubbed clean of any historical residues, ... extracted from any originary roots, especially any roots in violence or conflict."[35] Neoliberal fungibility detaches from the ontological realm and becomes "a semiotic virus," an embrace of "the multiplication of differences, not the differences themselves."[36] Victor Bascara has identified this "will to abstract" as a "hallmark of American imperialism," in which the fungibility of model-minoritized people justifies this neoliberal management of concrete, historical difference, which has always been "a necessary condition of reaping profit under the current, highest stage of capitalism."[37]

While noting that the process of rendering fungible can apply to any and all bodies, and that racial fungibilities themselves are now interchangeable with each other, I argue that the codability and transferability of race have taken on distinctly techno-Orientalist hues. Prior to this neoliberal intensification, Day's argument shows that, because "Chinese bodies are in nearly exclusive alignment with quasi-mechanized labor temporality" and "labor in the abstract," the romantic anticapitalism of white, concrete labor "performs an aesthetic function by giving Chinese shape to the unrepresentable: giving bodily form to the abstract, temporal domination of capitalism."[38] In techno-Orientalist procedures, the inverse is true: the Orientalist shape is precisely the subtraction of the body along with the informational logic that appears as a digital whitewashing. The foundation of this specific form of abstraction in "the colonial ontology of anti-black racism that grounds modernity, liberalism, and global capitalism" and its shift to digital regimes of calculation is an important genealogical trajectory for what Niu analyzes as "post-posthumanism," where the techno-Orientalist figure in cyberspace begins to merge with bionanotechnological constructions of the human.[39] The resulting erasure or suppression of the actual historical relationships between Asian countries and the production of various technologies are motivated by the anxiety around the increasing interpenetration of realms considered to be separate: the disembodied zone of the digital and the embodied arena of racialized labor and violence; weightless codable consciousness and the materiality of technological hardware; the posthuman control over human subjects as information patterns, and the post-posthuman ability to program reality at the molecular level. Thus, visions of a transracial future and the fear of not being able to discern racial identity follow from Asian recodability. Their association with and susceptibility to code generates not only the Orientalist signifiers in cyberpunk and elsewhere of "genetic engineering, artificial life, interplanetary corporate power, police-state surveillance ... [understood] as fundamentally Asian," but also the seeming unraced neutrality of code and the technological infrastructure that encourages both a belief in cyberspace as a raceless, disembodied space of freedom and the "identity tourism" that can be found there.[40]

Altered Carbon and Techno-Orientalist Fictions of Transracial Transfer

Altered Carbon—both the British writer Richard Morgan's novel (2002) and the American showrunner Laeta Kalogridis's television series (2018–2020)—is a cyberpunk fiction in which the Asian protagonist Takeshi Kovacs, originally from the planet Harlan's World "settled by a Japanese keiretsu using East European labor," finds himself "resleeved" hundreds of years later on earth in a futuristic San Francisco called "Bay City." In the *Altered Carbon* universe, alien technology left behind on Mars by advanced beings enables "the digitization of personality," which Pawel Frelik calls "the conceptual cornerstone of [canonical cyberpunk] narratives."[41] This technology allows human beings to record and store all of the experiences and memories that make up their identity in an implanted "cortical stack" from the age of one forward, to transmit this data through a "needle-cast" network running through state institutions such as prisons, morgues, and health clinics, as well as corporate travel agencies, and to resleeve or "cross-sleeve" their identities in tank-grown clones or in unclaimed, unoccupied bodies that differ dramatically from their birth sleeves. After serving a prison sentence in bodiless storage for much longer than an ordinary human lifespan, Takeshi is "spun up" by a 357-year-old "Meth" ("Methusalah") who desires his elite Envoy Corps military training to solve a crime. The premise of digitization leads to the show's primary sight gag: the Asian Takeshi—played in brief flashback scenes by Korean American Will Yun Lee, even more briefly by the Hong Kong actor Byron Mann, and in a single musical appearance by South Korean singer Jihae Kim—enjoys the entirety of season 1 in the sleeve of white police officer Elias Ryker (played by Joel Kinnaman), and most of season 2 in a Black sleeve (played by Anthony Mackie), sometimes alongside Takeshi Prime (Will Yun Lee), during scenes of simultaneous double-sleeving (Figure 4.4). Underscoring this movement of transracial transfer, the opening sequence of the show represents the title's transformation from Japanese characters to letters of the alphabet.

The narrative premise allows Richard Morgan in the novel and the five non-Japanese actors in the show to inhabit the identity of Takeshi, and for the

FIGURE 4.4 The character Takeshi Kovacs "sleeved" in differently raced and gendered bodies in Netflix's *Altered Carbon* (2018–2020). (Source: Netflix.)

character Takeshi to occupy a range of differently raced and gendered bodies. *Altered Carbon* has built into its plot the "postcorporeal subjectivity" that, as Lisa Nakamura has argued in her long-standing contributions to the critique of race-free, transnational "cybernetic tourism," allows users to occupy the role of the privileged "Western subject" who roams unencumbered by material constraints and coordinates amid souvenirs of difference.[42] LeiLani Nishime notes that science fiction films are a zone in which one frequently encounters the "pernicious linking of whitewashing to stories of racial progress," "future worlds where technologies make it possible to replace biological bodies with synthetic ones."[43] She connects this touristic whitewashing to yellowface in a longer history of Asian racial masquerade, in which white performers play up Chinese servility and foreignness within the context of the nineteenth-century use of cheap Asian labor and anti-Chinese, anti-immigration efforts. As Nishime argues, the condition for the possibility of yellowface is the detachability of Asianness from the bodies of Asians—the performance of "race, ethnicity, sexuality, and nationality rendered as disembodied aesthetic spectacle" projected by white actors.[44] Building on Nishime's insights that cultural responses to economic threats from Asia produce the reduction of "Asianness to a series of detachable aesthetic traces," disconnected from specific bodies and the specificities of bodies," I argue that *Altered Carbon* symptomatically stages the anxiety that Asianness now resides in the detachability of identity from bodies itself, and that the techno-Orientalist fantasy is operating in the projection of a propagating code or digital information that allows for transracial transfer.

In many ways, one could simply read the series and book as an allegory about the difficulties of identifying others on the internet and what people do with their anonymity. Nakamura, Beth Kolko, and Gilbert Rodman have noted that the absence in online worlds of "visual and aural cues that serve to mark people's identities IRL [in real life]" encourages race and gender masquerading.[45] Race in cyberspace gets reconfigured into the belief that historically situated and embodied identities are purely a matter of "constructing an identity for yourself."[46] The show *Altered Carbon* plays the potential for switcheroo for laughs, and the potential for postracial self-fashioning for science fiction wonderment, by presenting a number of other cross-sleevings besides Takeshi's. For instance, the minor character Eva Elliott, the hacker spouse of Black character Vernon Elliott (played by Ato Essandoh), is spun back up from a prison sentence as a white man. More prominently, the Mexican American police officer Kristen Ortega (played by Martha Higareda), whose neo-Catholic family is against the practice of resleeving the dead in new bodies, spins up her deceased grandmother in the body of a white supremacist prisoner for her family dinner on Dia de los Muertos (Figure 4.5). Lars Schmeink regards the jarring contrast between mind/identity and body in this scene as an example of the show's resistance to the "hypercapitalist commodification of bodies," and he sees the transhuman fantasy of disembodied identity as showcasing the dissonance generated by seeing a

FIGURE 4.5 Officer Ortega's deceased grandmother is "spun up" in the body of a white supremacist prisoner for the Ortega family dinner on Dia de los Muertos in *Altered Carbon*. (Source: Netflix.)

loved one reembodied in "a totally opposite body image" diegetically.[47] About the novel's literal transhumanism as a whole, Frelik makes the case even more strongly: Morgan's novum of the cortical stack does not devalue meat but rather "explicitly serves embodiment," and the novel presents "nothing short of a return of the human.... The stack and sleeve technologies may make Kovacs posthuman but in his soul/stack he remains like us."[48] But it is hard to imagine these visual gags and the tone in which they are rendered to be functioning as a critique of its own plot premise that what we take as masquerade is simply how humans exist in the world of datafied and commodified identities—with only the most minor complications of sleeve shock.

The narrative in the book and show dodges the representational politics of whitewashing and other forms of Asian erasure by means of a premise that race and gender are now immaterial, a matter of transferable encoded memories. The universal platform for cortical data transfer normalizes racial and gendered masquerade and encourages the transhumanist transcendence of embodied, material, and historically sedimented realities through its focus on the techno-biology of the human. But that is precisely where techno-Orientalist fantasy and a peculiar form of Asian racialization is on full display. In their excellent introduction to the first volume of *Techno-Orientalism*, the editors remind readers that "Asian American studies has always attended to constructions of culture, race, and the body partly because U.S. techno-Orientalist imagination has its roots in the view of the Asian body ... as a form of expendable technology, a view that emerged in the discourse of early U.S. industrialization."[49] That body in its variability has been as susceptible as other racialized bodies to atavistic or futurist "racebending" and "racial masquerades" that willfully ignore the distinctive economic, social, and global relations structuring it. But what might be unique in techno-Orientalist contexts is the way Asian bodies and even their absent presence can figure the general transhumanist desire to reduce the complex specificities of racializations to a simple state of universal codability. The fiction of transracial transfer depends on the separation of racial code from racialized bodies located in specific historical and socially embodied contexts. The peculiar fungibility associated with Asians under neoliberalism has mutated into a fully frictionless scenario in which one kind of difference is the same as any other difference, and the proliferation of differences no longer anchored in the material is both cause for celebration and for a kind of racial epidemiological anxiety, a fear of the spread of unreadable racial code beneath the skin and new uncertainties about identification." As Elizabeth Wissinger has argued, in our current Fourth Industrial Revolution, following the move "from mechanical to electrical to digital forms of production" and finally to a biotechnological situation, the reality of "datafied bodies" has emerged in contemporary practices of health-tracking and self-optimization, corporate data mining and surveillance, and the willing surrender of all life processes to "the whirring hoppers of corporate profit-seeking."[50] And that avatar, profiled so relentlessly, simultaneously contains a major question mark—"could be anyone in there," as Takeshi puts it. This more general datafied milieu, as Wendy Chun has said of the internet, has "been made oriental."[51]

Notes

1 Sherryl Vint, "Long Live the New Flesh: Race and the Posthuman in Westworld," in *Reading Westworld*, ed. Alex Goody and Antonia Mackay (Cham, Switzerland: Palgrave, 2019), 144.
2 Lars Schmeink, "Embodiment in Altered Carbon," in *Sex, Death and Resurrection in Altered Carbon: Essays on the Netflix Series*, ed. Aldona Kobus and Łukasz Muniowski (Jefferson, NC: McFarland, 2020), 68; Hans Moravec, *Mind Children: The Future of Robot and Human Intelligence* (Cambridge, MA: Harvard University Press, 1988), 4, 177.
3 See the film essay *Inhuman Figures: Robots, Clones, and Aliens*, written by Michelle Huang, directed by C. A. Davis (Washington, DC: Smithsonian Asian Pacific American Center, 2021). https://smithsonianapa.org/inhuman-figures/.
4 Betsy Huang, *Contesting Genres in Contemporary Asian American Fiction* (New York: Palgrave Macmillan, 2010), 112.
5 Stephen Hong Sohn, "Introduction: Alien/Asian: Imagining the Racialized Future," *MELUS* 33, no. 4 (2008): 10.
6 Stacy Alaimo, "New Materialisms," in *After the Human: Culture, Theory and Criticism in the 21st Century*, ed. Sherryl Vint (Cambridge: Cambridge University Press, 2020), 178–179; Greta Aiyu Niu, "Techno-Orientalism, Nanotechnology, Posthumans, and Post-Posthumans in Neal Stephenson's and Linda Nagata's Science Fiction," *MELUS* 33, no. 4 (2008): 73–96.
7 Alaimo, "New Materialisms," 177.
8 Richard K. Morgan, *Altered Carbon* (London: Gollancz, 2002; repr., New York: Del Rey, 2006), 255; Laeta Kadogridis, *Altered Carbon* (Netflix Streaming Services, 2018), season 1, episode 5.

9. Christopher T. Fan, "Asian/American (Anti-)Bodies: An Introduction," *Post45*, November 23, 2015, https://post45.org/2015/11/asianamerican-anti-bodies-intro/.
10. Fanchen, https://www.youtube.com/@Fanchen/about. See Fanchen, "Guy thinks I can't play Rush E and calls me a liar," posted June 24, 2021, YouTube, 1 min., 44 sec., https://youtu.be/on4IoQ2MQ7M?si=LnxEuWs5zlXKSawW.
11. As of 2025, Fanchen's video now has over seventeen million views, and Sheet Music Boss has over ninety-four million.
12. Huang, *Contesting Genres*, 115.
13. Kingsley Marshall, "Music as a Source of Narrative Information in HBO's Westworld," in Goody and Mackay, *Reading Westworld*, 104. Kingsley is citing an unpublished interview with Nolan: "Jonathan Nolan, Co-Creator, HBO's Westworld," interview with Kingsley Marshall, conducted September 3, 2018. However, the idea is something that journalists have reported Nolan saying in public settings. See Kim Renfro, "Here's Why Modern Songs Play on the Saloon's Piano in 'Westworld,'" *Business Insider*, October 11, 2016, https://www.businessinsider.com/westworld-piano-songs-2016-10; and Rachael Durkin, "Westworld's Player Piano Is the Great Character That Keeps Getting Overlooked," *The Conversation*, April 18, 2018, http://theconversation.com/westworlds-player-piano-is-the-great-character-that-keeps-getting-overlooked-95238.
14. Lou Cornum, "What Are We Really Getting Out Of 'Westworld'?" *BuzzFeed News*, June 21, 2018, https://www.buzzfeednews.com/article/loucornum/westworld-season-two-native-american-fantasy.
15. Vint, "Long Live the New Flesh," 147.
16. See "What is Body Worlds?," FAQ, Bodyworlds.com, Körperwelten. Accessed January 25, 2025, https://bodyworlds.com/about/faq/.
17. *20/20*, "The Business of Bodies" (ABC, February 15, 2008).
18. Vint, "Long Live the New Flesh," 144.
19. Vint, "Long Live the New Flesh," 151; see Alexander Weheliye, *Habeas Viscus: Racializing Assemblages, Biopolitics, and Black Feminist Theories of the Human* (Durham: Duke University Press, 2014), 3, 43.
20. David L. Eng, "Transnational Adoption and Queer Diasporas," *Social Text* 21, no. 3 (2003): 10, https://doi.org/10.1215/01642472-21-3_76-1.
21. Eng, "Transnational Adoption and Queer Diasporas," 11.
22. Karin Evans, *Lost Daughters of China* (New York: J. P. Tarcher/Putnam, 2000), 72.
23. Evans, *Lost Daughters of China*, 172.
24. Aaron Bady, "'Westworld,' Race, and the Western," *New Yorker*, December 9, 2016, https://www.newyorker.com/culture/culture-desk/how-westworld-failed-the-western; Dustin Abnet, "Escaping the Robot's Loop? Power and Purpose, Myth and History in Westworld's Manufactured Frontier," in Goody and Mackay, *Reading Westworld*, 225.
25. Jennifer Bisset, "Creator of The Matrix Code Reveals Its Mysterious Origins," *CNET*, October 19, 2017, https://www.cnet.com/culture/entertainment/lego-ninjago-movie-simon-whiteley-matrix-code-creator/.
26. Michelle Huang, *Inhuman Figures: Robots, Clones, and Aliens* (Smithsonian Asian Pacific American Center), 4, accessed January 29, 2023, https://smithsonianapa.org/inhuman-figures/.
27. Iyko Day, *Alien Capital: Asian Racialization and the Logic of Settler Colonial Capitalism* (Durham, NC: Duke University Press, 2016), 44.
28. Day, *Alien Capital*, 42, 46.
29. Day, *Alien Capital*, 45.

30. Seb Franklin, *Control: Digitality as Cultural Logic*, Leonardo Books (Cambridge, MA: MIT Press, 2015), xviii.
31. Seb Franklin, *The Digitally Disposed: Racial Capitalism and the Informatics of Value*, Electronic Mediations (Minneapolis: University of Minnesota Press, 2021), 118.
32. Saidiya Hartman, *Lose Your Mother: A Journey along the Atlantic Slave Route* (New York: Macmillan, 2008), 31.
33. Saidiya Hartman, *Scenes of Subjection: Terror, Slavery, and Self-Making in Nineteenth-Century America* (New York: Oxford University Press, 1997), 21.
34. Shannon Winnubst, "The Many Lives of Fungibility: Anti-Blackness in Neoliberal Times," *Journal of Gender Studies* 29, no. 1 (January 2020): 21, 107, https://doi.org/10.1080/09589236.2019.1692193.
35. Winnubst, "The Many Lives of Fungibility," 109.
36. Winnubst, "The Many Lives of Fungibility," 107, 103.
37. Victor Bascara, *Model-Minority Imperialism* (Minneapolis: University of Minnesota Press, 2006), 15, 11.
38. Day, *Alien Capital*, 46.
39. Niu, "Techno-Orientalism," 73.
40. Timothy Yu, "Oriental Cities, Postmodern Futures: 'Naked Lunch,' 'Blade Runner,' and 'Neuromancer,'" *MELUS* 33, no. 4 (2008): 59; Lisa Nakamura, "Race in/for Cyberspace: Identity Tourism and Racial Passing on the Internet," in *CyberReader*, ed. Victor J. Vitanza (New York: Allyn Bacon, 1999), 444.
41. Pawel Frelik, "Woken Carbon: The Return of the Human in Richard K. Morgan's Takeshi Kovacs Trilogy," in *Beyond Cyberpunk: New Critical Perspectives*, ed. Graham J. Murphy and Sherryl Vint (New York: Routledge, 2010), 174.
42. Lisa Nakamura, *Cybertypes: Race, Ethnicity, and Identity on the Internet* (New York: Routledge, 2002), 88.
43. LeiLani Nishime, "Whitewashing Yellow Futures in *Ex Machina*, *Cloud Atlas*, and *Advantageous*: Gender, Labor, and Technology in Sci-Fi Film," *Journal of Asian American Studies* 20, no. 1 (February 1, 2017): 30, https://doi.org/10.1353/jaas.2017.0003.
44. Nishime, "Whitewashing Yellow Futures," 33.
45. Beth Kolko, Lisa Nakamura, and Gilbert Rodman, introduction to *Race in Cyberspace*, ed. Beth Kolko, Lisa Nakamura, and Gilbert Rodman (New York: Routledge, 2000), 4.
46. Kolko, Nakamura, and Rodman, introduction, 6.
47. Schmeink, "Embodiment in Altered Carbon," 76, 77.
48. Frelik, "Woken Carbon," 187, 188.
49. David S. Roh, Betsy Huang, and Greta A. Niu, "Technologizing Orientalism: An Introduction," in *Techno-Orientalism: Imagining Asia in Speculative Fiction, History, and Media*, ed. David S. Roh, Betsy Huang, and Greta A. Niu (New Brunswick, NJ: Rutgers University Press, 2019), 10.
50. Elizabeth Wissinger, "The Sociology of Self-Tracking and Embodied Technologies: How Does Technology Engage Gendered, Raced, and Datafied Bodies?" in *The Oxford Handbook of Digital Media Sociology*, ed. Deana A. Rohlinger and Sarah Sobieraj (New York: Oxford University Press, 2020), 323.
51. Wendy Hui Kyong Chun, "Introduction: Race and/as Technology; or, How to Do Things to Race," *Camera Obscura* 24, no. 1 (May 2009): 34.

5

Outsiders Within

The Indigenous/Minority
Question and Techno-Orientalist
Gaze in India

M. IMRAN PARRAY

The binary oppositions of East/West and Orient/Occident form the framework around which the discourse of Orientalism (and now techno-Orientalism) is articulated and theorized. Edward Said thought of the Orient as a cultural contestant of Europe and "one of its deepest and most recurring images of the Other."[1] As the West's negative alter ego, the East has been projected as "alluring and exotic, dangerous and mysterious, always the Other."[2] David Roh, Betsy Huang, and Greta Niu define techno-Orientalism—a framework that reimagines traditional Orientalism—as "the phenomenon of imagining Asia and Asians in hypo- or hypertechnological terms in cultural productions and political discourse."[3] As is evident from the existing scholarship, much of the theorizing in this direction has been focused around bigger geocultural constructs such as East and Asia, though Japan (and now more increasingly China) occupies what David Morley and Kevin Robins term "a threatening position in the Western imagination."[4]

This approach of addressing the East or Asia as a single entity in debates of both classical Orientalism and techno-Orientalism, I contend, is problematic. It limits the possibilities of locating granular operations of techno-Orientalism within a nation-state such as India, a postcolony of multiple sociopolitical and cultural fault lines, where techno-Orientalist moves as forms of Othering emerge from within against minorities and the Indigenous. In reimagining the phenomenon of techno-Orientalism from its grand theoretical vision, this chapter places the geopolitical entity of the nation-state (in this case India) at the center of the techno-Orientalist equation by arguing that the nation's techno-Orientalist gaze projects its Indigenous/minority communities as a technological-physical threat to the country and its Hindu majority community.[5] This "fear of small numbers" as Arjun Appadurai succinctly explains with respect to the category of "minority" produces an "anxiety of incompleteness" among the numerical majority and activates among certain majoritarian groups predatory and ethnocidal tendencies toward the "small numbers."[6] At the level of cultural representation, such anxieties crystallize in multiple configurations of myth, prejudice, and hate—both online and offline—directed at populations and communities living at the margins and peripheries of the country that form the minority.[7] The core argument thus is that Muslims are Orientalized (and techno-Orientalized) by the majoritarian interests of the Indian state—"the country"—or what Lisa Lau and Ana Cristina Mendes term "Hindu Orientalism," a recasting of Western Orientalism in a re/neocolonial India that works against minorities and the marginalized such as Dalits, tribes, and Muslims.[8]

Focusing on seasons 1 and 2 of *The Family Man*, an Amazon Prime web series created by Indian filmmakers Raj Nidimoru and Krishna DK, this chapter shows how members of marginalized communities and those who share a conflicted relationship with the Indian state are often projected as outsiders within, individuals who possess both premodern and hypermodern traits and pose great threats—physical and technological—to the country's Hindu majority population and the de facto socioeconomic order.[9] In illustrating this argument, the chapter explores the representation of the Muslim and Adivasi body—in the form of a Malayali Muslim man, a Kashmiri Muslim man, and a Sri Lankan Tamil woman fighter—who are projected as technologically and strategically dangerous and need to be eliminated at all costs for the security of the state. I argue that these characters are metonymies of the condition of the two most marginalized sections in India, which at the same time lays bare a re-Orientalist gaze that works from within the confines of the nation-state. This chapter thus contextualizes the on-screen imaginary with off-screen reality, thereby folding the critique of an emergent genre into a wider socioeconomic and political order of the time. Although speculative fiction has been shown to be the most fitting genre for expressing techno-Orientalism, it is proposed that the genre of web series with futuristic impulses also exposes deep racial and technological anxieties. I argue that *The Family Man* and other such web series running on Indian OTT (over-the-top) platforms augment the already-existing prejudices against minorities and the Indigenous communities by projecting them as inherently violent and as a perpetual threat to national security. This techno-Orientalist rendering of the minority body that activates among the majority a sense of incompleteness has, it is posited, following Appadurai, the tendency to "drive majorities into paroxysms of violence against minorities," a cause of concern in India in view of lynch mobs running amok and a democracy (a nation-state) looking the other way.[10]

Carol Breckenridge and Peter van der Veer argue that nationalism "is the avatar of orientalism in the later colonial and postcolonial periods."[11] There are oppressive potentialities in all nationalisms and hence the nation-state is "inherently violent and exclusionary."[12] Roh, Huang and Niu alert us of the danger of internalization of techno-Orientalist patterns by Asian creators, who might imitate without any critical scrutiny the same dehumanizing model that the West deployed on them.[13] Lisa Lau proposes re-Orientalism as the "process of Orientalism by Orientals." Re-Orientalism, or what has been also described as "new orientalism," fundamentally relates to the perpetuation of Orientalism(s) by the "Orientals."[14] Irfan Ahmad identifies the phenomenon of re-Orientalism in India, which is crystallized in the form of structural marginalization and Othering of the country's largest minority, Muslims. Stigmatization of Muslims' deprivation is executed via tropes such as "Islamic terrorism," "fanaticism," "Muslim appeasement," "separatism," and "disloyalty to the nation."[15] If, as Toshiya Ueno explains, the "basis of Orientalism and xenophobia is the subordination of Others through a sort of 'mirror of cultural conceit,'" nowhere is such subordination more explicitly visible than in today's India where the Indigenous and minority communities, particularly Muslims, are subjected to widespread abuse and various other forms of socioeconomic and political Othering.[16] The perpetuation of abuse and Othering against Muslims does not only take place in real life; the reel life too demonstrates these impulses. As Mita Banerjee argues, two kinds of Orientalism operate in Bollywood, the Indian film industry, an enormous site of popular cultural production. Firstly, Bollywood films project India as a feast of color to a Western observer; secondly, the film as an apparatus celebrates the "triumph of Hindu India, in all its colorful exuberance, over a Muslim minority which it views as both colorless and immoral."[17] Recent Bollywood productions such as *The Kashmir Files* (2022) and *The Kerala Story* (2023) are new entries in a long list of films that seek to inject into the Muslim minority body—both real and imagined—a sense of immorality.

The phenomenon of Orientalizing and techno-Orientalizing the Muslim community in India has a deep and complex history. Across political and intellectual spectrums, Muslims are often assumed to be an undesirable and foreign entity while the identity of India is projected as inherently and purely Hindu, a place that belongs to the majority Hindu population alone.[18] Another major reason behind Orientalizing/Othering of Muslims is the partition of the subcontinent in 1947 and the creation of two new nation-states: India and Pakistan. The partition triggered one of the largest displacements of human populations in the twentieth century. Despite the creation of a new nation-state for Muslims, millions chose to stay in India, considering it as their homeland. But

Indian Muslims have been systematically marginalized by the postcolonial Indian nation-state.

Constructing the Muslim "Other" as Technological Threat

Released on September 20, 2019, *The Family Man*, season 1, starts with three men on board a hijacked fishing boat in the Arabian Sea, en route from Lakshadweep, a tropical archipelago, to Kerala, the southernmost state in India. As the boat nears the coast of Kochi, it is intercepted by the Coast Guard. The men are arrested for trying to illegally enter Indian territory. The investigation is handed over to Threat Analysis and Surveillance Cell or TASC, a covert intelligence agency that works in close coordination with other elite agencies such as the National Investigation Agency, Research and Analysis Wing, and Force One. Senior analyst Srikant Tiwari (played by Manoj Bajpayee) and his associate JK Talpade (played by Sharib Hashmi), the main protagonists of the narrative world, begin investigating the case. "Moosa Rehman from Kasaragod, aged twenty-your years. IIT Kanpur," Tiwari slowly reads from the file, perplexed, while looking at Talpade. The case file introduces the other two characters: "Moinuddin, alias Asif. Trivandrum bus blast, Mysore High Court blast. A suspect in both cases," and Aboobacker, a mechanical engineer. As shall unfold in the narrative, the men were sent on a mission—"Mission Zulfikar"—hatched by agents of Pakistan's Inter-Services Intelligence (ISI) and terrorist groups affiliated to Islamic State of Iraq and Syria (ISIS) to launch a nerve gas attack in India's capital state of Delhi. One of the three, Moosa (played by Neeraj Madhav), a chemical weapons expert trained in the ISIS camps, together with a Kashmiri Muslim man, Sajid Ghani (played by Shahab Ali)—a "horse gone berserk" and a "lone wolf" who is "overwhelmingly passionate" and "does jihad too, in his own inimitable style"—would push India to the edge with a vicious plan (Figure 5.1). As the makers of *The Family Man* introduce the arrested Muslim men—by invoking their religious identity, educational/technological competence, and direct links with ISIS, other terror groups, and the ISI, and projecting them as a threat to India's security—the first techno-Orientalist expressions with regard to the country's largest minority are rendered explicit that anchor the rest of the narrative.

Since "techno-Orientalist imaginations are infused with the languages and codes of the technological and the futuristic," the projection of and emphasis on the Indian Institute of Technology (IIT) as Moosa's alma mater is the first sign of his technological capability.[19] Apart from being an IIT graduate and a chemical weapons expert, Moosa's character demonstrates a strong technological fluency. He is shown to share an intimate relationship with technology. The show reflects a visual aesthetics that techno-Orientalizes both Moosa and Sajid. When Moosa and Sajid take control of Orion Chemicals, a Delhi-based chemical factory, to release a toxic gas with a plan to poison the Delhi population, Sajid

FIGURE 5.1 A Kashmiri Muslim man, Sajid Ghani, holding a nerve gas canister inside a hideout in Kashmir. (Source: Amazon Prime.)

compliments Moosa by saying, "You sound like a kid in a toy store." Such is the technological fluency of Moosa that it is child's play for him to navigate and appropriate systems that enable them to turn a factory into a ticking time bomb. By executing a system override at Orion Chemicals, Moosa transforms a normal chemical factory into a disaster site. What makes Moosa and Sajid part of the same gaze are their premodern and hypermodern capabilities. Moosa is referred to as "Al-Qatil" or "the murderer," who is "just like ISIS, brutal. Killing infidels is his only goal in life." On the other hand, Sajid is "dangerous and intelligent ... a highly motivated individual." He possesses extensive training in improvised explosive devices (IEDs). Together these characteristics transform Moosa and Sajid into unpredictable demons. They are shown escaping the gaze of India's billion-dollar security grid as if they are supernatural. This power of easy mobility attributed to Moosa and Sajid is a manifestation of techno-Orientalizing of the minority body. In real life, however, minorities' lives are defined by layers of checkposts and interrogations, the symbols of state violence.

There are a number of emblematic web series running on Indian OTT platforms such as *Crackdown*, *Kathmandu Connection*, and *Special Ops* that help us to understand this Orientalizing and techno-Orientalizing of the Muslim minority in India. In *Crackdown* (2020), Muslim characters are shown to be in league with foreign terror organizations—mainly Pakistan based—desperate to wage biological warfare against the Indian state and its citizens. The series' tagline is "eliminate the threat before it eliminates us." The "threat" is sourced to a Muslim man, woman, child, and household, satisfying the visual pleasure of an average Indian Hindu viewer. Homes of Muslims are shown to be bomb-making garages while working Muslims are projected as overground workers for terrorists ready to aid the detonation of bombs at airports, the massacre of innocent civilians, and the destruction of public infrastructure. In the very first scene of *Kathmandu Connection* (2021), the Muslim call for prayer, the *Adhan*, forms the diegetic music for an antiterror operation carried out by the Indian counterterrorism police to neutralize members of Hizb-ul-Mujahideen—a militant organization operating from Kashmir—who are hiding in a room in Delhi's Mehrauli area.

In *The Family Man*, a paradoxical discourse on Muslim technological access and danger runs concurrently, which becomes explicit when Tiwari frowns at Moosa's educational qualification. This triggers in Tiwari a sense of trepidation since an average Indian Muslim must remain, from the standpoint of Hindu Orientalism, a member of a backward community, bereft of any intellectual or technological knowledge and access. Otherwise, the Muslim man, like Moosa, will transform this knowledge and technological access into a bomb. Therefore, backward they must stay in order for the country to grow and remain safe from the Muslim danger. The "backwardness" of Muslims in India was a highlight of an official panel, the Sachar Committee Report of 2006, which found Indian Muslims to be victims of widespread injustices and thus recommended their inclusion in the national mainstream.[20] Unlike a re-Orientalized sketch of an average Muslim as a backward Indian, as the story unfolds in *The Family Man*, Moosa presents a huge challenge to the Indian security establishment and confounds them with his skills and knowledge.

Juxtaposing Tech with Islamophobic Imagery

A number of web series, including *The Family Man*, superimpose technology and Islamophobic imagery to establish spatiotemporal relationships between incongruent symbolic systems that forge a direct connection between Islam and violence. When Sajid prepares a scooter bomb, he is shown wearing a black skull cap that bears the Islamic symbols of the crescent and star. Such a link is also established when Major Sameer of the ISI and Moosa are shown discussing "Mission Zulfikar" in a room filled with containers of highly toxic chemicals, a satellite communication system, and the calligraphic name of the Prophet Muhammad on the front wall (Figure 5.2). This sort of representation is deeply Islamophobic, implicates Islam as a violent religion/ideology, and frames its adherents as savages and killers on one hand and as technologically dangerous on the other. Throughout its two seasons, *The Family Man* invokes such an imaginary of Islam and its followers. In *Kathmandu Connection*, a police raid on a house bearing the name "Islam's" to nab at-large Muslim terrorists is suggestive of the Orientalizing of Muslims via this Islamophobic imagery.

This Islamophobic imagery is not new or specific to the genre of web series alone. Instead, it builds on the already existing prejudices in mainstream film industries such as Hollywood and Bollywood that have been shown to racialize

FIGURE 5.2 Pakistani spy Major Sameer meeting ISIS militant Moosa Rehman at an ISIS camp in Iraq amid barrels of chemicals, telecommunication systems, and Quranic words scribbled on the wall. (Source: Amazon Prime.)

Muslim identities in service to Islamophobia.[21] Hindi cinema, in particular, is reductionist, projecting Muslims as "a synecdoche to signify a terrorist, religious extremist, Pakistan loyalist, anti-Hindu and a traitor."[22] As Kalyani Chadha and Anandam P. Kavoori show, Bollywood at once exoticizes, marginalizes, and demonizes Muslims as the "Other."[23] In Indian filmmaker Neeraj Ghaywan's view, the demonization of Muslims takes place also via the technical resources of cinema such as color palette, production design, camera angles, and so on. "You look at the color palette, or how the production design is done while portraying Muslims. It's mostly black, it's mostly dimly lit, and it's shot at a low level to have a demonizing effect. But as you show the other side, it's all flowery and bright lights and shot at an eye level," Ghaywan noted in a 2020 discussion, "The Quint's Films & Politics Roundtable," hosted by *The Quint*, a web-based Indian news organization.[24] The genre of web series perpetuates these popular-hegemonic frames and amplifies them in a new ecosystem of producer-consumer engagement propelled by private capital and the mushrooming growth of web television platforms. This mix of private capital, the emerging genre of web television, and lax regulations on web content implicates the state in enabling such programming on OTT platforms. Instead of regulating this aspect of OTT platforms, the state has assumed what Vibodh Parthasarathi calls "considered silence," a "persistent lack of regulatory response on certain challenges."[25]

Between Good and Bad Muslims

By Orientalizing and techno-Orientalizing Indian Muslims, *The Family Man* produces a binary opposition of good Muslim and bad Muslim through the characters of Imran Pasha, the Force One officer, and Zoya, the TASC junior analyst, a female Muslim from Lucknow. This good/bad Muslim narrative is explicitly drawn in many other espionage series running on Indian OTT platforms. In *Crackdown* (2020), top Muslim intelligence officers Riyaz Pathan and Zorawar Kalwa are shown hunting a network of Muslim terrorists run by Zaheer and Hamid, two Kashmiri Muslims working at the behest of Pakistan's army and the ISI. The technologically fluent character of Qadir Qazi, a Muslim terrorist, infiltrates into India after escaping an Afghan prison, dons a new identity, and aids terrorists in hijacking an airplane from the highly guarded Delhi airport. As Hilal Ahmed points out, the good/bad Muslim binary "is inextricably linked to the debates on postcolonial Muslim identity." As Ahmed further observes, "A number of different phrases are used interchangeably to describe good Muslims—secular Muslims, cultural Muslims, nationalist Muslims and so on—to counter the bad guys or communal Muslims, separatist Muslims, pro-Pakistan elements and, more recently, the terrorist Muslims."[26] While Pasha, Zoya, Pathan, and Zorawar represent the trope of "good Muslim" in India, the ones who are on the side of the state and fight for the nation, the likes of Zaheer, Hamid, Moosa and Sajid—technologically dangerous "antinational" elements—represent the "bad Muslim" trope.

The Euro-Western political regimes, particularly the antiterrorism alliance represented by George W. Bush and Tony Blair, have for long sought to distinguish between good and bad Muslims, and, as Mahmood Mamdani highlights, the implication of such a binary is that "Islam must be quarantined and the devil must be exorcized from it by a civil war between good Muslims and bad Muslims."[27] That is exactly what transpires in the context of *The Family Man* and other web series as loyal Muslim officers are shown to be in constant civil war with the likes of Moosa and Sajid, their coreligionists gone rogue.

Trauma and Memory

As has been posited in the sections above, Muslims are Orientalized and techno-Orientalized by the majoritarian interests of the Indian state as a replication of the very Orientalist strategies that the British colonial state imposed on them. In between this process of Orientalism and techno-Orientalism as a rite of passage, *The Family Man* unravels deep and traumatic experiences that the Muslim characters, such as Moosa as an Othered Indian citizen, have had to endure. His plan to join ISIS stemmed from the 2002 Gujarat riots in which far-right Hindu activists killed his family members. In an emphatic tone, he narrates his scars from the past:

> At that time, my uncle used to live in Baroda. Mom, dad and my sister were visiting them. They didn't step out for two days. They were hoping that the riots would end soon. But they didn't. Two days later, a mob entered the

house. My mother and sister hid under the bed. They witnessed it all. They watched the mob slaughter my father and my uncle. My sister screamed involuntarily. Those bastards dragged them out and thrashed them like animals! My mom wasn't lucky enough. She survived. Since then she wakes up screaming every single night. My father, sister and uncle appear in her dreams every night.

The Gujarat riots started on February 27, 2002, after a coach of a Sabarmati Express train was set on fire in a Muslim neighborhood of the Godhra area, which led to the death of fifty-seven persons, including twenty-five women and fourteen children. On board the train were nationalist Hindu activists coming back from Ayodhya, where Ram Temple was to be built at the site of the demolished Babri Masjid.[28] The "Hindu backlash" followed and an anti-Muslim campaign of violence and terror was launched with the "not-so-hidden complexity of the state machinery and the ruling party."[29] At the time of the riots, Narendra Modi, the current prime minister of India, was the state's chief minister. He is widely criticized for having turned a blind eye to the anti-Muslim violence. A report by the United Kingdom government has remarked that the event, supported by the Gujarat government, had "all the hallmarks of an ethnic cleansing."[30] The spectacular case of Bilkis Bano demonstrates the right-wing brutality that was unleashed on the minority body. Bilkis was twenty-one years old and five months pregnant when she was gang-raped by Hindu men during the violence while fourteen of her family members, including a three-year-old daughter, were brutally murdered. In 2022, on the eve of India's seventy-fifth year of independence from the British colonial yoke, the Gujarat government released all eleven convicts on August 15.

The scars of the characters Moosa and Bano fold the gap between the reel and real life of minorities in India. If in reel life, the subjects are perpetually Othered in discourse, these discourses and projections are put to practice in real life, rendering the minority body vulnerable to all kinds of misfortunes. Sajid's brother too was a victim of this misfortune. "He was fourteen when these bastards shot him," he shares with a terrorist recruiter/handler. "He was returning from school. He saw a procession. People were raising slogans. He joined them. It was a peaceful procession from Lal Chowk to Jama Masjid. No stone-pelting, no furor, nothing." Personal experiences of Moosa and Sajid thus illustrate the kind of violence Muslims in the plains of India and the Himalayan territory of Kashmir face, where an enduring political conflict has brutalized the local population for decades now.

In response to this shared grief that they witness every day at the hands of the state and majoritarian Hindu groups, *The Family Man* projects Muslims as thirsty for revenge. This kind of portrayal predicated on revenge reduces them to a community of revenge seekers and takes away from them the agency of citizens equally contributing to the society. The processes of re-Orientalism and techno-Orientalism are thus foregrounded in Othering the Muslim minority in terms of viewing them either as second-class citizens, children of a lesser god, innately backward or antiquarian, or as dangerous social elements, lacking morality, and hence in need of being completely purged from the nation and society.

Techno-Orientalizing the Tamil Minority Body

Premiered on June 4, 2021, *The Family Man*, season 2, builds on the Orientalizing and techno-Orientalizing of the Muslim minority and deploys a somewhat similar treatment on the Adivasi/Indigenous body—another minority—represented, among others, by Raji (played by Samantha Ruth Prabhu), a Sri Lankan Tamil liberation fighter. The web show creates an imaginary axis between communities at the periphery and their alliance with international and regional terror outfits. Thus, the ISI, together with brainwashed Indian Muslims, is shown to be in league with Sri Lankan Tamil Eelam fighters in spreading terror across South Asia. The context of this techno-Orientalizing of another minority is the long fight of the Liberation Tigers of Tamil Eelam (LTTE), which ran a more than two-decade-long military operation against the Sinhala majority–dominated government that was engaging in systematic persecution of minority Tamils in Sri Lanka.[31] By foregrounding the Eelam threat and its links with the ISI and Indian Adivasi constituencies, *The Family Man* lays bare the anxieties of the state with regard to internal secessionist movements in the Punjab, Kashmir, the northeast, Chhattisgarh, and south India. Raji and her affiliates represent this imaginary threat that emanates from the Indigenous and Adivasi communities.

The framing of this unlikely partnership among the ISI, Indian Muslims and Tamils, and Eelam leaders has historical antecedents. Sinhala history has portrayed Muslims as "pirates" and blended this stereotype with that of Tamils, the traditional "other" in Sri Lanka; both have been perceived as common enemies.[32] So, when Raji and Sajid are shown fighting for the same cause, it is a recirculation of the old stereotype that both are the common enemies who pose great threats to the state—both India and Sri Lanka—and must be eliminated at all costs. The character of Raji, brutalized by the Sri Lankan army, represents the Adivasi community and their resilience. While the displacement of tribal communities in India is partly attributed to urban development, it is deeply rooted in the extractivist vision of the Indian state. This vision legitimizes the state to wage an endless war against its own citizens in locales collectively tagged as the "red corridor," locations that are rich in natural resources such as bauxite and iron ore but are projected as hotbeds of Naxalite and Maoist terrorism. The violence that is mounted against minorities and Indigenous peoples is embedded in what Dickens Leonard calls "the social category of spatial power relations, bodily (dis)locations and displacements."[33] It is this bodily dislocation and displacement that Raji fights against, echoing the desire of those who have been driven out of their homes. Together with Sajid,

FIGURE 5.3 Sajid Ghani and Selvarasan, a Tamil militant commander, assembling C4 explosives on a minijet. (Source: Amazon Prime.)

Raji's character brings to the fore what film critic Tanul Thakur calls "the sorrows of geographical orphans craving their homelands."[34]

In trying to establish this connection, *The Family Man* Orientalizes and techno-Orientalizes Raji and her affiliates, akin to how it treats the Muslim minority body through Moosa and Sajid. Raji is at once homeless, a highly trained pilot, and a guerilla fighter, while her handler, Selvarasan, is an expert in aviation engineering, having received his education from an Irish university. The web series projects them as savages, experts in bomb-making and combat training. The bomb that they build from C4 explosives at Tigris Aviation, an abandoned flying school in the southern Indian city of Chennai, is meant to be placed on a very light jet and rammed into the convention center where a bilateral meeting between the prime ministers of India and Sri Lanka is scheduled (Figure 5.3). Violence and terror are the only attributes Raji and others have been identified with, an example of re-Orientalization at work. The state violence on Adivasis is briefly alluded to as a juxtaposition of the enduring tensions between the two. Like Moosa, who represents the off-screen reality of the Indian Muslim minority, Raji sums up the struggles and scars of the Adivasi populations in South Asia: "My father was a schoolteacher. That day, our new textbooks had arrived. My father had gone to school to distribute them. The Lankan Army carried out an air raid on the school and we never found my father's dead body. When I was fifteen years old, the army opened fire on my brother and his friends and killed them. He was nine years old. After that, three soldiers captured me. Then they carried me off to their base and there . . . one after another. So many soldiers, I lost count." Raji's experiences can be contextualized within the framework of "rape as a weapon of war" that has remained central to state policy globally against the marginalized and Indigenous to crush dissent and erase difference. This sentiment is best enunciated in the context of India by Mahasweta Devi in her powerful collection of short stories, *Breast Stories*. In her short story "Draupadi," a tribal woman, Dopdi, is captured by the army and brutally raped. After the rape, she is told by the rapists to cover herself up but she remains naked in defiance of their orders. Raji, sharing a deep homology with Dopdi, thus becomes what Gayatri Chakravorty Spivak calls the subject of "resistant rage."[35] As Ramachandra Guha observes, the sixty years of India's democracy since independence in 1947 were least rewarding for Adivasis who were at the receiving end of the nation-building project, peripheral and always the Other.[36] This enduring Othering of Adivasis and the failure of the state has given rise to Naxalist and Maoist insurgencies that seek to challenge the state and its extractivist policies laced with violence on the Indigenous populations and disregard of local traditions and faith systems. The Indian government, for instance, in its upcoming census will not count faiths practiced by the Adivasi population as separate religions. In states like Jharkhand, Chhattisgarh, and Odisha, Adivasis worship nature. Local communities across India have been campaigning for long to have their faiths—the Sarna faith for example—recognized as separate religions.[37] In Sri Lanka, the struggles of minority Tamils continue to be sifted through the prism of terrorism and secessionism. The Tamil National Alliance remains agitated in view of Colombo's failure to address Tamils' urgent concerns. Some of their core concerns are "release of political prisoners, answers to families of [forcibly] disappeared persons, and the persisting land grabs in the north and east."[38] But with India and Sri Lanka entering a new phase of diplomatic and military partnership to counter China's rising influence in the South China Sea, the minorities and the Indigenous on either side are less likely to have their voices heard, while new spells of suffering might be unleashed on them given India's and Sri Lanka's past histories of racial and antiminority violence.

"There Is the Truth. And Then, There Is the Country"

In contrast to its treatment of characters that represent the Muslim and Adivasi communities as Others, *The Family Man* and a host of other web shows featuring on Indian OTT platforms take pains to mellow the mechanistic and automatic brutality of the state by humanizing characters that symbolize it. When an innocent Kashmiri Muslim, Kareem Bhat, is gunned down in cold blood in *The Family Man*, the TASC and Force One officers express remorse over glasses of whiskey. This exchange of thoughts between Tiwari and Pasha is telling.

> PASHA It doesn't matter whether it was a mistake or not. We killed them. Now we're defaming them. Everybody's calling them terrorists.
> TIWARI There is the truth. And then, there is the country.

The "country" here becomes a catchphrase used by representatives of the state to justify violence on citizens in general and on minorities and the Indigenous in particular. The tendency of the state and its representatives toward oppression coincides with the diagnosis of the state by Russian revolutionary agitator Mikhail Bakunin, who provocatively remarks, "There is no horror, no cruelty, sacrilege, or perjury, no imposture, no infamous transaction, no cynical robbery, no bold plunder or shabby betrayal that has not been or is not daily being perpetrated by the representatives of the states, under no other pretext than those elastic words, so convenient and yet so terrible: "*for reasons of state.*""[39] In *The Family Man*, the trigger-happy tendency of state forces toward violence becomes brutally explicit when Talpade tells Tiwari on the latter's transfer to Kashmir, "Don't stroll around all alone in Srinagar. Always carry your gun. When in doubt, shoot first." "When in doubt, shoot first" defines the creed of the Armed Forces Special Powers Act, a draconian law that gives unbridled powers to Indian armed forces in conflict regions of Kashmir and the northeast. Under the law, the forces can shoot anyone merely based on suspicion and thus enjoy impunity from being tried in a court of law. Such laws and the culture of impunity—despite chilling evidence of widespread and grave human rights violations—together make evil banal, as Hannah Arendt would say, with representatives of the state like the armed forces having lost the capacity to think and act in a humane manner.

The self-reflections by state officers in *The Family Man* are at best melodramatic rhetoric to mellow down the extent of evil in the eyes of viewers and citizens. The expression "There is the truth. And then, there is the country" thus normalizes terror in the name of the nation even as agents of the state locate themselves between the fragile banalities of who they think they are and what they think they are doing. The Indian armed forces are accused of committing a number of crimes in the region mainly under the vanguard of the Armed Forces Special Powers Act, which offers them impunity from being subjected to legal trial. Rights groups have for long been fighting for "a process of accountability for institutional criminality" in Kashmir as evinced by Parvez Imroz and his colleagues in the landmark report *Alleged Perpetrators*. The report documents five hundred individuals from the army, paramilitary forces, and police as perpetrators of crime. According to the report, the "institutional culture of moral, political and juridical impunity has resulted in enforced and involuntary disappearance of an estimated 8000 persons [as of November 2012], besides more than 70,000 deaths, and disclosures of more than 6000 unknown, unmarked and mass graves."[40] Nissim Mannathukkaren speaks about violence against minorities becoming banal in India and the dangers of such banality to the human condition and democracy. He argues that the rhetoric of development serves as the biggest tool in this banalization of violence and evil.[41]

Conclusion

By foregrounding a multitude of racial, religious, and ethnic anxieties and fault lines within the Indian nation-state, the genre of web series symbolized by *The*

Family Man is fast becoming an ideological apparatus that at once Orientalizes and techno-Orientalizes the country's minorities and the Indigenous. This cinematic rite of passage from Orientalism to techno-Orientalism—by portraying the minorities, particularly Muslims, and Indigenous peoples as both socioeconomically backward and technologically dangerous—makes for a new reading about the perpetuation of racial and ethnonational anxieties in the new media genre of web television. As has been shown, the prejudicial, skewed, and banal intertwining of terrorism, Islam, nation, and minorities is not specific to *The Family Man* alone. Drawing on the already-existing prejudices and radical acts of Othering against the minorities and Indigenous, dozens of such web shows, streamed on various Indian platforms such as Voot, Zee5, and Sony Liv, have given rise to a violent imaginary in what I call a techno-Orientalist gaze that projects the country's Indigenous and minority communities as a technological and physical threat to the country and its Hindu majority community. In what Koichi Iwabuchi terms "a paradoxical combination of traditionalism . . . and high-technology," *The Family Man* and other web shows cast a techno-Orientalist gaze on minorities, interpreting them at once as both barbaric and technologically dangerous.[42] As a consequence, such web shows and the overall format of the genre have the tendency to produce new social pathologies that serve the majoritarian Hindu interests commanded by the right-wing institutions such as the ruling Bharatiya Janata Party (BJP) and its ideological fountainhead, Rashtriya Swayamsevak Sangh. And the evidence of such new social pathologies can be found everywhere in the "new India"—from the streets to the Parliament.

Notes

1 Edward Said, *Orientalism* (New York: Pantheon, 1978), 1.
2 Lisa Lau and Ana Cristina Mendes, "Introducing Re-Orientalism: A New Manifestation of Orientalism," in *Re-Orientalism and South Asian Identity Politics: The Oriental Other Within*, ed. Lisa Lau and Ana Cristina Mendes (New York: Routledge, 2011), 1.
3 David S. Roh, Betsy Huang, and Greta A. Niu, "Technologizing Orientalism: An Introduction," in *Techno-Orientalism: Imagining Asia in Speculative Fiction, History, and Media*, ed. David S. Roh, Betsy Huang, and Greta A. Niu (New Brunswick, NJ: Rutgers University Press, 2015), 2.
4 David Morley and Kevin Robins, *Spaces of Identity: Global Media, Electronic Landscapes, and Cultural Boundaries* (London: Routledge, 1995), 147.
5 By Indigenous, I refer to populations and communities that "have become marginalized as an aftermath of conquest and colonization by the people from outside the region [and those who] govern their life more in terms of their social, economic and the cultural institution than the laws applicable to the society or the country at large." Virginius Xaxa, "Tribes as Indigenous People of India," *Economic and Political Weekly*, December 18, 1999, 3590. In the context of India, the term "Adivasi" connotes communities whose indigeneity is related to aspects of marginalization.
6 Arjun Appadurai, *Fear of Small Numbers: An Essay on the Geography of Anger* (Durham, MC: Duke University Press, 2006), 8.

7   I use "peripheral" and "marginal" to denote three sociopolitical conditions in India; see M. Imran Parray, "Choking the 'Periphery': Pride and Prejudice in India's Globalizing Internet Imaginary," *Internet Histories* 5, no. 3–4 (2021): 8. Firstly, they refer to regions within the Union of India that witness perpetual violence, locations that bear, to invoke Laura Guimarães Corrêa, "interlocked oppressions" and where citizens "experience life from the standpoint of an *outsider-within*, a familiar stranger with an *oppositional gaze*": Corrêa, "Intersectionality: A Challenge for Cultural Studies in the 2020s," *International Journal of Cultural Studies* 23, no. 6 (2020): 823, emphasis in original. Secondly, peripheral and marginal refer to alternative politico-cultural aspirations—contrary to the ones projected by the state and market forces—and regional communities that seek to break free of the interlocked oppression. Thirdly, those populations and areas could be termed peripheral and marginal that are either marginalized or grossly misunderstood or misrepresented in mainstream Indian popular media and scholarship.
8   Lau and Mendes, "Introducing Re-Orientalism," 12; Irfan Ahmad, "Hindu Orientalism: The Sachar Committee and Over-representation of Minorities in Jail," in *The Politics of Muslim Identities in Asia*, ed. Iulia Lumina (Edinburgh: Edinburgh University Press, 2021), 115-144.
9   A number of communities share a conflicted relationship with the Indian state, including, among others, Adivasis, Dalits, Kashmiris, and Nagas. While some conflicts are rooted in the Indian state's denial of constitutional rights to its citizens such as Adivasis and Dalits, others are territorial in nature such as in Kashmir and Nagaland. Also see note 1.
10 Appadurai, *Fear of Small Numbers*, 8.
11 Carol A. Breckenridge and Peter Veer, eds., *Orientalism and the Postcolonial Predicament: Perspectives on South Asia* (Philadelphia: University of Pennsylvania Press, 1993), 12.
12 Étienne Balibar, "Racism and Nationalism," in *Race, Nation, Class: Ambiguous Identities*, ed. Étienne Balibar and Immanuel Wallerstein (New York: Verso, 1991), 45–46. See also Alina Sajed, "Between Algeria and the World: Anticolonial Connectivity, Aporias of National Liberation and Postcolonial Blues," *Postcolonial Studies* 26, no. 1 (2023): 13.
13 Roh, Huang, and Niu, "Technologizing Orientalism: An Introduction," 7.
14 Lau and Mendes, "Introducing Re-Orientalism," 1.
15 Ahmad, "Hindu Orientalism," 115.
16 Toshiya Ueno, "Japanimation and Techno-Orientalism," in *The Uncanny: Experiments in Cyborg Culture*, ed. Ed Bruce Grenville (Vancouver: Arsenal Pulp Press, 2002), 223. Also see Lindsay Maizland, "India's Muslims: An Increasingly Marginalized Population," Council on Foreign Relations, last modified March 18, 2024, https://www.cfr.org/backgrounder/india-muslims-marginalized-population-bjp-modi.
17 Mita Banerjee, "More than Meets the Eye: Two Kinds of Re-Orientalism in Naseeruddin Shah's What If?" in *Re-Orientalism and South Asian Identity Politics: The Oriental Other Within*, ed. Lisa Lau and Ana Cristina Mendes (New York: Routledge, 2011), 124.
18 Ahmad, "Hindu Orientalism," 117.
19 Quote from Roh, Huang, and Niu, "Technologizing Orientalism: An Introduction," 2.
20 The Sachar Committee was a high-level commission constituted by the Government of India "to prepare a report on the social, economic and educational status of

the Muslim community of India." The committee, composed of its chairman, Justice Rajender Sachar, and six other members, released its report in 2006.
21 Maheen Haider, "The Racialization of the Muslim Body and Space in Hollywood," *Sociology of Race and Ethnicity* 6, no. 3 (2020): 382–395.
22 Sanjeev Kumar, "Metonymies of Fear: Islamophobia and the Making of Muslim Identity in Hindi Cinema," *Society and Culture in South Asia* 2, no. 2 (2016): 233.
23 Kalyani Chadha and Anandam P. Kavoori, "Exoticized, Marginalized, Demonized: The Muslim 'Other' in Indian Cinema," in *Global Bollywood*, ed. Anandam P. Kavoori and Aswin Punathambekar (New York: New York University Press, 2008), 131–145.
24 The Quint, "Bollywood's Role in India's Heated Political Environment," *The Quint's Films & Politics Roundtable*, January 2, 2020, https://www.youtube.com/watch?v=N8M4BgaTm38.
25 Vibodh Parthasarathi, "Between Strategic Intent and Considered Silence: Regulatory Contours of the TV Business," in *The Indian Media Economy: Industrial Dynamics and Cultural Adaptation*, ed. Adrian Athique, Vibodh Parthasarathi, and S. V. Srinivas (Oxford: Oxford University Press, 2018), 1:145.
26 Hilal Ahmed, "The Good Muslim–Bad Muslim Binary Is as Old as Nehru," *The Print*, November 6, 2018, https://theprint.in/opinion/the-good-muslim-bad-muslim-binary-is-as-old-as-nehru/145770/.
27 Mahmood Mamdani, "Good Muslim, Bad Muslim: A Political Perspective on Culture and Terrorism," *American Anthropologist* 104, no. 3 (2002): 766.
28 Christophe Jaffrelot, "Communal Riots in Gujarat: The State at Risk?" *Heidelberg Papers in South Asian and Comparative Politics* 17 (2003): 3.
29 Asghar Ali Engineer, "Gujarat Riots in the Light of the History of Communal Violence," *Economic and Political Weekly* 37, no. 50 (December 14, 2002): 5047.
30 Hartosh Singh Bal, "UK Govt Inquiry Says VHP Planned to 'Purge Muslims' in 2002 Riots, Acted with Guj Govt's Support," *The Wire*, January 24, 2023, https://thewire.in/communalism/full-text-bbc-documentary-gujarat-riots-modi-uk-report.
31 Joanne Richards, "An Institutional History of the Liberation Tigers of Tamil Eelam (LTTE)," Graduate Institute of International and Development Studies (Geneva: Geneva Graduate Institute, 2014), 6.
32 Alexander McKinley, "Merchants, Maidens, and Mohammedans: A History of Muslim Stereotypes in Sinhala Literature of Sri Lanka," *Journal of Asian Studies* 81, no. 3 (2022): 527.
33 Dickens Leonard, "Towards a Caste-Less Community," *Economic and Political Weekly* 54, no. 21 (2019): 47.
34 Tanul Thakur, "Review: Family Man 2 Is a Worthy Sequel, Even If It Lacks Intimate Dilemmas of First Season," *The Wire*, June 5, 2021, https://thewire.in/culture/review-family-man-season-2.
35 Gayatri Chakravorty Spivak, Introduction to *Imaginary Maps*, ed. Mahasweta Devi, trans. Gayatri Chakravorty Spivak (New York: Routledge, 1997), viii.
36 Ramachandra Guha, "Adivasis, Naxalites and Indian Democracy," *Economic and Political Weekly* 42, no. 32 (August 11, 2007): 3305.
37 Vijaita Singh, "Only Six Religion Options Make It to Next Census Form," *The Hindu*, May 26, 2023.
38 Meera Srinivasan, "No Action from Govt. on Tamils' Concerns, Says TNA," *The Hindu*, January 5, 2023, 15.

39 Quoted in Noam Chomsky, *For Reasons of State* (New York: Penguin Books, 2003), iii, emphasis in original.
40 Parvez Imroz et al., *Alleged Perpetrators: Stories of Impunity in Jammu and Kashmir* (Srinagar: IPTK, 2012), 7.
41 Nissim Mannathukkaren, "The Banality of Evil," *The Hindu*, May 19, 2016.
42 Koichi Iwabuchi, "Complicit Exoticism: Japan and Its Other," *Continuum* 8, no. 2 (1994): 69.

6

On Forms of the Black Box

―――――――――――――――◆◇◆―――――――――――――――

Race and Technology
in STS and Global Critical
Race Studies

CLARE S. KIM AND
ANNA ROMINA GUEVARRA

In 2017, the National People's Congress enacted the Cybersecurity Law of the People's Republic of China, which established a legal framework for digital sovereignty and data regulation without outlining the processes for reviewing the cybersecurity of digital services. Soon after, major U.S. news publications and scientific journals criticized the adverse effects that the information technology (IT) legislation would have on technoscientific development and on the global knowledge economy by employing the figure of the "black box," a technical artifact that conventionally is conceptualized as "just performing its function, without any need for, or perhaps any possibility of, awareness of its internal workings on the parts of users."[1] For instance, *Slate* highlighted U.S. policymakers' and industry experts' frustrations over the legislation enabling a regime of "'black box' review" such that "U.S. companies in China experienced an unfair playing field: cybertheft, pressure to turn over technology and intellectual property to Chinese partners, and the laws and regulations that advantage Chinese companies."[2] Likewise, a Policy Forum article for *Science* warned that an "arms race" was imminent over "proprietary 'black box' technologies."[3] More recently, a 2023 *Wall Street Journal* piece declared that China was "creating a black box around information on the world's second-largest economy."[4] Recalling frequent tropes that connect Asia(ns) with machines, such reporting forms an increasingly pervasive discourse on the shape of U.S.-China relations and the racialized dynamics of information control.

If the U.S. media's imagery of black boxes casts China as secretive and a potential threat to technology transfer and the future of the knowledge economy, then Chinese authorities have drawn on the same language to celebrate their ingenuity and effectiveness. "Do you know why the Chinese are so naturally good at deep learning?" asked Dinglong Huang, Chinese CEO of a tech company specializing in artificial intelligence (AI), in 2018. It is "because the black box has been part of Chinese society and culture since the very beginning. Chinese medicine. There is an input, some herb or infusion. You have no idea how it works, but it does. All you can do to get a different result is enter a different input."[5] In the context of technoscientific development, the figure of China as potential threat and perpetrator of intellectual theft runs parallel to Huang's framing of China's productivity and hard-working qualities as rooted in a distinctly Chinese past.

One way of understanding these varied castings of Asia(ns) is through the prism of race and technology articulated by the analytic apparatus of techno-Orientalism in critical global race studies. In this framing, such figurations of the black box invoke a long-standing practice of describing and imagining Asia(ns) as racialized Other(s) in technological terms. But just as technology evolves, so do the form and application of the black box. While acknowledging the historical and racialized visions of race and technology that previous accounts of techno-Orientalism have illuminated, this chapter aims to show that an alternative formulation of the race-technology dyad, which cuts orthogonally across the contextual approach, offers a more comprehensive model for situating the connections between Asia(ns) and technoscience and the sociopolitical worlds in which they move.

In this chapter, we trace the "black box" as a concept, a technical device, and a metaphor that we hope will model a closer engagement among science and technology studies (STS), the history of science, and global critical race studies. We argue that this engagement provides a useful intervention that illuminates the historicity and particular manifestations of techno-Orientalism in STS. We do this by tracing the varied ways black boxes have been used as material and metaphorical icons of opacity in computing, information processing, and racial narratives. Likewise, we examine their relevance to the construction of Asians and Asian-ness as inscrutable or incapable of reason and intelligence. Our approach is neither chronological nor comprehensive but taxonomic. To this end we first explore how black boxes have been racialized as Asian/Other in technical designs and representations, and how racialized assumptions and other forms of difference have underwritten the production and imaginaries of black boxes. We then show how the black box as an abstract concept and metaphor has shaped sociohistorical accounts of race and difference (Orientalism) and the collective lives and politics of Asia(ns). We do so in an attempt to take stock of what history can do for descriptions and accounts of techno-Orientalism.

In excavating this history, we show that accounts of black boxes inherit an overlapping history of technical, political, and abstract metaphorical messaging. Our examination not only reveals the racialized discourses between black boxes and Asia(ns), but also encourages further reflection on how humanistic and social-scientific scholars are imbricated in this co-constitutive process of knowledge production. The possibilities for overcoming hegemonic racial discourses, we suggest, remain constrained so long as we fail to situate critical theories and frameworks in existing histories, hierarchies, and valuations of race and technology.

Black Boxes and Difference/Differentiation

Racialized conceptions of difference informed the designs and descriptions of digital technologies and information machines throughout the twentieth century. In "The Onlife Manifesto: Philosophical Backgrounds, Media Usages, and the Futures of Democracy and Equality," media theorist Charles Ess offers electronic media as an originary example and explains how they came to be filtered through racialized differences. He writes, "In 'Western' societies, the affordances of what [Marshall] McLuhan and others call 'electric media,' including contemporary ICTs [information and communication technologies], appear to foster a shift from the modern Western emphases on the self as primarily *rational, individual*, and thereby an ethically *autonomous* moral agent towards greater (and classically 'Eastern' and pre-modern) emphases on the self as primarily *emotive*, and *relational*—i.e., as constituted exclusively in terms of one's multiple relationships, beginning with the family and extending through the larger society and (super)natural orders."[6] In this telling, such transformations in racialized differences were coeval with the development of telephone networks, digital media, and other devices that perform intricate functions but whose internal mechanisms are not so easily understood. For Ess, these changes in technological forms also signified their potential as agents of relationality, with such demarcations grounded in racialized imaginaries of social relations and premodern Oriental societies of the East.

Since the mid-twentieth century, critics, scientists, and technical experts have attached the label of "black box" to these new technological forms. But techno-scientific inquiry into the concept of the black box has a longer history preceding this mid-century coinage. In this section's historical survey of the racialization of black boxes, we treat the black box as an "epistemic space," by which we mean "the broader realm in which a [techno]scientific concept takes shape" under multiple names, contexts, and modes of inquiry that "nevertheless elude full and final representation."[7]

An early precursor of the black box can be traced to the seventeenth- and eighteenth-century designs of automatons. In continental Europe, mechanical constructions such as clocks and human-like machines engendered mechanistic explanations that conceived of human bodies as complex machinery.[8] Mechanical parts such as cogs and pistons were equated with bones, muscles, and other organs. In "Enlightened Automata," historian of science Simon Schaffer argues that debates about the bodily, mental, and social mechanisms of automata were debates about the proper relations of human labor to emerging industrial capitalism. Enlightenment and mechanical philosophers understood themselves as those who could stand apart and understand the mechanical principles of nature, bodies, and society, whereas workers and artisans, in their philosophy, were like machines—devoid of sentiment or formal reason—to be understood and managed.[9] Here, Schaffer attends to the automaton as a figure that could differentiate the human from the merely mechanical. At the same time, the designs of machines that appeared to operate on their own pointed to a wider array of black-box-like devices whose concealment of their inner workings "both invited and foreclosed speculation into the inventor's methods."[10]

Closer scrutiny of automata designs intimates how visions of the properly human (read: white) were parsed from the less than human with reference to ethnic alterity. One of the most popular automata to circulate across Europe and North America was a chess-playing machine called the Mechanical Turk. Constructed by Wolfgang von Kempelen in the seventeenth century, the device was a counterfeit clad in Orientalist attire and whose cabinet of whirring gears concealed a skilled human chess player who would direct the play of the Turk (Figure 6.1). Automata like the Mechanical Turk were designed to elicit awe and wonder from the audiences to which they were presented.[11] Bernard Dionysius Geoghegan writes that the ethnic alterity framing "the interface provided a perfect distraction from the real mechanisms of human labor powering the machine."[12] While the automaton's interior relied upon a mechanistic design, its presentation also depended upon a racial imaginary that was grafted onto the machine's surface and obfuscated the invisible labor that made it work.

If conceptions of ethnic alterity were grafted onto the surface of seventeenth- and eighteenth-century automata, then the introduction of black boxes in the mid-twentieth-century marked a moment when such features were removed but whose racial logics remained embedded and enshrouded in secrecy. During World War II at the MIT Radiation Laboratory, the term "black box" referred to the use of black-speckled boxes to encase radar electrical equipment such as amplifiers, receivers, and filters.[13] Two differing trajectories followed from this technical form. On the one hand, its material components were drawn upon to produce alternative configurations that in turn preceded the development of flight data recorders and other discrete devices to which the term "black box"

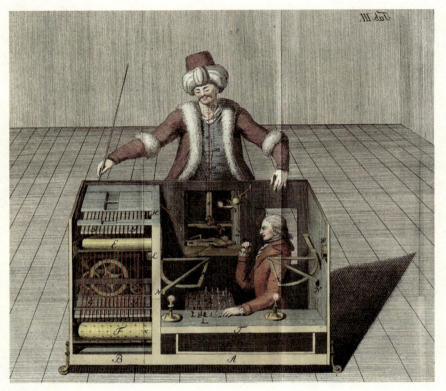

FIGURE 6.1 The chess-playing Mechanical Turk. (Source: University Library of the Humboldt-Universität zu Berlin.)

was most commonly applied. On the other hand, those same features also facilitated its use for information processing. A black box, as mathematician Norbert Wiener used the term, referred to a unit defined to perform a function before one knew how it worked in terms of "inputs," "outputs," "messages," and "feedback." By contrast, what he called "white boxes" designated devices whose inner mechanisms were specified and understood to the user.[14] For Wiener, this distinction was crucial for explaining how an entity sustained itself and how it related to or clashed with other entities.

Black boxes also became significant for articulating a new field of scientific study called cybernetics. Derived from the Greek word for steersman and coined by Wiener in 1947, cybernetics aimed to study communication and control processes in both machines and complex living organisms. This new field analyzed the circulation of messages and information to explain how a human and/or nonhuman entity sustained itself. The design of these technological forms, as Peter Galison has shown, was premised on a vision of the enemy pilot so merged with machinery that (his) human-nonhuman status was blurred.[15] The resulting servomechanical enemy was fashioned in opposition to an enemy Other that could be identified in the Pacific theater.

Influenced by the field of cybernetics, black box devices acquired an explicitly racialized and militarized ontology at midcentury. In the killing frenzy of World War II, one version of the enemy Other for the United States was barely human and human-machine-like. An example can be seen in the anti-Japanese propaganda circulating at the time. In a 1942 poster stamp titled "This Is the Enemy," Harley Melzian depicted a Japanese soldier not only as a monstrous animal but also as one that threatens the white racialized order (or the white woman in particular). In other propagandic films produced by the U.S. Department of Defense such as *Know Your Enemy* (1945), the notion of the Japanese as human automatons by nature (and thus as extensions of the abstracted black box) gained traction. As Long Bui elaborates in *Model Machines*, the representation of Japan or the Japanese as a "war machine" was predicated on "fixating on Japanese people's 'natural' docility toward authority" and "exaggerating Japan's cultural backwardness" with "people treated as little more than rice-cultivating 'human machines.'"[16] Militarism undergirded black box designs and was supported by a whiteness and masculinity that animated a racialized enemy Other.

Yet black boxes, like race, are malleable things. The inscrutability of black boxes gathered an immaterial form in the mid-to-late-twentieth-century traditions of logic, behaviorism, and early AI research. Abstracting from earlier recourse to the physical body, technical experts came to view the mind as a black box and computers as thinking machines. Cyberneticians, who had originally defined the black box as a system whose inner workings were unknown or unknowable, further developed and expanded on the concept as a tool and metaphor for different kinds of scientific research. As Elizabeth R. Petrick has shown, W. Ross Ashby, Stafford Beer, and W. Grey Walter, building on the works of Norbert Wiener, drew on the black box as a "way to understand and manipulate complex systems across a wide range of scientific fields: biology, neurology, psychology, electrical engineering, and physics." Operating simultaneously as a mathematical theory, a device, and a metaphor, black boxes were utilized to help "build an electronic computer that would model the human brain."[17] For instance, in the 1950s, Ashby moved from working on a mathematical black box to a physical one and constructed his most famous device: the homeostat. It was a feedback machine consisting of both a box of switches that controlled electrical resistance and a magnet on top. The homeostat responded to disequilibrium by altering its own resistance values—inputted as "information"—until the magnet on top was centered and achieved stability.[18] In this way, it was intended as a model of how the brain responded to feedback from its environment to course-correct some part of the body. Abstracted away from the concerns of wiring and vacuum tubes, black boxes here were figured as learned and conditioning machines akin to the human mind.

Building on cybernetics and a longer history of mathematical logic aimed at formally describing human thinking, one approach to Artificial Intelligence (AI) research dominated in the early postwar period: treating digital computers and human minds as symbolic information processing systems. According to this AI model, symbolic information served as inputs for both minds and computers that, in turn, produced solutions, decisions, and other "intelligent" outputs.[19] Researchers believed AI could be achieved by first identifying the formal rules that human minds follow, and then translating those rules into instructions—or programs—that the computer could execute. Of particular significance to researchers was an understanding of intelligence as both universal and abstract. Whether a brain, a body, a machine, an organization, or a hybrid system of all, the material form from which rules were to materialize and be enacted did not matter. But as historians of science have shown, such presumptions belied a predisposition for privileging white, European and American men while deeming others—women, people of color, colonized and enslaved people—as irrational or incapable of exhibiting intelligence.[20]

The work of philosopher John Searle offers one example where this conceptualization has structured thinking about tests for artificial intelligence in relation to a racialized Chinese Other. In "Minds, Brains, and Programs" (1980), Searle presents his so-called Chinese Room Argument as a black box counterargument to the Turing Test conceptualized by the mathematician Alan Turing (1950).[21] Turing argued that since consciousness is subjective, the only way we can know if a computer is intelligent is to ask it questions. If the computer's answers cannot be distinguished from those of a human, the computer has to be considered capable of thought and, by extension, sentient. Searle counters Turing with a hypothetical situation of himself locked inside a room and tasked with translating Chinese characters. He does not recognize them as such but perceives them to be black squiggles, as inputs processed into outputs using predefined instructions. Though he is like a computer program, an effective processor of messages such that his responses are intelligible to Chinese speakers, he still does not understand the squiggles he is manipulating. Thus Searle's Chinese Room Argument makes the case about the complexity of human thought, which cannot be easily simulated by or manipulated through artificial intelligence. However, the coherence of this framing belies and depends upon a valorization of the Chinese language as enigmatic and unintelligible. Creating a paradoxical construction of alterity in which the inscrutability of Chinese characters serves as a resource for a seemingly universal vision of artificial intelligence, Searle ultimately reinscribes a Western one.

The late twentieth-century rise of predictive analytics, machine learning, and data-driven computational practices precipitated a statistical turn that redirected the focus of AI and computing work away from simulating symbolic and linguistic behavior towards acquiring data for algorithmic processing and information management. In other words, these new approaches displaced a logical and rules-based formulation of the black box for one valued for its storage capacity and containment utility. This development has proceeded in tandem with the deployment of a wide range of informatic techniques, engendering new forms of algorithmic control, such as the commodification of personal data within emergent regimes of risk and surveillance, as well as the proliferation of standardized data management practices across domains ranging from design and architecture to governance and statecraft.[22] Consequently, researchers and technical experts increasingly aim to design automated systems that bypass the capacity of human thinking and behavior.[23] Some examples include the development of machine learning or automated game-playing systems that obtain new strategies by playing against themselves rather than training against human experts or utilizing human-inspired heuristics.

Despite efforts to design complex and objective black box systems that draw upon the capacity of computing technologies rather than human behavior, racialized logics and other notions of difference continue to suffuse their forms.[24] When Google's DeepMind system AlphaGo sparred with and won against Korean professional Go player Lee Sedol in 2016, AI researchers' reactions to the neural net's decision-making process revealed their enlistment of a racialized imaginary in algorithmic form. Researchers were surprised by the moves made by AlphaGo, which had trained on datasets produced by itself to make statistical predictions. AlphaGo had utilized "nonhuman" moves, they exclaimed, which made it "very difficult to find out why [it] made a particular decision."[25] Part of their surprise also stemmed from the selection of Go as the game to be played. "If Chess is the quintessential European game, Go is the quintessential Asian game," commented software developer Jeff Atwood. "Go requires a completely different strategy. Go means wrestling with a problem that is essentially impossible for computers to solve in any traditional way."[26] The researchers' shock at the opaque and unaccountable nature of the algorithmic system undergirding AlphaGo paralleled their indifference to racializing Go, as a non-Western game, as similarly impenetrable and inscrutable.

Where might this particular historical trajectory of black boxes—from their Enlightenment precursors to their abstracted, encoded (and implicitly racialized) forms—lead to? Many places. One contemporary material manifestation is the case of EngKey (Figure 6.2), a telepresence robot that was created to teach English in South Korean elementary schools.[27] Behind EngKey's innovation is the laboring body of a Filipina teacher working remotely in the Philippines whose identity and presence are occluded by a white avatar. It is not only a cost-effective measure for addressing the demand for such instructors, but also allows a labor-exporting state like the Philippines to keep its laborers "in place." Collectively, the design and representation of EngKey instantiates the historical legacies of black box designs and multiple processes of invisibility, whereby visions of black boxes and completely automated machines often rely on the deliberate masking of the racialized and gendered aspects of imaginaries and labor that figure into globally networked capitalism.[28]

Collectively, the historical sketch we have drawn here reveals a tension in the changing material forms and racialized imaginaries of black boxes between their

FIGURE 6.2 EngKey. (Source: The Korean Institute of Science and Technology.)

descriptions—ranging from the mechanical and logical to the informatic and statistical—and claims about what they enable or explain about Others. The shuttling back and forth between the technical and racialized forms of black boxes underscores that ambiguity and has been contoured by notions of materiality associated with each conceptualization of the black box.

Asians and/as Black Boxes

Not only have black boxes been designed and represented through technical and symbolic forms that often racialized them as Asians or Others, but the metaphor of the black box has also contoured sociohistorical and popular accounts of the collective identity and politics of Asia(ns). What happens when black boxes as material and technical entities are seemingly left behind, manifesting as mobile metaphors for thinking about identity, mobility, and/or social agency? Do figurations of difference from other contexts follow them, or are these undone and redone? We turn now to surveying the functions and effects of adopting the black box as an analytic within humanistic and social scientific scholarship. We do so not only to interrogate the assumptions that underwrite analyses, but also to reflect on how scholars, including us, have been participants—unacknowledged or not—in the technical and racialized discourses we seek to disentangle. By mapping out the at once diverse and circumscribed contexts in which the black box metaphor has taken on meaning, we also gain insight into when, whether, and for whom this framing took place. The proliferation of "black box" metaphors points to transformations in both understanding and inhabiting forms of difference such as race, particularly in connection with informatics, work, migration, and economic independence in the context of U.S. empire.

In the 1980s, many technically trained Asians and Asian Americans found their lives transformed by the potential for upward socioeconomic mobility, which was partially a result of the increasing demands for labor in the production of consumer electronics and was facilitated by immigration reforms that favored technical skill. U.S.-based commentators worried that the proliferation of Asian-owned manufacturing and media electronics corporations—primarily from Japan—risked casting Asians as complex figures of economic and technological integration on the one hand, and as figures of cultural and national exclusion on the other. Since the 1990s, STS scholars seeking to unpack the sociotechnical dynamics of this development have utilized the phrase "opening up the black box" as a metaphorical stand-in for an approach to studying the social aspects of science and technology. Popularized by Trevor Pinch and Bruno Latour, the "black box" metaphor has pervaded accounts of the social construction and politics of technology-related issues, being understood as a system one can apprehend, as Langdon Winner puts it, "solely in terms of its inputs and outputs."[29] In the conclusion to his *Confronting Nature: The Sociology of Solar-Neutrino Detection*, Trevor Pinch invoked the term to explain how technologies could appear uncontroversial and undisputed. It would be when the use of a device—in his case, a solar-neutrino telescope—became "black boxed" and any subsequently extracted data were understood as matter-of-factly speaking for "Nature." As he elaborated, "It is as if the social struggle over a piece of knowledge has become embedded in a piece of apparatus. Black-boxed instruments are the carriers of social relations."[30] Here, the analytic of the black box functioned to signify the closure of technoscientific debate.

Similarly, in *Science in Action*, Bruno Latour promoted the figure of the "black box" as a productive analytic to examine the legitimacy of science and technology as a social process. Borrowing from cyberneticians' understandings of the "black box," Latour elaborated on a black box as a complex network of enrolled actors all acting as one single automaton.[31] In this articulation, which is more commonly referred to as Actor-Network-Theory, scientific facts and technological artifacts can be treated the same: both require the gathering of networks of enrolled actants together such that they are assembled into a thing that acts as one and serves to exert power over other things.[32] Latour's framework demonstrates how seemingly neutral and stable technoscientific objects like minicomputers could be alternatively figured as unsettled and unstable points of inquiry that depended upon each other and the networks that supported them.[33] Later, in *Pandora's Hope: Essays on the Reality of Science Studies,* Latour elaborated on the process of "blackboxing" as a generative lens by which to uncover the myriad ways that "technical work is made invisible by its own success," to the extent that technology carries now a "fundamental opacity."[34] Collectively, these framings offer alternative explanations for why technoscientific objects like automata could be rendered consistently as holistic, universal technologies without presuming their inherent neutrality.

The turn of the twenty-first century saw a state-led rescripting of Asian identity in line with an increasingly securitized migration discourse following a post-9/11 world. Throughout the 2000s and 2010s, the U.S. government and corporations increasingly turned to algorithmic processes and data-driven surveillance regimes to identify, determine, and structure the financial and social lives of those under and moving through their purview. The resulting association of Muslims as security threats and increased scrutiny on them was not lost on media scholars. Legal scholar Frank Pasquale has denoted the emergence of a "black box society" to foreground the technical complexity of algorithmic systems and the access to their wide surveillance capacities due to the scale of their applications, the technical expertise required to engage their code, and their proprietary nature. In Pasquale's view, understanding algorithmic systems and big data technologies—including those used for reputation and search—as black boxes denotes asymmetrical power relations, highlighting the ways in which the United States or tech companies can operate under "interlocking technical and legal prohibitions."[35]

If previous STS scholarship has drawn on the black box metaphor to denote issues of closure related to the politics of technoscience, subsequent work has mobilized the metaphor to identify problems of transparency and accountability. More recently, scholars of globalization, Asia, and migration have adopted the black box as an analytic to understand migrant networks and infrastructures of labor across Asia. As Johan Lindquist, Biao Xiang, and Brenda S. A. Yeoh write in their essay on the changing political economy in Asia: "Although much is known about why migrants leave home and what happens to them upon arrival, considerably less is known about the forms of infrastructure that facilitate their mobility. By infrastructure we mean the institutions, networks and people that move migrants from one point to another. This is what we identify as the "black box" in migration research. . . . A focus on brokers, we argue, is one productive way of opening the black box of migration research."[36] In this case, the focus on the brokering process allows for the identification of the actors and the in-between processes that govern how people are moved through the migration process. In the context of Asian migration, opening up the black box of migration provides a window for understanding the assemblage of infrastructures that enable the recruitment of migrants through various pathways. This shift allows for a distinctive focus on the infrastructures (e.g., brokers) that anchor and facilitate the migration process rather than the lived experiences of migrants.

This plays itself out differently in different contexts. For example, the Philippines's infrastructure consists of a highly coordinated system of actors working together to produce a labor-exporting economy—the Philippine state, the private recruitment agencies, technologies, and nursing schools, among many others.[37] In Sri Lanka, the fate of the workers depends on the *khafala* ("sponsorship") system in the Middle East, which monitors the everyday structure of their mobility by tying their employment and residency status to their *kafeel* (employer or sponsor).[38] Xiang Biao's work on the infrastructure of a global labor management system called "body shopping" constitutes another example.[39] These "body shops" are facilitated by consultants and brokers who play the critical intermediary role of recruiting and placing information technology workers from India to various sites around the world. Across all these cases, these intermediaries not only define but constitute the very core of the black box of migration. On their own terms, the attention to migration and knowledge infrastructures renders a gnarly knot of performances, labor hierarchies, sexualities, racial formations, vectors of desire, and more, with no definitive foundation. What is notable about these scholarly invocations of the black box is their dive into technical imagery to illuminate the role of material infrastructures, positionality, and power relations. They are recuperative moves not unlike other scholarly attempts to move beyond a singular, self-contained subject and a generalized or unified position for analysis and additional inquiry.

However, nonreflexive and uncritical deployments of black boxes depend upon accepting a particular relationship between technology and its varied material forms or materiality: the ability of technologies to transform into and function as epistemological figures that coordinate, suspend, or illuminate difference. Consequently, a vast array of social, material, and institutional arrangements disappears and are replaced by hermeneutics and language. Moreover, we implicitly condone the assignment of metaphors and other analytic categories as exempt from history and periodization. The continued reliance on this assumption constrains scholars' abilities to identify differential power relations and properly situate Asia(ns) within/across time and space, and thereby risks reinscribing their opaqueness and inscrutability in humanistic and social scientific analyses. This problem will persist so long as metaphors continue to be presumed or imagined as immaterial. What would an account of techno-Orientalism and black boxes look like that does not take for granted the contexts, networks, and materialities to which black boxes have been invoked and utilized? Can we have an analysis that also acknowledges social science and humanities scholars as situated participants within the same infrastructure(s) as politicians, technologists, and scientists?

Rethinking Race and Technology in STS and Global Critical Race Studies

Insofar as this chapter is about black boxes, it has sought to amplify analyses of race and technology by considering not only form and structure, but also agency. As we have shown, black boxes as material-formal entities have been shaped and reshaped by rhetorics of race and other forms of social difference. Asians have figured as nonhuman, mechanistic laborers, with such castings opposing putatively hetero-white normative principles as reason, rationality, and innovation. At other times, Asians have been cast as embodiments of the antithesis of artificial intelligence, as in the examples of the Mechanical Turk and the Chinese Room Argument. Closer scrutiny of the rhetorical relays between race and technology in these accounts of black boxes reveals how the trope of the Asia(n) as machine builds on nested relations and histories in the crosscurrents of labor migration, geopolitics, Enlightenment philosophy, and scientific epistemology.

We have revisited the racialization and Othering of Asians at a moment when the social sciences and humanities have increasingly drawn different framings of materiality and materialisms to make sense of digital labor and migration, the "posthuman" and "nonhuman," and the circulation and politics of knowledge and technology.[40] Scholars in Asian and Asian American studies have deployed the theoretical framework of techno-Orientalism to illuminate the myriad ways that the Asian body has functioned "as a form of expendable technology" across literary and cultural media.[41] In doing so, they have risked treating the materiality of technology as self-evident, which in turn flattens the functions that technoscience achieves in accounts of the world, no matter at what scale or level of abstraction. Whether in the form of a mechanical device or an abstracted technical process, the role of technologies through the lens of techno-Orientalism invokes two effects concerned with the production of racial difference. It operates at the register of bodily experience as well as allows for the conflation of culture with embodied racial difference. Stated otherwise, whether addressed toward the domain of embodiment and biopolitics or that of geopolitics, technologies always function as a neutral medium. This raises two problems for scholarly analyses of technology and race. First, this approach risks reducing a broad range of racial representations and power relations to always already being effects of technological intervention, thereby eliding hierarchies of difference, labor, and agency. Second, it obfuscates the ways in which racial differences and social imaginaries have in turn contoured technologies. Important to keep in mind, however, is not just the role of technoscientific concepts in formatting what scientists and other actors take to be material, but also the historicity of the analytic categories that humanistic and social scientific scholars adopt.

Therefore, in this chapter, we offer a methodology that can strengthen the techno-Orientalist critique by not only introducing an encounter between critical race theory and science and technology studies to analyses of techno-Orientalism, but also calling into question the notion of encounter itself, asking how we can make sense of our own entanglements and partial engagements with technologies and other theories that furnish our world. This method follows on the calls of feminist STS scholars Sophia Roosth and Astrid Schrader to "suspend divisions between doing science, making sense of science, and simply getting on in the world," with the understanding that writing about technology and "writing *about* science [are] never separable from the work of science itself."[42] Neither objects of study nor modes of inquiry can be contained in disciplinary discourses when subjects and objects encompassing techno-Orientalism get involved with one another, leaving inherently indeterminate who engages with whom. Upending the distance drawn between critical race theory and science and technology studies, we underscore how the dynamics and historical production of techno-Orientalism are a function of shifting entanglements of theories and sociotechnical arrangements.

While we cannot offer a fully realized overview of all possible avenues, as such a project must await further study and elaboration, we wish to highlight one direction that might incorporate productive cross-engagement between STS and the allied fields of global critical race studies. Donna Haraway and other feminist and critical race STS scholars remind us that metaphors can be understood as "descriptive technologies" that open up new possibilities for analysis without naively disavowing "the cultural significance, material contexts, or sacredness" of the concept.[43] In this framing, black boxes can simultaneously be read and reread as things material and formal, concrete and conceptual, and as operating across domains of knowledge ranging from the humanistic to the technical in different times and places. For example, the proliferation of black box metaphors and technologies in relation to U.S.-Chinese relations, referenced at the beginning of this chapter, can be reread and newly understood as instantiating a form of what Iván Chaar López calls "'networked asymmetries'—the uneven associations and differential arrangements through which actors are enrolled in shifting relations."[44] Understanding the differential arrangements that produce and are produced of China as black boxes through networked asymmetries allows for a nuanced understanding of the privileged yet discriminated experiences of Asians and Asian Americans, while at the same time highlighting instances of disposability and nonessential knowledges that are elided at that level of description. Such cross-engagement offers another way of drawing out the unimaginable expanse of global capital and interimperial formations at work.

Overlaying this STS approach with the existing toolkit of Asian and Asian American studies not only helps bring these fields into more productive conversation with each other, but also invites further reflection and scrutiny into other unacknowledged assumptions that limit our ability to unsettle sedimented narratives and representations. We can turn previous articulations of techno-Orientalism back in on themselves, asking after the technoscientific concepts that have contoured understandings of difference. In their early description of techno-Orientalism, for instance, David Morley and Kevin Robins contended that the 1980s' "association of technology and Japaneseness now serves to reinforce the image of a culture that is cold, impersonal and machine-like, an authoritarian culture lacking emotional connection to the rest of the world."[45] Tellingly, their analysis of cultural identity drew upon the metaphor of the black box. A robust "theoretical understanding of cultural identity" was unnecessary, they averred, since "cultural identity is a black box" and "attack from the exogenous forces of 'foreign' communications empires" posed the only real threat to the expression.[46] Their emphasis on exteriority and spatial logics casts the production of difference as politically symptomatic rather than as analytically trenchant.

Closer engagement with STS and feminist, critical race, and post/decolonial scholarship can draw attention to the politics of technological forms at work in analyses of techno-Orientalism, making explicit the discursive connections drawn between critical (race) theory and technoscientific accounts of the world. Following feminist STS scholars like Ruha Benjamin, we can observe the ways in which race and racism get encoded. "We can conceptualize race *itself* as a kind of technology," she reminds us, as "one of the many conduits by which past forms of inequality are upgraded."[47] Benjamin's invocation of "race as technology" enables her the ability to excavate how seemingly neutral black box technologies have been rooted in anti-Blackness. She posits that the "anti-Black box links the race-neutral technologies that encode inequity to the race neutral laws and policies that serve as powerful tools for white Supremacy."[48] Where might Asia(ns) figure in this dynamic, and how would they be discerned? Considering this, future scholarship can routinely take stock of the language and conceptual tools undergirding analyses of techno-Orientalism to avoid naturalizing and essentializing the difference between race and technoscience.

Notes

We would like to thank Justin Phan, Gayatri Reddy, anonymous reviewers, and the editors and contributors of this volume for their invaluable feedback on different stages of this manuscript.

1. Donald MacKenzie, *Inventing Accuracy: A Historical Sociology of Nuclear Missile Guidance* (Cambridge, MA: MIT Press, 1990), 26.
2. Samm Sacks and Graham Webster, "The Trump Administration's Approach to Huawei Risks Repeating China's Mistake," *Slate*, May 21, 2019, https://slate.com/technology/2019/05/u-s-china-huawei-executive-order-foreign-adversary-national-security.html.
3. Jesse M. Keenan, "A Climate Intelligence Arms Race in Financial Markets: Public Policy Grapples with Private 'Black Box' Models," *Science* 365, no. 6459 (September 20, 2019): 1240.
4. Lingling Wei, Yoko Kubota, and Dan Strumpf, "China Locks Information on the Country Inside a Black Box," *Wall Street Journal*, April 30, 2023, https://www.wsj.com/articles/china-locks-information-on-the-country-inside-a-black-box-9c039928.
5. Dinglong Huang, quoted in Bruno Maçães, "China's Black Box Superiority," *Politico*, November 12, 2018, https://www.politico.eu/blogs/the-coming-wars/2018/11/china-black-box-superiority-cybersecurity-artificial-intelligence-ai/.
6. Charles Ess, "The Online Manifesto: Philosophical Backgrounds, Media Usages, and the Futures of Democracy and Equality," in *The Online Manifesto: Being Human in a Hyperconnected Era*, ed. Luciana Floridi (New York: Springer, 2014), 98.
7. Staffan Müller-Wille and Hans-Jörg Rheinberger, *A Cultural History of Heredity* (Chicago: University of Chicago Press, 2012), xi–xii.

8   For a more thorough survey of the automaton in the Western imagination from antiquity to the Industrial Age, see Minsoo Kang, *Sublime Dreams of Living Machines: The Automaton in the European Imagination* (Cambridge: Harvard University Press, 2011).

9   Simon Schaffer, "Enlightened Automata," in *The Sciences in Enlightened Europe*, ed. William Clark, Jan Golinski, and Simon Schaffer (Chicago: University of Chicago Press, 1999), 126–165.

10 Roger Moseley, *Keys to Play: Music as a Ludic Medium from Apollo to Nintendo* (Oakland: University of California Press, 2016), 159.

11 Lorraine Daston and Katherine Park, *Wonders and the Order of Nature, 1150–1750* (New York: Zone, 2001).

12 Bernard Dionysius Geoghegan, "Orientalism and Informatics: Alterity from the Chess-Playing Turk to Amazon's Mechanical Turk," *Ex-Position*, no. 43 (June 2020): 48.

13 David Mindell, *Between Human and Machine: Feedback, Control, and Computing before Cybernetics* (Baltimore: Johns Hopkins University Press, 2002).

14 Arturo Rosenblueth, Norbert Wiener, and Julian Bigelow, "Behavior, Purpose, and Teleology," *Philosophy of Science* 10, no. 1 (1943): 23–24.

15 Peter Galison, "The Ontology of the Enemy: Norbert Wiener and the Cybernetic Vision," *Critical Inquiry* 21, no. 1 (1994): 228–266.

16 Long T. Bui, *Model Machines: A History of the Asian as Automaton* (Philadelphia: Temple University Press, 2022), 89.

17 Elizabeth R. Petrick, "Building the Black Box: Cyberneticians and Complex Systems," *Science, Technology, & Human Values* 45, no. 4 (2019): 576–577.

18 W. Ross Ashby, *An Introduction to Cybernetics* (London: Chapman & Hall, 1957). See also Norbert Wiener, *Cybernetics: Or, Control and Communication in the Animal and the Machine*, 2nd ed. (Cambridge, MA: MIT Press, 1961); Stafford Beer, *Cybernetics and Management* (New York: John Wiley, 1959); and W. Grey Walter, *The Living Brain* (New York: W. W. Norton, 1953).

19 Allen Newell and Herbert A. Simon, *Human Problem Solving* (Englewood Cliffs, NJ: Prentice Hall, 1972). See also Hunter Crowther-Heyck, "Defining the Computer: Herbert Simon and the Bureaucratic Mind—Part 1," *IEEE Annals of the History of Computing* 30, no. 2: 42–51; Hunter Heyck, *Herbert Simon: The Bounds of Reason in Modern America* (Baltimore: Johns Hopkins University Press, 2005.)

20 See Alison Adam, *Artificial Knowing: Gender and the Thinking Machine* (New York: Routledge, 1998); André Brock, *Distributed Blackness: African American Cybercultures* (New York: New York University Press, 2020); Lorraine Daston, *Rules: A Short History of What We Live By* (Princeton, NJ: Princeton Univ. Press, 2022); Stephanie Dick, "Of Models and Machines: Implementing Bounded Rationality," *Isis* 106, no. 3 (2015): 623–34; Sanjay Seth, *Beyond Reason: Postcolonial Theory and the Social Sciences* (Oxford: Oxford University Press, 2020).

21 John Searle, "Minds, Brains and Programs," *Behavioral and Brain Sciences* 3, no. 3 (1980): 417–457; Alan Turing, "Computing Machinery and Intelligence," *Mind*, n.s. 59, no. 236 (October 1950): 433–460.

22 Dan Bouk, *How Our Days Became Numbered: Risk and the Rise of the Statistical Individual* (Chicago: University of Chicago Press, 2015); Josh Lauer, *Creditworthy: A History of Consumer Surveillance and Financial Identity in America* (New York: Columbia University Press, 2017); Reinhold Martin, *The Organizational Complex: Architecture, Media, and Corporate Space* (Cambridge, MA: MIT Press, 2005); Shoshana Zuboff, *The Age of Surveillance Capitalism: The Fight for a Human Future at the New Frontier of Power* (New York: Public Affairs, 2019).

23. Luciano Floridi, *The Fourth Revolution: How the Info Sphere Is Reshaping Human Reality* (Oxford: Oxford University Press, 2016).
24. For a refutation of algorithms as objective "black boxes," see Angèle Christin, "The Ethnographer and the Algorithm: Beyond the Black Box," *Theory and Society* 49, no. 5 (2020): 897–918. More broadly, scholars note that machine learning systems can introduce ethically and socially problematic decision-making practices that reproduce inequalities or entrench differences. See Julia Angwin, "Machine Bias," Pro Publica, accessed 12 July 2022, https://www.propublica.org/article/machine-bias-risk-assessments-in-criminal-sentencing; Virginia Eubanks, *Automating Inequality: How High-Tech Tools Profile, Police, and Punish the Poor* (New York: Macmillan, 2017); Safiya Noble, *Algorithms of Oppression: How Search Engines Reinforce Racism* (New York: New York University Press, 2018); Cathy O'Neil, *Weapons of Math Destruction: How Big Data Increases Inequality and Threatens Democracy* (New York: Broadway Books, 2016).
25. Alan Winfield, quoted in Ariel Belcher, "Demystifying the Black Box That Is AI," *Scientific American,* August 9, 2017, https://www.scientificamerican.com/article/demystifying-the-black-box-that-is-ai/.
26. Jeff Atwood, "Thanks for Ruining Another Game Forever, Computers," Coding Horror: Programming and Human Factors (blog), 25 March 2016, https://blog.codinghorror.com/thanks-for-ruining-another-game-forever-computers/.
27. Anna Romina Guevarra, "Mediations of Care: Brokering Labour in the Age of Robotics," *Pacific Affairs* 91, no. 4 (December 2018): 739–758.
28. For an extended overview of how robots and artificial intelligence have served to reinforce the logics and structures of colonial and capitalist social order in histories of human labor, see Neda Atanasoski and Kalindi Vora, *Surrogate Humanity: Race, Robots, and the Politics of Technological Futures* (Durham, NC: Duke University Press, 2019).
29. Langdon Winner, "Upon Opening Up the Black Box and Finding It Empty: Social Constructivism and the Philosophy of Technology," *Science, Technology, & Human Values* 18, no. 3 (1993): 365.
30. Trevor J. Pinch, *Confronting Nature: The Sociology of Solar-Neutrino Detection* (Dordrecht, Netherlands: D. Reidel, 1986).
31. Bruno Latour, *Science in Action: How to Follow Scientists and Engineers through Society* (Cambridge, MA: Harvard University Press, 1987), 131.
32. Bruno Latour, "On Recalling ANT," *The Sociological Review* 47, no. 1 (1999): 15–25.
33. Latour, *Science in Action*.
34. Bruno Latour, *Pandora's Hope: Essays on the Reality of Science Studies* (Cambridge, MA: Harvard University Press, 1999), 304.
35. Frank Pasquale, *Black Box Society: The Secret Algorithms that Control Money and Information* (Cambridge, MA: Harvard University Press, 2015), 8.
36. Johan Lindquist, Biao Xiang, and Brenda S. A. Yeoh, "Opening the Black Box of Migration: Brokers, the Organization of Transnational Mobility and the Changing Political Economy in Asia," *Pacific Affairs* 85, no. 1 (2012): 8–9.
37. Anna Romina Guevarra, *Marketing Dreams, Manufacturing Heroes: The Transnational Labor Brokering of Filipino Workers* (New Brunswick, NJ: Rutgers University Press, 2010); Guevarra, "Mediations of Care"; Yasmin Ortiga, *Emigration, Employability and Higher Education in the Philippines* (New York: Routledge, 2019).
38. Frank Eelens and J. D. Speckmann, "Recruitment of Labor Migrants for the Middle East: The Sri Lankan Case," *International Migration Review* 24, no. 2 (1990): 297–322.

39 Biao Xiang, *Global Body Shopping: An Indian Labor System in the Information Technology Industry* (Princeton, NJ: Princeton University Press, 2008).
40 N. Katherine Hayles, *How We Became Posthuman: Virtual Bodies in Cybernetics, Literature, and Informatics* (Chicago: University of Chicago Press, 1999); Wendy Hui Kyong Chun, *Control and Freedom: Power and Paranoia in the Age of Fiber Optics* (Cambridge, MA: MIT Press, 2006); Karen Barad, *Meeting the Universe Halfway: Quantum Physics and the Entanglement of Matter and Meaning* (Durham, NC: Duke University Press, 2008); Bui, *Model Machines*; Leslie Bow, *Racist Love: Asian Abstraction and the Pleasures of Fantasy* (Durham, NC: Duke University Press, 2022).
41 David S. Roh, Betsy Huang, and Greta A. Niu, "Technologizing Orientalism: An Introduction," in *Techno-Orientalism: Imagining Asia in Speculative Fiction, History, and Media*, ed. David S. Roh, Betsy Huang, and Greta A. Niu (New Brunswick, NJ: Rutgers University Press, 2015), 11.
42 Sophia Roosth and Astrid Schrader, "Feminist STS Out of Science: An Introduction," *Differences* 25, no. 3 (2012): 2.
43 Constance Penley, Andrew Ross, and Donna Haraway, "Cyborgs at Large: Interview with Donna Haraway," *Social Text*, no. 25/26 (1990): 8; T. L. Cowan and Jas Rault, "Introduction: Metaphors as Meaning and Method in Technoculture," *Catalyst: Feminism, Theory, Technoscience* 8, no. 2 (2022): 8.
44 Iván Chaar López, "Latina/o/e Technoscience? Labor, Race, and Gender in Cybernetics and Computing," *Social Studies of Science* 52, no. 6 (2022): 831.
45 David Morley and Kevin Robins, *Spaces of Identity: Global Media, Electronic Landscapes and Cultural Boundaries* (London: Routledge, 1995), 169.
46 Morley and Robins, *Spaces of Identity*, 71.
47 Ruha Benjamin, *Race after Technology: Abolitionist Tools for the New Jim Code* (London: Polity, 2019), 149. For alternative formulations of the relation between race and technology, see the special issue of *Camera Obscura* edited by Wendy Hui Kyong Chun, especially her contribution entitled "Introduction: Race and/as Technology; or, How to Do Things to Race," *Camera Obscura* 70 (2009), no. 24: 7–35.
48 Benjamin, "Introduction," 35.

Part III
Sinofuturism

7

Infrastructure and/as Mediation

China 2098's
Tempro-Affective Politics

IAN LIUJIA TIAN

This chapter examines infrastructure in the making of Asian futures by exploring *China 2098*, a series of thirty computer graphics created by artist Fan Wennan.[1] Since its first appearance on the social media platform Weibo, *China 2098* has drawn praise from Chinese nationalist commentators who call it "infrastructural punk": an alternative to cyberpunk and techno-Orientalist representations of Chinese technology and bodies as "threats."[2] Critiques of Chinese ethnonationalist government see Fan's work as ideological propaganda that justifies authoritarianism.[3] This chapter situates *China 2098* within broader discussions about the politics of infrastructure. It moves beyond a political-economic lens by turning attention to how the representation of infrastructure refashions the politics of affect and constructs a lineal temporality, informed by China's nationalism and capitalist development.[4] This "infrastructural optimism" combines future advanced technology with the past and despair with hope. Such a mediative process reveals that Chinese triumph over technological development might continue to reproduce, rather than contest, empire-building and expansion.

Infrastructure has in recent years become the key focus in debates over Asia's place in the global economy. With China's emergence as a global hegemonic investor, infrastructural projects have become promises of prosperity, advancement, and profits.[5] In China's Belt and Road Initiative, a manifestation of maritime and land-based global supply chain capitalism by another name, infrastructure takes center stage in mediating powerful interests, relations, and forms of practices. In the United States, infrastructure has also resurfaced as an important political project. After decades of disinvestment and months of cross-party negotiations, President Joe Biden in 2021 announced the $1.2 trillion Bipartisan Infrastructure Investment and Jobs Act that promises to "upgrade" America.[6] Across the Asia-Pacific, from former Philippine president Rodrigo Duterte's "Build! Build! Build!" initiative and Indonesia's establishment of a new capital city to future high-speed train projects in Thailand, one could characterize the so-called Asian century as the "infrastructure century."

Scholars have argued the rush to build infrastructure does not necessarily guarantee improved mobility and prosperity for all. Infrastructure is rather about the disruptions on the way, which enable speculations of price and value in the finance capitalist era.[7] While it is critical to investigate, following critical geographers, how infrastructure entangles with the circulation of goods in supply chain capitalism, it is also important to explore how infrastructure affects people's imagination of new modes of being and living.[8]

Situating *China 2098* in Chinese Science Fiction

While scholars of Chinese science fiction (SF) have considered the plots of several contemporary SF novels as evidence of non-Western understanding of technology, this chapter highlights the central role of infrastructure in mediating a new mode of imagining the future.[9] *China 2098* emerged both from China's recent "SF fever," such as the popularity of Hao Jingfan's and Liu Cixin's works, and from the Chinese leadership's push to tell a better China story on the world stage. But SF as a literary genre has a much longer history in China. In the late Qing period, SF was first introduced by reformists such as Liang Qichao, who translated Jules Verne's *Two Years' Vacation*, which was the first time the genre became available to Chinese-speaking readers. During the Republic era, writing about the future was a form of social critique of the semicolonial and semifeudal conditions in China.[10] After the success of the communist revolution, a selection of Soviet SF writers was commonly read until the Cultural Revolution, when tales about utopias, industrialization, and space travel were considered "nonrevolutionary."

Since the Reform and Opening-Up period, especially since the 1990s, SF has once again emerged as an alternative literary genre in China's literary scene. A new sense of social justice and critique has also accompanied the revival of SF. Han Jingfan's *Folding Beijing* and Han Song's *2066: Red Star Over America* are just two of the prime examples. *China 2098* differs from these works in substantial ways. It is not intended to be social commentary but presents an "objective"

FIGURE 7.1 Screenshot of the ArtStation website featuring Fan Wennan's work. (Source: ArtStation.)

singular national future. As Virginia L. Conn argues, contemporary SF in China embodies the photographesomenon, a time loop where "an objective national past becomes always-already written by and understood through the lens of a future still to come."[11] *China 2098* perhaps best proves Conn's point through its panoramic illustrations of infrastructure.

Published via the artist Fan Wennan's own account on Weibo (the Chinese counterpart of Twitter) as well as on a website called ArtStation in 2020, *China 2098* is a series of thirty computer graphics depicting various infrastructures with a socialist aesthetic. Fan drew these pictures first as sketches, and later utilized computer software such as Blender to turn them into digital illustrations. Fan's style creates an encounter between two distinct aesthetic forms: socialist tropes during the Maoist era (such as slogans, young women in red uniforms, muscular men, and images of Karl Marx; see Figure 7.1) and modernist architecture (emphasis on volume, limited ornamentation, clean and asymmetrical). A short explanation accompanies each piece of artwork, detailing what the infrastructure does and how it is constructed. Narrated in a matter-of-fact tone by a space station mechanic called Laoqi (老七), these explanations form a coherent and linear story about global events taking place from the 2030s till 2098, including climate catastrophe induced by rising sea level, China-U.S. conflict, and space voyages.[12] It is important to note that these explanations are only in Mandarin Chinese and serve as a textual mediation on how readers should interpret the images. While *China 2098* is aimed at the domestic Chinese readership, Fan also aims for an international audience. The decision to upload his works on ArtStation, a digital artists' platform, is perhaps a strategy for broader outreach. For non-Mandarin readers, then, the visuality of infrastructure might be the key appealing aspect. But if read together with the texts, *China 2098* contains more nuanced contexts and interpretations.

This chapter attempts to offer one such interpretation by unpacking *China 2098*'s broader ramifications on the infrastructural politics of the Asian future. I suggest that *China 2098*'s use of futuristic infrastructure with socialist era (1949–1979) motifs requires contextualized critical readings beyond simply seeing it as an alternative "infrastructural punk." If techno-Orientalism depicts Asian subjectivities being subsumed by machines, *China 2098* attempts to flip the script by depicting human mastery of infrastructure, machines, and nature. Is simply reversing the narrative, however, sufficient to combat techno-Orientalism? What role does infrastructure play in Fan's reinterpretation of socialist realism? What are the connections between infrastructure, politics, and affect?

The rest of the chapter first draws on social sciences and humanities scholars' critical examinations of infrastructure. They have demonstrated how the representation of infrastructure produces affective attachments and lineal temporal narrative in national state-building projects. While it is important to conceptually distinguish infrastructure as a physical object from the representation of it, in this chapter, I treat infrastructure not solely as an object but as relations and forms of practices that structure the way we live. Methodologically, separating the two as distinctive units of analysis is not beneficial, because the representation of infrastructure is integral and dialectically constitutive of infrastructure as a concept/object.

Thus, I argue that infrastructure is more than invisible, taken-for-granted "objects" that facilitate movements of goods and provide shelter.[13] In *China 2098*, infrastructure is not just something that we build and utilize, or something that simply overpowers. Rather, it produces a positivist feeling about a future where needs, comfort, and satisfaction thrive after ecological collapse. I will demonstrate, through my readings of *China 2098*, that Fan's reliance on visual and linguistic depictions of futuristic infrastructure can be interpreted as an "infrastructure sublime," much like David Nye's famed "technological sublime," that structures our imaginations about the future convergent with a modernist belief in technocratic governance.[14] That is, infrastructure mediates affective contradictions by offering "scaling up" as the subliminal response to contemporary crisis: huge projects provide consistency, survival, and thriving amid climate crisis, inequalities, and geopolitical tension. Further, socialist signs in the artwork make evident how infrastructure renders the socialist past as an allegory of the capitalist present and a promise of the postapocalyptic future. This temporal and affective politics has wider application to examine the future of Asia, which is haunted by past colonial histories, and where future reinvention is promised by infrastructure's ability to reclaim space, sovereignty, and the future.

Infrastructure and Mediation

Infrastructure is traditionally associated with urban planning and engineering and refers to the subterranean networks that support the operation of a given structure.[15] However, anthropologists and human geographers have argued against the presumed neutrality of physical infrastructure (e.g., roads, buildings, and pipes) in economic and scientific analysis.[16] According to geographers, infrastructure's significance emerges when it ceases to function.[17] Further, scholars have also highlighted the role of infrastructure in empire-making and nation-building.[18] In North America, for example, oil pipelines are common but contentious infrastructure that highlights the violence of settler colonialism on Indigenous land.[19] Another intervention, proposed mainly by anthropologists and literary critics, is the expanded definition of infrastructure as including both technological and "soft" language, cultural systems, affect, and religious learnings.[20] The emphasis on institutionalized structure that facilitates and binds goods, cultural flows, people, and collectivities is central to this new definition. It allows for a different way of thinking about totality and structure. Indeed, Lauren Berlant has persuasively proposed viewing nonsovereign social relations as a type of infrastructure that binds people within a common.[21]

This chapter focuses on the first definition of infrastructure as technologized objects that facilitate the "flow of goods in a wider cultural as well as physical sense."[22] I start from the premise that infrastructure is always political. From its design to construction, infrastructure produces optimism about connectivity, prosperity, and a livable future. This positivity sutures diverse populations, histories, and contradictions, a process that I call mediation. How does infrastructure mediate these differences? Or put differently, what kinds of beliefs, feelings, and actions does infrastructure intend to produce?

To answer these questions, I need to clarify my use of "mediation," which is informed by media studies and anthropology scholars.[23] In media studies, mediation describes the mutual adoptability and complexity of informational content and media technology.[24] For social theorists, mediation is a necessary social process that depends on certain repetitive everyday practices with or without media technology, but its immediacy might appear externally imposed with a change of perspective.[25] For example, rituals and customs may appear to be "natural" mediations of social relations to those who are embedded within communities, but "arbitrary" to those who are unfamiliar with said rituals. Informed by social theory, I take a historical materialist approach to mediation, viewing it as a constitutive social process of reconciliation between opposing sets of social entities within capitalism, such as modes of production and political power.[26]

Infrastructures take on mediative capacities when, for example, postcolonial modern nation-states deploy infrastructural projects as signs of progress. In China as well as in other Asian countries, roads, dams, and other technological systems have been hypervisible.[27] These projects are not just institutionalized systems that enable flows of goods and people, but also provide people with structures of imagination, fantasies, dreams, and desire.[28] This is what I call infrastructural mediation between the reality and the possible, the past and the future, and the discontent and the promise.

Physical objects are signs within a historicized system of meaning; hence, it is perhaps more productive to ask how some objects and/or technological systems *become* infrastructure.²⁹ Exploring the historical and social conditions is just as important as investigating their efficiency and duality. Visual and discursive representations provide entry points to understand just how infrastructure invigorates and structures what is possible and desirable even before its completion. In other words, by focusing on the representation of infrastructure in SF, we might see the mediative process between futuristic imaginations and infrastructural optimism that teaches readers or viewers about what kinds of built future are livable. By conceptualizing infrastructure as mediation, I examine how, through texts and pictures, *China 2098* reveals the dialectic connection between the objective existence of infrastructure and the subjective effects of its representation.

Narrating Infrastructural Future

Readers who access *China 2098* online will first encounter Fan's illustrations of various infrastructure. A textual-based explanation appears when one clicks on the picture. I choose to start my analysis with the narrative before examining the visuals because it provides a comprehensive understanding of how disparate infrastructure in *China 2098* are connected. More than being supplementary to the visual, these explanations form an important part of the affective politics of infrastructure: they mediate how readers should interpret and attach affects to infrastructural projects. My analysis argues that the ways in which Fan intertwines narratives of infrastructure reveal how infrastructure mediates triumphs and despair, optimism and pessimism.

The story begins with Laoqi (老七), the narrator, who tells readers of a series of disastrous events retrospectively that led to his deployment to the Yellow River State space station near Earth in 2098. To summarize the plot, in the 2030s, a war breaks out between China and "Western imperialist countries." The imperialists lose the war and a revolution, led by racialized people, takes place in North America, where a Peoples' Union of America (PUA) is formed in 2068.

During the war, there are significant advancements in nuclear fusion technology, enabling the defossilization of energy production and consumption. With the use of stable and consistent energy through nuclear fusion, the UN (United Nations) starts to build a ring that consists of several huge space stations around the Earth for human survival due to the rising sea level in the 2050s. However, this project fails as multinational capitalists contracted to build the stations are unable to deliver the infrastructure. The governments of China, Japan, the EU (European Union), the PUA, and Russia take on the half-completed stations and continue building the ring around Earth. But it is too late. By 2056, all icebergs melt, resulting in a significant rise in sea level, inundating parts of Japan, Korea, China, and Southeast Asia.

The failure of the UN space station project is in contrast with the success of three main infrastructural initiatives built by the Chinese Communist Party and its people. A housing complex is built for climate refugees under the slogan "Everyone has a place to live, every house has an occupant" (人人有房住, 房房有人住). From 2058 to 2066, the East China Sea Dam is completed with the fastest speed. The story ends with Laoqi's elaborations on Red Flag (红旗), a digital infrastructural system that collects real-time information to make political and economic decisions efficiently and scientifically. It reduces bureaucracy by calculating the best next step for each level of government using a complex algorithmic program. The explanation provided by Fan through the protagonist is somewhat ambiguous for the readers. Fan sees Red Flag as an objective, algorithm-based decision-making "core" that directs workers. The brain, Red Flag, directs workers and bureaucrats as if they were organs of the body.

Fan's SF is deeply shaped by the optimistic belief in and promise of infrastructure to further the welfare of ordinary people, against the pessimism induced by large-scale disasters. To adapt Lauren Berlant's use of optimism as a normative affective politics in the United States, infrastructures in *China 2098* might provide the material terms upon which people believe in the future offered by the Chinese state. What matters is not if these projects benefit the masses but the affective attachment to and the experiential differences felt by the communities impacted by the infrastructures. By weaving the triumph over ecological crisis with the fulfillment of well-being through completions of infrastructures, Fan mobilizes optimism as a central affect in *China 2098*. Most of the texts that accompany each print are Laoqi's elaborations on how these projects work: the machinic logic and technological advancement that drive them. While these narratives may be viewed as explanatory supplements to the visual, Laoqi's explanations may also demonstrate the affective quality of infrastructure.

The East China Sea Dam, housing complex, and Red Flag computational system are three major projects that move the narrative arc of *China 2098*. They orient the despair over ecological crisis toward the triumph of infrastructural optimism. For example, when introducing the first illustration, Laoqi comments on the scale and the accomplishment of the East Asia Dam at length:

在黃河站可以輕松看見這座「東北亞人民的奇跡」: 大堤從廈門開始修建, 穿過台灣海峽, 琉球群島和種子島, 一直延伸到千葉附近; 在北面, 津輕海峽, 宗谷海峽和韃靼海峽被封死, 形成了一個龐大的「內海」, 接下來就是向大堤外抽水直到恢復洪水前的海岸線; 不過其中一小部分海水在淡化後被送入西北大盆地, 重塑當地的生態環境, 這就是東水西引。再後來, 人們開始在大堤內填海造陸, 不到10年就成了現在的樣子。(You can see the "miracle of Northeastern Asian people" from the Yellow River station in space. The south side of the dam starts from Xiamen, through the Taiwan Straight, Okinawa, Tanega Shima to Chiba. The north section connects Tsuguru, Soya, and Tartary Straights. An "inner sea" is formed so that water within the dam is pumped out. Some seawater is purified and transported to Tarim Lake in Central Asia to create wetlands as ecological reservations; this is the "Water transportation from East to West project." As water level decreases, people start land reclamation. It has taken less than ten years for the project to look like this in 2098.)[30]

Much like the Great Wall that both forged a barrier against the "barbaric" and constructed territorial boundaries of the idea of the "Middle Kingdom" (i.e., China), in *China 2098*, the climate disaster happening in the region (Northeastern Asia) is solved by the scaling-up of the dam in the "Great Wall against Seawater." The dam mediates and participates in the social processes of forging a sense of collective survival and affective reorientations to regional identification. To refer to the dam as a "miracle" suggests affective appreciations and triumphant optimism. It may be necessary to consider Fan's identification with and imagination of the region through the binary of collectivism/individualism and East/West alongside East Asia's contemporary realpolitik. If current divisions are the result of Cold War ruins, the Pax Americana, and the "New Cold War," then there is *China 2098*, which suggests the "end" of history, the eventual overcoming of East Asian divisions because of the climate crisis and the inevitable need for cross-border, large-scale projects to ensure provision of utilities.[31]

Further, my reading highlights the cheerful and appreciative tone in which Laoqi talks about the dam. This tone suggests that infrastructure is not just the products of labor but objects that interact with the subjective dimensions of people's lives. For example, Laoqi takes time to list all the geographical names to emphasize the dam's scale; he also stresses how fast it has been built: "不到10年就成了現在的樣子" (less than ten years). One would assume Laoqi's emphasis is to praise the power of human labor, yet the word "labor" does not appear in the explanation of the dam. With Laoqi's appreciative tone, we are left with awe and fascination of infrastructure. Such affective responses hint at what I have been trying to suggest in the chapter: infrastructure is a mediator of the affective through the concrete.

Another example is how Laoqi comments on one of the gates on the dam:

'它的建築風格和當時普遍的「安置點審美」差別很大，原因是在那個年代不僅中國人民需要一個標志，全人類都需要—— 一個戰勝洪水的勝利標志。(The aesthetics of the gate is different from that of the settlement housing complex for climate refugees; the reason is that at that time (rising seawater that floods half of the world), Chinese people, maybe all humankind, need a symbol to represent the victory of overcoming seawater.)

This commentary shows how the gate, as a part of the infrastructure of the dam, evokes feelings of comfort, pride, and exuberance. As an object whose main feature is to stop, purify, and transport seawater, the dam functions also as an optimistic monument: a symbol of survival and thriving after an ecological disaster where the only solution is the state. The Chinese Communist Party's politics appears to be a land/waterscape politics in which people find themselves. The gate does not simply stand for the Party's power; rather, Laoqi's narration suggests that Fan intends to use the gate to mediate the pessimistic feelings of the people after various disasters and wars (e.g., sorrow, anger, pain) through the concrete, producing a belief in the eventual overcoming of the climate crisis and an end of history in China.

Visualizing Infrastructural Nostalgia

One cannot deny the omniscient presence of dams, buildings, bridges, and space rings in Fan's drawings. Yet, *China 2098* also does more than present a top-down, state-driven picture of infrastructure; it redeploys signs of a rosy socialist past to gesture toward an optimistic future. In this way, the past becomes relevant only through the future's rewriting and reinvention. Some may consider this nostalgic fusion of socialist realist aesthetics with a science fiction future to be a visually novel experiment. However, I contend that the aesthetic choice made by Fan reveals how infrastructure mediates the temporal inconsistency between the glorious past (the socialist era), the crisis-ordinary capitalist present, and the hopeful future.

These illustrations resemble extreme long shots in cinematography, which create a distance between the viewer and the scene, typically showing the subject on a massive scale. Fan's incorporation of cinematic techniques invites audience members to "view" the future at a distance and, thus, intends to engender a subjective "impression" of the "objectivity" of the future.[32] Such an objective view further invites viewers to see themselves as figures within the panorama.[33] Fan's *China 2098* provides objective observations of infrastructure by suturing the past onto the surface of the future. For instance, in the image showing the mass scale of the main gate on the East Asia dam, it is clearly visible that it is named "Pioneer" (先鋒), which was a common phrase to name production units, streets, or newborn children during the socialist era. Other slogans in the narrative resemble early socialist-era phrases that consist of one or two seven-word sentences, sometimes rhymed at the end. These slogans were commonly printed on building surfaces during the Great Leap Forward and the Cultural Revolution. For instance, Laoqi uses the slogan "Carbon dioxide is treasurable, synthesized starch is indispensable" (二氧化碳是個寶, 合成澱粉少不了) to discuss the success of synthesized starch; he also deploys the phrase "Promise to turn dirt into greenery" (誓把黃土變青山) as he describes the dedication of workers in the land reclamation project.

These transtemporal fixings of socialist slogans on infrastructures create a time loop. What has existed in the past (the socialist era) and what exists in the present are reinterpreted as and merged into what will become of the future.

FIGURE 7.2 The ninth illustration of the *China 2098* series. Courtesy of Fan Wennan. (Source: ArtStation.)

Infrastructure, therefore, serves as a mediator of incongruent temporalities into an asynchronistic future where technocratic governance is the only solution to the problem of contemporary crisis. Fan's illustrations are similar to what ordinary Chinese citizens observe on a daily basis. Even after the official end of Maoist socialism, murals and slogans continue to be an important part of state ideological work.[34] For example, one of the drawings, titled "The Buildings that Never Fall," features several buildings in progress, with a slogan that says "Everyone has a place to live, every house has an occupant" (人人有房住, 房房有人住) attached to the scaffolding on red tiles (Figure 7.2).

This is a typical scene in contemporary China's real estate construction zones, where socialist and/or safety slogans take advantage of the height of commercial apartment complexes for wider reception. This visual binding of past, present, and future provides a regime of temporal consistency. In this linear narrative, aesthetic elements are rescued from the past to have a new life. The slogans and murals from the socialist era resupply infrastructural advancements with new intentions and futures. They mediate what kinds of dreams and lives are possible and livable, and perhaps how one might achieve them. I suggest that, by transposing nostalgic representations of socialism onto futuristic infrastructures, Fan renders contemporary problems absent from its visual legibility, eliding the contradiction between contemporary China's claim of socialism and the socioeconomic inequality acutely felt by the poor, and by ethnic and gender minorities.[35] The use of nostalgic socialist elements in *China 2098* brushes aside crucial questions about the socialist state's contradictory legacies (e.g., the Cultural Revolution); the violent transformation of China's economy since the 1980s, such as the privatization of social institutions; current repression of independent unions, labor strikes, and queer/feminist advocacy; and ordinary young Chinese's creative online activism about the impossibilities of surviving in a cutthroat authoritarian

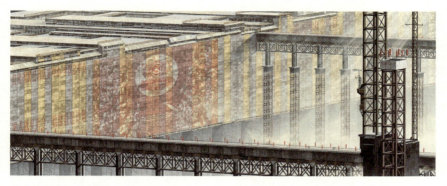

FIGURE 7.3 The sixteenth illustration of the *China 2098* series. Courtesy of Fan Wennan. (Source: ArtStation.)

capitalist society.[36] While there is limited space to discuss these pressing issues in this chapter, the use of socialist slogans by Fan blocks off inquiries regarding whose future, whose past, and whose present matter. It concels the imporant truth: that there might not be one unified trajectory of progress.

If, during the socialist period in China, slogans conveyed a radical anticapitalist reading of the modern as much as they served as ideological propaganda, the use of socialist slogans by Fan in *China 2098*, then, is more symbolic than a complete integration of socialist principles in the visual representation.[37] I propose that aesthetically socialist infrastructure serves as a temporal and affective mediator of China's assimilation into global, technomanagerial governmentality processes. The arrival of a futurist and technocratic infrastructure that organizes society is symbolized by the computational system Red Flag. Its headquarters is depicted by Fan as warehouse-style concrete buildings with steel walkways (Figure 7.3). One of the slogans on the bottom of the left side reads "Needs are commands, data is plan" (需求就是指令, 数据就是计划).

There is also a large mural on the wall of the Red Flag building that depicts people with delighted faces surrounding Tiananmen Square and the hammer and sickle, the communist symbol. Fan depicts Red Flag's technocratic modernity by appropriating socialist aesthetics to express an affective state of happiness and satisfaction. From the smiling faces to the slogans promising utility provision, the headquarters is not only a place for objective calculations for the economy, but also a place where politics can be felt. The headquarters does not simply represent the state; rather, the affective facial expressions on the wall imply a seamless relationship between what technomanagerial governmentality (Red Flag) will offer and how people will experience this technocratic rule in the future. Fan's depiction of technological progress (e.g., Red Flag) corresponds with the Chinese Communist Party's push for innovation during the China-- U.S. techno-cold war. *China 2098* is a commentary on this hegemonic rivalry, promising a Chinese exceptionalist future as the end of history. Infrastructure

FIGURE 7.4 The eighth illustration of the *China 2098* series. Courtesy of Fan Wennan. (Source: ArtStation.)

provides the material grounds on which technology politics are mediated, embodied, and felt.

While most of Fan's illustrations depict labor sites, people are rarely depicted. In an email communication between Fan and the author of this chapter, Fan states that the reason for the lack of "humans" in *China 2098* is his training in background design. He had to create a landscape first before it could be filled with humans by specialized artists in human figure creation.[38] While this might partially explain the invisibility of human bodies, Fan does include workers in his current artwork. But they are usually small compared to the size of the infrastructure in the background. Size matters here, in the sense that the human bodies serve as the sites of identification for the audience vis-à-vis the infrastructure. I suggest that viewers are, maybe unintendedly, invited to identify not with the human figures but with the objectivity and universality of infrastructure (Figure 7.4).

Not only are humans subsumed by the technological advancements of the future, so is the natural environment. We might recall the idea of the photographesomenon that I invoke above in the essay. In *China 2098*, human subjects are subsumed by the gaze of an objective "viewer" who makes no distinction between organic and inorganic matters, between objects and subjects. Viewers are invited, instead, to appreciate the universality of infrastructures and technological developments, such as buildings, bridges, or dams, which are the products of labor. Hence, even though Fan uses slogans that contain "workers" or "the masses" as visual cues of a mass-based consensus, his uncritical embrace of technology gives *China 2098* a dystopian quality, perhaps unintentionally. The invocation of "the masses" perhaps lends discursive legitimacy to a future of technocratic governmentality. Fan's lack of critical thinking on data, artificial intelligence, and the smartness of AI prevents him from breaking out of the technodeterminant mode of thinking that puts faith in technological systems and, perhaps, from envisioning a different vision of socialism that centers workers and their ability to struggle.

To summarize, Fan's use of slogans, murals, and names as visual and discursive signs is more than just a nostalgic homage to socialist realism. Beyond creative fusion, infrastructures with socialist realism combine contradictory feelings and produce a linearity between the past, present, and future, effectively foreclosing the possibility of a different vision of socialism.

Conclusion

Alexander Bogdanov's socialist science fiction novel *Red Star* (1908) depicted a socialist utopian Mars free of class, private property, and divisions of labor. This world, however, is far from perfect. Bogdanov warns socialists, through the main narrator Leonid, that social, political, and racial issues continue to haunt the new world. From depleting resources to colonizing other planets, *Red Star*'s utopia is far from a place where everyone can live happily ever after. Bogdanov perhaps knew in the 1900s, when the novel was first completed, that social and political tensions will not disappear in a socialist utopia.

China 2098 echoes descriptions of socialist utopia in *Red Star*, but it seems to stop short at reflecting upon the challenges ahead. In this chapter, I have tried to suggest how infrastructure takes central stage in mediating the contemporary capitalist contradictions globally. I propose that while *China 2098* offers a different interpretation of the relations between human and machine than techno-Orientalism, it falls short at critically engaging with the challenges and consequences of technology. Instead, Fan provides a modernist belief in and promise of infrastructural advancement as the antidote to global capitalism-induced climate crisis. Fan's symbolic use of socialist slogans and keywords (e.g., masses, people) does not imply a genuine engagement with socialist ideas and principles. Rather, infrastructure with socialist realist slogans neutralizes Chinese citizens' anxiety and desire amid contemporary affective contradictions: 1) pessimism amid growing inequality versus the state's propaganda of an optimistic socialist future; and 2) the temporary consistency among the socialist past, the capitalist present, and the postapocalypse future. I further argue that, in *China 2098*, infrastructure is not just something workers build and utilize; it shapes how people imagine a life worth living. As vital actors of political processes, big projects such as dams and spaceships provide the material terms upon which people believe in the enchanting future offered by the so-called socialist state.[39] Furthermore, infrastructure with socialist slogans and murals visually produces a linear schema from the past (the socialist era), the crisis-ordinary capitalist present, and the utopian future.

China 2098's emphasis on objects/materiality is a distorted reading of historical materialism. The challenge to techno-Orientalism may not simply be to champion technology. Returning to the questions I proposed in the beginning of the chapter, in an era when Asian future is becoming synonymous with infrastructural development, mastery of technological systems in Asian nations is not the antidote to techno-Orientalism. As my analysis of *China 2098* suggests, infrastructures mediate affective and temporal dissonances, obscuring socialist critiques of the contemporary contradictions and crises in which ordinary Chinese citizens, including Fan, find themselves in. In the Asia-Pacific region, large-scale projects are more than the material grounds for the circulation of capital; they sustain fantasies about a life without despair, and a temporal loop where the meaning of colonial pasts and the crisis-ridden present gets emptied out for an archaistic future that is always-already here.

Notes

1 See Fan's Artstation website: https://www.artstation.com/nangesfg.
2 See, for example, a review of *China 2098* in the state-affiliated news outlet *Global Times*: "Young Artist Paints Epic Sci-fi Future of China in 2098," *Global Times*, May 22, 2022, https://www.globaltimes.cn/page/202205/1266254.shtml.
3 Albert Werner Cheng, "China 2098: An Unintended Satire," *Medium*, September 6, 2020, https://medium.com/@albertwernercheng_92086/china-2098-an-unintended-satire-c4c6a178a525.
4 Affect theorists such as Brain Massumi have referred to "the autonomy of affect." Eric Shouse has suggested that affect consists of nonconscious and nonsignifying forces disconnected from the subject situated within the world of meanings. While affect seems to be automatic, preindividual, and devoid of social contexts, in this chapter, I suggest, following Ruth Leys, that affect has not been independent of signification and meaning in which the subject finds themselves. My use is closer to Raymond Williams's concept of "structures of feelings" to explain how infrastructure is lived and felt and the relations between these feelings and the more systematic belief. See Massumi, "The Autonomy of Affect," *Cultural Critique* 31 (1995): 83–109; Shouse, "Feeling, Emotion, Affect," *M/C Journal* 8, no. 6 (2005), https://doi.org/10.5204/mcj.2443; Leys, "The Turn to Affect: A Critique," *Critical Inquiry* 37, no. 3 (2011): 443; and Williams, *Marxism and Literature*, Oxford: Oxford University Press, 1977, 132.
5 Xiao Liu and Shuang Shen, "Introduction: Infrastructures as An Inter-Asia Method," *Interventions* 25, no. 3 (2023): 297–305.
6 Jonathan Ponciano, "Everything in the $1.2 Trillion Infrastructure Bill: New Roads, Electric School Buses and More," *Forbes*, November 15, 2021, https://www.forbes.com/sites/jonathanponciano/2021/11/15/everything-in-the-12-trillion-infrastructure-bill-biden-just-signed-new-roads-electric-school-buses-and-more/?sh=66723020161f.
7 Timothy Mitchell, "Infrastructures Work on Time," *E-Flux*, 2020, https://www.e-flux.com/architecture/new-silk-roads/312596/infrastructures-work-on-time/.
8 Tim Summers, "China's 'New Silk Roads': Sub-National Regions and Networks of Global Political Economy," *Third World Quarterly* 37, no. 9 (2016): 1628–1643; Deborah Cowen, *The Deadly Life of Logistics: Mapping Violence in Global Trade* (Minneapolis: University of Minnesota Press, 2014).
9 Yen Ooi, "Chinese Science Fiction: A Genre of Adversity," *SFRA* 50, no. 2–3 (2020): 141–148; Mingwei Song, "After 1989: The New Wave of Chinese Science Fiction," *China Perspectives* 1, 2015, 7–14.
10 Adrian Thieret, "Society and Utopia in Liu Cixin," *China Perspectives*, no. 2015 (2015): 33–39.

11 Virginia L. Conn, "Photographesomenonic Sinofuturism(s)," *SFRA Review* 50 (2020): 79–85.
12 The explanations only appear after clicking each picture. During data collection, I compiled the texts and realized these explanations are not just about the infrastructures; they also form a coherent storyline.
13 Partrik Svensson, "From Optical Fiber To Conceptual Cyberinfrastructure," *Digital Humanities Quarterly* 5, no. 1 (2011), http://www.digitalhumanities.org/dhq/vol/5/1/000090/000090.html.
14 David Nye, *American Technological Sublime* (Cambridge, MA: MIT Press, 1996).
15 United Nations Office for Project Services (UNOPS), "Infrastructure," 2022, https://www.unops.org/expertise/infrastructure.
16 Susan Leigh Star, "The Ethnography of Infrastructure," *American Behavioral Scientist* 43, no. 3 (1999): 377–391; Brian Larkin, "The Politics and Poetics of Infrastructure," *Annual Review of Anthropology* 42 (2013): 327–343.
17 Penny Harvey, Caspar B. Jensen, and Atsuro Morita, *Infrastructure and Social Complexity: A Companion* (London: Routledge, 2017); Martin Coward, "Hot Spots/Cold Spots: Infrastructural Politics in the Urban Age," *International Political Sociology* 9, no. 1 (2015): 96–99;
18 Christina Schwenkel, "Spectacular Infrastructure and Its Breakdown in Socialist Vietnam," *American Ethnologist* 42, no. 3 (2015): 520–534; Manu Karuka, *Empire's Tracks: Indigenous Nations, Chinese Workers, and the Transcontinental Railroad* (Berkeley: University of California Press, 2019); Beng Huat Chua and Koichi Iwabuchi, eds., *East Asian Pop Culture: Analysing the Korean Wave* (Hong Kong: Hong Kong University Press, 2008).
19 Anne Spice, "Fighting Invasive Infrastructures: Indigenous Relations against Pipelines," *Environment and Society* 9 (2018): 40–56.
20 Brian Larkin, *Signal and Noise: Media, Infrastructure and Urban Culture in Nigeria* (Durham, NC: Duke University Press, 2008); Ara Wilson, "The Infrastructure of Intimacy," *Signs: Journal of Women in Culture and Society* 41, no. 2 (2016): 247–280.
21 Lauren Berlant, "The Commons: Infrastructures for Troubling Times," *Environment and Planning D: Society and Space* 34, no. 3 (2016): 393–419.
22 Larkin, *Signal and Noise*.
23 Joshua Neves, *Underglobalization* (Durham, NC: Duke University Press, 2020); Lila Abu-Lughod, Brian Larkin, and Faye Ginsburg, *Media Worlds: Anthropology on New Terrain* (Berkeley: University of California Press, 2002); Martin Lister et al., *New Media: A Critical Introduction* (New York: Routledge, 2008).
24 Nick Couldry, "Mediatization or Mediation? Alternative Understandings of the Emergent Space of Digital Storytelling," *New Media & Society* 10, no. 3 (2008): 373–391; Joanna Zylinska and Sarah Kember, *Life after New Media: Mediation as a Vital Process* (Cambridge, MA: MIT Press, 2012).
25 William Marzzarella, "Culture, Globalization, Mediation," *Annual Review of Anthropology* 33 (2004): 345–367.
26 W.J.T. Mitchell and Mark B. N. Hansen, eds., introduction to *Critical Terms for Media Studies* (Chicago: University of Chicago Press, 2010), vii–xxii; Aeron Davis, *The Mediation of Power* (London: Routledge, 2007).
27 Alessandro Rippa, Galen Murton, and Matthaus Rest, "Building Highland Asia in the Twenty-First Century," *Verge* 6, no. 2 (2020): 83–111.
28 Hannah Knox, "Affective Infrastructures and the Political Imagination," *Public Culture* 29, no. 2 (2017): 363–384.

29 Susan Leigh Star and Karen Ruhleder, "Steps toward an Ecology of Infrastructure," *Information Systems Research* 7, no. 1 (1996): 111–134.
30 All translations are by the author.
31 Naoki Sakai, *The End of Pax Americana: The Loss of Empire and Hikikomori Nationalism* (Durham, NC: Duke University Press, 2022); Lisa Yoneyama, *Cold War Ruins: Transpacific Critique of American Justice and Japanese War Crimes* (Durham, NC: Duke University Press, 2016); Ho-Fung Hung, *Clash of Empires: From "Chimerica" to the "New Cold War"* (Cambridge: Cambridge University Press, 2022).
32 Chelsea Birks, "Objectivity, Speculative Realism, and the Cinematic Apparatus," *Cinema Journal* 57, no. 4 (2018): 3–24.
33 Winfried Pauleit, "Video Surveillance and Postmodern Subjects: The Effects of the Photographesomenon, an Image Form in the Futur Anteriéur," in *Ctrl [Space]: Rhetorics of Surveillance from Bentham to Big Brother*, ed. Thomas Y. Levin, Irsula Frohne, and Peter Weibel (Cambridge, MA: MIT Press, 2002).
34 Evans, Harriet. "Ambiguities of Address: Cultural Revolution Posters and Their Post-Mao Appeal," in *Red Legacies in China*, ed. Jie Li and Enhua Zhang (Leiden, The Netherlands: Harvard University Asia Center, 2016), 85-114.
35 Ho-fung Hung, "Labor Politics under Three Stages of Chinese Capitalism," *South Atlantic Quarterly* 112, no. 1 (2013): 203–212; Shana Ye, "'Paris' and 'Scar': Queer Social Reproduction, Homonormative Division of Labour and HIV/AIDS Economy in Postsocialist China," *Gender, Place and Culture* 28, no. 12 (2021): 1778–1798; Uradyn Bulag, "From Yeke-Juu League to Ordos Municipality: Settler Colonialism and Alter/Native Urbanization in Inner Mongolia," *Provincial China* 7, no. 2 (2002): 196–234.
36 "Lying flat" (躺平) is an online slogan that gained attraction in China's social media in 2021. Mostly young people participated in this online movement by refusing to overwork, stopping forced socializing, and doing the bare minimum of what is required at work. The movement reflects the social pressures and affective dissonances of young people in a cutthroat, authoritarian capitalist system. Jing Gong and Tingting Liu, "Decadence and Relational Freedom among China's Gay Migrants: Subverting Heteronormativity by 'Lying Flat,'" *China Information* 36, no. 2 (2021): 200–220; Laikwan Pang, "China's Post-Socialist Governmentality and the Garlic Chives Meme: Economic Sovereignty and Biopolitical Subjects," *Theory, Culture & Society* 39, no. 1 (2022): 81–100.
37 Benjamin Kindler, "Maoist Miniatures: The Proletarian Everyday, Visual Remediation, and the Politics of Revolutionary Form," *Modern China* 48, no. 5 (2022): 911–947.
38 Wennan Fan, personal communication, February 26, 2024.
39 Penny Harvey and Hannah Knox, "The Enchantments of Infrastructure," *Mobilities* 7, no. 2 (2012): 521–536.

8

Techno-Orientalist Deflections

How Documentaries Frame China's AI Threat

GERALD SIM

In a 2018 speech about America's China policy, Vice President Mike Pence criticized China's "social credit score" as "an Orwellian system premised on controlling virtually every facet of human life." He alluded repeatedly to the Chinese Communist Party (CCP)'s "control and oppression" and highlighted its use of artificial intelligence (AI) to censor and control China's population. China's State Council had released a "Planning Outline for the Establishment of a Social Credit System" in 2014, then the "Next Generation Artificial Intelligence Development Plan" in 2018, to build a $150 billion industry and lead the world in AI by 2030. Few experts questioned that China is a surveillance state, but many criticized simplistic Western media portrayals of the social credit system.[1] Contrary to popular accounts, the actual system is disjointed, inefficient, and not very punitive. Doubts among knowledgeable figures in China about its efficacy further hamper intragovernmental information sharing.[2] And Sesame Credit, a private and voluntary rewards platform that successfully marketed itself as a precursor to the national social credit program, is not even part of the legal system.[3]

For Jeremy Daum of Yale Law School's Paul Tsai China Center, the narrative that Pence parroted reflects "global fears about new applications of information technology," corporate and state control over data, and "behaviorist conditioning."[4] He adds that dystopian portrayals normalize surveillance elsewhere: "Anything less invasive than our imagined version of social credit seems sort of acceptable, because at least we're not as bad as China."[5] To wit, these ideological representations resolve psychosocial contradictions and buttress political power. This resonates with Edward Said's understanding that Orientalism enacts political, economic, and military might. He too describes psychic projections: "European culture gained in strength and identity by setting itself off against the Orient as a sort of surrogate or even underground self."[6] Along those lines, science fiction author Ted Chiang thinks "that most fears about A.I. are best understood as fears about capitalism... about how capitalism will use technology against us," especially when technology and capitalism are so closely intertwined as to be indistinguishable.[7]

But when critics like Daum refer primarily to journalistic reporting, they advance a case that is at the same time too restrictive and overly broad. The critique underexamines powerful media such as film, dwells on Chinese state power, and overlooks depictions of how its citizens are interpellated. By comparison, recent AI documentaries, namely *In the Age of AI* (2019) and *Coded Bias* (2020), reveal much more about the cultural fantasies being triggered. These influential films visualize China as a strange place where people acquiesce to surveillance, as if complicit with state power. Thus constructed, America's adversary is more than a political regime, it is a 1.4 billion–strong horde.

Representations of the ethnopolitical threat are meaningfully driven by neoliberal, if not libertarian, interests within Silicon Valley. To be clear, I want to neither efface nor excuse the transgressions of China's authoritarian regime. But in the context of what Daum, Said, and Chiang collectively imply are projections, transferences, or sublimations, I craft a remit for this chapter that pertains less to what China is doing, than to how its endeavors are understood in the domestic American context. Culturally, it behooves us to recognize techno-Orientalist syntax in all its forms. Politically, I scrutinize how these discursive imaginings service the antiregulatory dispositions of Big Tech. Its writ against antitrust actions and government oversight pushes audiences to accept these companies as allies and as America's bulwark in this fast-moving AI war.

The "techlash" of the 2010s intensified concerns around data privacy, surveillance, social harms, and the business practices of behemoths like Google, Amazon, and Meta. Those feelings gathered bipartisan support for legislative oversight and antitrust enforcement. Subsequently, the Biden administration took meaningful steps to place major critics of Big Tech—namely Lina Khan, Tim Wu, and Jonathan Kanter—at vital posts. Those appointees started to take on an industry whose aversion to governmental intervention possesses philosophical roots traceable to cybernetician Norbert Wiener's view that society can be modeled after a self-regulating cybernetic system, and to countercultural ideals borrowed from self-governing communes.[8]

Silicon Valley's case against regulation found a useful cudgel when China charged into the race for AI supremacy. At a 2018 U.S. Senate hearing, Mark Zuckerberg responded to concerns about Facebook's size by saying that its Chinese competitors were "a real strategic and competitive threat that American technology policy should be thinking about."[9] In *Wired*, editor-in-chief Nicholas Thompson and political scientist Ian Bremmer raised alarms about "The AI Cold War that Threatens Us All."[10] The article criticized the liberal opposition within tech's workforce to Pentagon partnerships, and adhered to typical framings of competition with China, within which the U.S. effort is disadvantaged by American commitments to privacy and civil liberties. Likewise, political scientist Graham Allison and former Google CEO Eric Schmidt point out in a 2020 Harvard Kennedy School paper titled "Is China Beating America to AI Supremacy?" that without encumbrances like privacy protections, China can make more nimble and straightforward policy decisions. Allison and Schmidt affirm American exceptionalism, before folding it into a bargain for less regulation and more collaborations with the Department of Defense: "American AI faces serious headwinds, including a culture that values privacy over security, distrusts authority and is suspicious of government; IT companies wary of working with the U.S. Defense Department and intelligence agencies; dysfunctional public policies inhibiting recruitment and immigration; laws that make it difficult to compile big data sets; and the prospect of further regulations and antitrust action against the companies that are now America's national champions—and are driving American advances in this arena."[11] Schmidt has since continued to press the point. As he explained to the Senate Armed Services Committee the following year: "The private sector is America's great strength. We move faster and [more] globally than any government could . . . and we need global platforms or be forced to use the Chinese ones which is a disaster. . . . I think the government will need to help with some forms of funding, and we need to let the private sector build those things and make it successful."[12] Wu, Biden's special assistant for technology and competition policy, describes "the China argument" as Big Tech's version of "too big to fail" with "superficial nationalistic appeal." He warns about technology giants' underlying desire to preserve monopolistic control, which is contravened by the government's vigorous antitrust enforcement during the 1970s and 1980s that had helped to maintain the United States' innovation edge over Japan.[13]

Such admonishments have not deterred Silicon Valley. Techno-Orientalism infuses the cultural logic that it deploys to speak both to and for the state. While American companies and the government are adversarial on regulatory issues, their interests converge around the notion of China's external threat. The ethnocentric media narrative, in turn cultural, racial, and nationalistic, tends that common ground. And within the media landscape, documentaries that harbor public trust are potent expressions of that discourse. The pair of films that I examine in this chapter give voice to industry interests and may even serve as the discursive arm of the American state. These close readings further show how techno-Orientalism embeds itself in the sociotechnical imaginary.

Techno-Orientalist Tropes

Big Tech has mobilized decisively in the political arena. Since 2017, companies led by Google, Amazon, Facebook and Apple have quadrupled donations to research groups in Washington. The Center for Strategic and International Studies, the Center for a New American Security, the Brookings Institute, and the Hudson Institute received at least $625,000 in 2017–2018, a figure that the *Financial Times* estimates could have risen to as high as $2.7 million in 2019–2020. Along with the oil and gas industry, the technology sector has become a leading funder of these think tanks, part of a broader strategy to forestall regulation by reframing technology-related issues in terms of national security.[14] *Wired* reported that Facebook, Amazon, and Google collectively spent more than $47 million on lobbying and advocacy in 2020.[15] Moreover, these campaigns have international impact, because political wisdom in the United States has a habit of finding its way to Europe, where American technology companies also invest in lobbying against regulatory moves, particularly antitrust.

The United States is often said to be locked in an "AI Arms Race" with China, a term used most often for military applications of AI and the companies run by industry iconoclasts like Palmer Luckey and Peter Thiel.[16] The phrase has been criticized for being nationalistic, diplomatically obstructive, and empirically inaccurate.[17] Although that jingoistic fervor is associated with Luckey's and Thiel's profit motives and reactionary ideology, American technology companies also further the militaristic narrative that exploits "the specter of an AI race to head off user protections they see as burdensome."[18]

Overt examples of this messaging come from ostensibly nonpartisan lobbying groups like the American Edge Project (AEP) and the Taxpayers Protection Alliance (TPA). The more established TPA, which has received money from Google, opposes government spending and regulation broadly, while the technology-focused AEP launched in 2019 with a sole benefactor, Facebook.[19] AEP funds supportive academics and sponsors newsletters that target government insiders inside the Beltway.[20] Its assiduously bipartisan ads, which frequently marshal Cold War rhetoric to spread fear about adversaries like China, favor close-ups of the Chinese flag flapping ominously in slow motion, juxtaposed in one instance against an American boy in the back seat of a car holding up a miniature space shuttle against a bright sky.

In 2022, TPA ran a million-dollar campaign on national cable news and online in twenty states.[21] One ad features Pence's former national security advisor, Lt. Gen. Keith Kellogg. Against darkened images of China's flag, parading soldiers, and Chinese president Xi Jinping (Figure 8.1), Kellogg speaks before the Marine Corps War Memorial: "Generations of American soldiers have fought

FIGURE 8.1 The "China argument" in a political ad opposing tech industry regulation. (Source: Taxpayers Protection Alliance.)

to preserve our freedoms and standing in the world. The political attacks on our tech industry could squander their sacrifices and empower communist China to surpass the United States economically and militarily. The stakes are high. We cannot let China win the tech race. Tell Congress to stand up for American innovation and America's security." The striking ad recapitulates rhetoric common in recent documentaries about AI. Take *Do You Trust This Computer?* (2018), a film about the dangers of AI that is of lesser renown than its endorsement by Elon Musk would suggest. During its obligatory reference to foreign AI threats, tech founder Sean Gourley, whose company Primer holds major defense contracts, explains, "Today the Secretary of Defense is very very clear, we will not create fully autonomous attacking vehicles. Not everyone is going to hold themselves to that same set of values. And when China and Russia start deploying autonomous vehicles that can attack and kill, what's the move that we're going to make?" Under a montage of Cold War military imagery that jumps into the present with wide-angle shots from Chinese and North Korean military parades, author James Barrat identifies danger in artificial superintelligence being "surrounded by a bunch of people who are really just excited about the technology. They want to see it succeed but they're not anticipating that it can get out of control." The film ends on a silent dining room full of Zoomers—techpreneurs perhaps—faces buried in their laptops, in what could well be any Bay Area café at midday. After the image fades to black, closing intertitles warn that "the pursuit of artificial intelligence is a multi-billion dollar industry, with almost no regulations." The hollow coda neither cites nor suggests specific legislation. Rather, having framed all allusions to politics with nationalism, militarism, and Orientalism against an Eastern bloc, its message is more aptly understood as an appeal to fear and paranoia, voiced by Silicon Valley figures to cultivate helplessness and defeatism.

While nonprofits like AEP and TPA are legally compelled to reveal themselves in ad disclaimers, the ideologies of most AI documentaries go undeclared. Nonfiction films often carry patinas of factuality and objectivity. Some circulate with the benefit of vital imprimaturs. The documentaries discussed below aired on the Public Broadcasting Service (PBS), program distributor for public television stations in the United States and a major distribution avenue for nonfiction films. For many of these works, the gravitas of a broadcast premiere on PBS is financially invaluable. Media scholars have raised questions about PBS's complicated commitment to journalistic balance and public television's idealistic foundations in Jürgen Habermas's concept of enlightened democracy.[22] Nonetheless, PBS remains a venue for people who watch documentaries for edification, to become informed citizens. The impact that these films have on public discourse is thus reason to scrutinize their technopolitics, just as their high-mindedness and stated purpose necessitate holding them accountable.

Validated by PBS, many films find their way to academic libraries and educational settings. What are the language and cultural grammar of the conversations they facilitate? These films exemplify "public pedagogy"—modes of informal learning beyond educational institutions and traditions, facilitated instead by art, culture, and political discourse. Films are especially potent in this regard because the medium combines entertainment and politics with such alacrity that they circulate freely in the culture before settling in public memory.[23] Indeed, the texts discussed below assert a civic educational purpose by dint of aesthetic form, market positioning, and mode of production. They are tonally sober and eschew stylization, willing to be critical but never *self*-critical. MacArthur Fellows, academics, and experts with university affiliations populate their interviews, all seeking to elevate public understanding of AI.

In the Age of AI and *Coded Bias* typify the practice. *In the Age of AI*, coproduced by *Frontline* for PBS, addresses the transformations underway in the economy, the labor market, and state power. Shalini Kantayya's *Coded Bias* spotlights the fallibility and opacity of facial recognition technology, and the threat to privacy and civil liberties. It premiered at the Sundance Film Festival with the more buzzy *The Social Dilemma*. Two months after the latter started streaming on Netflix, *Coded Bias* had a limited theatrical release in New York City. This preceded a short run in virtual cinemas, followed by a television premiere on PBS's *Independent Lens* in March 2021. The documentary landed in Netflix's catalog a month later.

The Technopolitics of AI Documentaries

Coded Bias is an unreserved voice of the techlash. The film calls out companies by name and builds its advocacy on an "activist goes to Washington" narrative that centers on computer scientist Joy Buolamwini of the MIT Media Lab, an eminently likable figure with whom audiences easily identify. She is supported by an impressive lineup of writers, scholars, and activists in AI and data ethics, including prominent names like Meredith Broussard, Cathy O'Neil, Zeynep Tufekci, Amy Webb, and Safiya Noble, along with cameo appearances by computer scientist Timnit Gebru and sociologist Mona Sloane. Kantayya stated that during research for the documentary, she came to see "this canon inside of tech that was not being heard. The role of women and feminism as a force for change within Silicon Valley has been long underestimated."[24] She referred repeatedly to her cast as "brave," "brilliant," and "badass," stressing to *Variety* that "my mantra is we can't leave the tech bros alone on this. They need to hear and feel the groundswell from the public."[25]

Hence, *Coded Bias* presents a movement to hold the "tech bros" accountable. Early on, shortly after an interview with Broussard, we see Buolamwini walking to Harvard Book Store and sitting down with a copy of O'Neil's *Weapons of Math Destruction*. Her voiceover recounts O'Neil's book event at that location. We cut to O'Neil preparing for a speaking event in New York City. Later, Buolamwini boards the Boston subway and cracks open Broussard's *Artificial Unintelligence*. The film thus positions these women as counterpoints to Big Tech. Kantayya highlights Buolamwini's visit to a Black hair salon, where she recounts her childhood ambition of becoming a robotics engineer to her hairdresser. There is an irresistible optimism in her close-ups that highlight the colorful spectacles and Wu Tang Wakanda earrings framing Buolamwini's face. In that vein, O'Neil remembers a sixth-grade teacher diverting her from honors algebra because girls "would never need math." We then see her tutoring her daughter in geometry with a Spirograph. At the end of the film's middle act, the crew physically meets. Buolamwini, O'Neil, Broussard, and Sloane commiserate at a Chelsea restaurant. The scene of four women at a Manhattan brunch evokes the HBO series *Sex and the City* (1998–2004) along with its themes of affirmation and empowerment. After they raise a toast to data ethics on behalf of marginalized people, the film cuts away.

In *Weapons of Math Destruction*, O'Neil describes how algorithms exacerbate inequality through systems that make projections about recidivism and creditworthiness. Kantayya platforms that critique and interviews criminal justice activist Tonya LaMyers, a parolee who was undeservingly labeled as high risk by a recidivism prediction algorithm that overruled the opinions of a judge and parole officer. As such, we see how a social credit system functionally exists in America. It is another example of *Coded Bias*'s many helpful interventions that enhance the public conversation. But the film's avowed politics lead me to lament the subsequent recourse to techno-Orientalist tropes—"images and models of information capitalism and the information society" designed to preserve the West's identity as part of its material project to control the future, in which the Orient's economic and cultural transformations are seen as rapid and inexorable.[26]

Techno-Orientalism's contradictory discourse associates Asians with hypo- as well as hypertechnological figurations that elicit attraction and condescension from the West. It signifies Japan and China differently. Japan is a competitor in innovation, a culture on which the West projects its technological fantasies, and a modernity that can potentially replace Western modernity. China on the other hand competes as a manufacturing engine fueled by "a vast, subaltern-like labor force *and* . . . a giant consumer market whose appetite for Western cultural products, if nurtured, could secure U.S. global cultural and economic dominance."[27]

Recent reactions to the CCP's ambitions recycle those stereotypes. China's advantage in the AI arms race is put down to the massive population that generates limitless data sets to improve its neural networks. Interviewed for *In the Age of AI*, venture capitalist Kai-Fu Lee explains in the film that "AI is basically run on data and fueled by data. The more data, the better the AI works, more importantly than how brilliantly the researcher is working on the problem. So in the age of AI where data is the new oil, China is the new Saudi Arabia." Just as crude oil is refined, data sets must be "cleaned" in a process of identifying, rectifying, and removing inaccurate, duplicative, or incomplete data. On that score, Allison and Schmidt believe that "China wins by default due to the size of its population." "To the extent that the next decade is an era of implementation, the advantage lies with China. In implementation, the overwhelming competitive advantage is quantity of quality data. Both in collection and in having a cadre of grunts to clean the data, China wins."[28] Chinese labor is thus defined by immense quantity but inferior quality. *In the Age of AI* illustrates this with Chinese multitudes in urban centers, all holding mobile devices, shopping in modern malls and traditional wet markets peppered with QR codes, demonstrating a high rate of technological adoption matched by widespread willingness to live under surveillance (Figure 8.2).

In the West, this is the stuff of nightmares, explored safely in niche science fiction (*Black Mirror*, 2011–2023) and conspiracy narratives (*Mr. Robot*, 2015–2019, and *Humans*, 2015–2018). Western media have often mused about how the *Black Mirror* episode "Nosedive," in which a woman's life in a social credit dystopia takes a turn for the worse, presaged China's social credit system.[29] By contrast, *In the Age of AI* and *Coded Bias* show the Chinese submitting ever so dutifully, so strangely, to facial recognition scans just to buy fast food. These agoraphobic impressions mutually intensify the meganumerophobia triggered by technology and policy coverage of China's massive dollar investments in AI. Chinese masses are depicted as passive, subservient drones—data points, more objects than subjects—while unfathomable financial sums are the governmental corollary. Their success comes not from ingenuity and nous, but from how far they can turn the spigots of currency and data. *In the Age of AI* introduces us to emerging innovators in China, but they differ from the 1980s-era hypertechnological type associated with Japan. The Chinese are Orientalized by their amoral rejection of freedom, civil liberties, and democracy. Images of

FIGURE 8.2 The archetypal image of digital adopters among China's multitudes. *In the Age of AI* (2019). (Source: *FRONTLINE* and Five O'Clock Films.)

submissive multitudes further imply that what Arthur Kroker and Michael Weinstein call the "virtual class" or "techno-intelligentsia" enjoy freedom only if they serve party goals.[30] Under authoritarian capitalism, without Western values, these are utterly rapacious sociopaths.

Coded Bias Otherizes China as a malevolent surveillance dystopia expanding its social credit system. The CCP is "explicit" about it, O'Neil emphasizes, twice: "It's like algorithmic obedience training." We are introduced to a young Hangzhou resident, Wang Jia Jia, who buys produce, clothing, and soft drinks under watchful cameras everywhere. She is adorned paradoxically with antiestablishment trappings: cerulean streaks in her hair, prominent tattoos, a skateboard, and a Kangol beret. Those accoutrements underscore her submission to the authoritarian gaze on her movements: making purchases at the mall, a convenience store, and a subway station's vending machine. Then, as if she were a character in an episode of *Black Mirror*, Wang gives the system glowing marks. She avows that it is convenient and incentivizes good behavior. Those comments are calibrated to both startle and bewilder PBS watchers. The film puts her in contrast with Brooklyn residents organizing to remove facial recognition cameras from their housing complex, irate Londoners standing up to the Metropolitan Police, and Hong Kong's political protesters most viscerally of all, prodemocracy insurgents in helmets and gas masks resisting en masse, disabling cameras with spray paint and laser pointers. Wang is among Chinese people we see "surveilled" on low-resolution monitors, their visages behind digital grids mapping their faces (Figure 8.3). Visually menaced by these overbearing intrusions, the Western spectator is provoked to resist. Why then do people in China just go along?

FIGURE 8.3 Wang Jia Jia, supporter of China's social credit system. *Coded Bias* (2020). (Source: 7th Empire Media.)

In reality, they probably do not. Research shows that while Chinese citizens express high rates of social acceptance and trust in the reliability of facial recognition technology, domestic opposition is not de minimis. According to surveys, attitudes depend on specific local conditions and experiences with terrorism, the state, familiarity with the technology, and highly particular historical norms in regard to AI's effect on social relationships.[31] Absent such caveats, the idea of "China" is reduced to a Sinophobic meme.

In the Age of AI does offer balance through *Frontline*'s long-form journalistic style of documentary. PBS's public funding model also affords independence from corporate financing and editorial influence. But the film relies on the same cadre of experts: technologists, CEOs, venture capitalists with inevitable ties to Stanford and MIT, *Wired* editors, beat reporters for the *New York Times*, and the ubiquitous futurist Jaron Lanier. Most prominent among them is Kai-Fu Lee, former Apple engineer, then Microsoft and Google executive, turned venture capitalist. Bookended by discussions of China as the other pole in the current battle between two AI superpowers are segments about the future of work, corporate surveillance, and the threat to democracy. The film is visually and emotionally muted in accordance with *Frontline*'s house style, and circumspect if not ominous about our AI future. Lee is introduced as a leader in China's tech sector. He explains China's commitment and introduces some of its latest innovations. The account mirrors his new book, *AI Superpowers*, which the film mentions near the conclusion, during his lecture and book-signing event in New York City. The slippage between what is articulated in Lee's book and the film's "objective" statements narrated in *Frontline* narrator Will Lyman's baritone means the film defers to Lee's point of view, in which China is on the front foot and gaining. In that context it is notable that even with a diverse audience for the book event, we only witness him in conversation with deferential Americans. Allison and Schmidt had asked, "Is China Beating America to AI Supremacy?" Before that, a *New York Times* title wondered, "Is China Outsmarting America in A.I.?"[32] This film's dire answer appears to be "yes."

In The Age of AI begins as *AI Superpowers* does, with a computer program's pair of famous victories over the world's best exponents of Go, a Chinese board game considered significantly more complex than chess. AlphaGo, developed by Google subsidiary DeepMind, vanquished South Korean champion Lee Sedol in 2016 and China's Ke Jie the following year. The first chapter of Lee's book, "China's Sputnik Moment," describes the tremors felt throughout China afterward. "Sputnik moment," a reference to the Soviet Union's successful 1957 launch of a satellite into orbit that precipitated intense anxiety in the West about being surpassed, had found its way to AI discourse a few years before, when AI researcher Ben Goertzel used it agnostically to describe technological potential, while retaining the framing as a general contest between nations.[33] Lee recycles the Cold War metaphor that the *New York Times* technology reporter Paul Mozur used as well.[34] These days, the term politicizes China's intentions, militarizes the conflict, elevates it to a mortal threat, and Otherizes a rival. Later in the film, China is said to be "raising a bamboo curtain."

By centering China, *In the Age of AI* unfolds through Lee's perspicacious dicta. Between its first segment ("China has a Plan") and its fifth ("The Surveillance State"), we move through three stories from middle America. "The Promise" highlights the advancements that deep learning has made in autonomous vehicles and medical imaging, before "The Future of Work" takes stock of innovation's impact on the economy and the labor market. We meet Shawn and Hope Cumbee, married truckers from Beaverton, Michigan. Seated at her kitchen table, Hope relates the family's financial struggles. Hope and the couple's son are cross-edited with Shawn, who is waiting in a Tennessee garage for his truck to be repaired. Shawn is skeptical of the threat of self-driving vehicles: "I really ain't worried about the automation of trucks." The film switches immediately to a closeup of Hope being told that autonomous trucks are already running "exit-to-exit" on the interstate. The camera stays on her surprise as she quietly realizes the consequences for her family. Given *Frontline*'s normally measured tones, the moment is wrenching. Eight minutes of expert interviews later, we meet another casualty of industrial automation in nearby Saginaw. Harry Cripps, president of an auto workers' union, drives through an empty downtown street on a drab winter's day. Later, while packing boxes with produce for a food drive, he describes the job losses that automation has wrought. The film cuts to the bustling and pristine interior of a robotics plant in nearby Rochester Hills, before turning away to lament the sight of boarded homes and urban blight.

By construing the victims of AI exclusively through the lives of the "white working class," *In the Age of AI* underscores its own techno-Orientalism. Setting aside the absence on screen of Black or brown faces from majority-minority Saginaw, the film maps AI's effect on labor by peering only at industries external to it, thus overlooking the racially disparate labor and working conditions *within* AI. As Neda Atanasoski and Kalindi Vora have shown, whereas utopian imaginings of true freedom for the liberal subject through technology have always obscured "the uneven racial and gendered relations of labor, power, and social relations that underlie the contemporary conditions of capitalist production," in the post–civil rights era of racial liberalism, technoliberal futures continue to be structured by race even while claiming postracial ideals.[35]

In the Age of AI evinces that logic. First, it equates human obsolescence with white job loss, while disavowing racism and xenophobia. Over aerial footage of factory ruins, we hear audio from a Trump campaign rally: "We want to keep our factories here. We want to keep our manufacturing here. We don't want it moving to China, to Mexico, to Japan, to India, to Vietnam." Are the economically anxious heartlanders we just met susceptible to this protectionist rant? It tees up a counterpoint from an economist who explains that automation, not offshoring, is more responsible for layoffs. Atanasoski and Vora contend that tech industry leaders often make similar moves to absolve outsourcing or reject Trump's xenophobic isolationism.[36] But foisting blame onto automation fashions another alibi for the unremitting relevance of race in the technoliberal imaginary.

Second, the film excludes China's population from the fourth chapter's account of "The Surveillance Capitalists." Harvard professor Shoshana Zuboff makes a requisite appearance to synopsize her tome of the techlash, *The Age of Surveillance Capitalism*. She differentiates the titular era from how "industrial capitalism claimed nature . . . for the market dynamic to be reborn as real estate. . . . Industrial capitalism claimed work for the market dynamic to be reborn as labor." Surveillance capitalism, however, claims "private, human experience," which she enunciates twice, slowly, for good measure. Paradoxically, therefore, we are made to see that it is the liberal subject who is threatened while the illiberal subject thrives. By drawing narrative partitions and ideological distinctions, *In the Age of AI* walls China off from its understanding of surveillance capitalism, whose hazardous effects are in effect portrayed as an American experience. We must note that Zuboff does not subscribe to this distinction. She writes that the Chinese are victims as much as we are: "As distinct as our politics and cultures may be or have been, the emerging evidence of the Chinese social credit initiatives broadcasts the logic of surveillance capitalism and the instrumentarian power that it produces. Sesame Credit doubles down on every aspect of surveillance capitalist operations, with hundreds of millions of people caught in the gears of an automated behavioral modification machine and its bubbling behavioral futures markets dispensing perks and honors like Pokemon fairy dust in return for guaranteed outcomes."[37] *In the Age of AI* elides this when it depicts China as the threat. The final segment returns to China for another montage of urban life: pedestrians and commuters on surveillance

FIGURE 8.4 Uyghurs in Xinjiang Province, victims of the Chinese surveillance state. *In the Age of AI* (2019). (Source: *FRONTLINE* and Five O'Clock Films)

monitors, overlaid with digital markers to signify that they are being biometrically identified. They still seem nonplussed. Then, to illustrate a "total surveillance state," the film cuts to Xinjiang with shots of Uyghur masses being involuntarily surveilled. They are darker complexioned, mustachioed, clad in Muslim headwear, distinct from the urban shoppers we just saw (Figure 8.4). Unlike Uyghurs living in the "open air prison" of Xinjiang, the urban denizens neither want nor deserve sympathy and protection. They have at best reconciled themselves to authoritarianism.

At worst, they reject democracy, relinquish civil rights, and consider privacy a Western whim. We cannot fathom this, interpellated instead by what Jodi Melamed describes as "racial capitalism," the "forgetting of interconnections, of viable relations and of performances of collectivity that might nurture greater social wholeness, but are deactivated for capital and state management."[38] During the Cold War, American liberalism conflated Soviet ideology and technology. Soviet robots were "correlated with the mechanization of totalitarianism," against which American freedom and liberal values were defined.[39] When that peril receded, technoliberal modernity shifted its preoccupation onto race. The current paranoia is now about Chinese humans, not because they are robots of productivity (as Japanese autoworkers were perceived before), but because they seem like mechanized drones of authoritarianism.

For its part, China is leveraging these American projections for its own messaging objectives. The state-run China Global Television Network (CGTN) proffers a strategic narrative about AI that interprets U.S. fear as a crisis of confidence: perceptions of China's threat reflect American weakness. CGTN programming takes pride in China's global stature, articulates its technological aspirations, and affirms its geopolitical position to an international audience.[40] Such variances in content between American and Chinese media remind us that the countries' media apparatuses and technology sectors are different. The Chinese government controls both, whereas American companies can be outwardly adversarial to the state.[41] It also suggests how we can track future shifts in the techno-Orientalist narrative, should China miss its lofty targets or if Big Tech's regulatory resistance recalibrates after the end of Lina Khan's term as chair of the Federal Trade Commission, whose energies may be redirected under the second Trump administration. As these situations progress, will American interests persist with the China argument? Will Silicon Valley adjust its position or strategy on regulation? Techno-Orientalist discourses may abate, but are just as likely to take new form.

Notes

1 Jamie Horsley, "China's Orwellian Social Credit Score Isn't Real," *Foreign Policy*, November 2018, https://foreignpolicy.com/2018/11/16/chinas-orwellian-social-credit-score-isnt-real/; Louise Matsakis, "How the West Got China's Social Credit System Wrong," *Wired*, July 29, 2019, https://www.wired.com/story/china-social-credit-score-system/; Shazeda Ahmed, "Credit Cities and the Limits of the Social Credit System," in *Artificial Intelligence, China, Russia, and the Global Order*, ed. Nicholas D. Wright (Montgomery, AL: Air University Press, 2019), 55.
2 Ahmed, "Credit Cities," 60.
3 Horsley, "China's Orwellian Social Credit Score"; Matsakis, "How the West."
4 Jeremy Daum, "China through a Glass, Darkly," *China Law Translate*, December 24, 2017, https://www.chinalawtranslate.com/en/china-social-credit-score/.
5 Matsakis, "How the West."
6 Edward Said, *Orientalism* (London: Routledge and Kegan Paul, 1978), 3.
7 Ted Chiang, "Transcript: Ezra Klein Interviews Ted Chiang," *The Ezra Klein Show*, March 30, 2021, https://www.nytimes.com/2021/03/30/podcasts/ezra-klein-podcast-ted-chiang-transcript.html.
8 Fred Turner, *From Counterculture to Cyberculture: Stewart Brand, the Whole Earth Network, and the Rise of Digital Utopianism* (Chicago: University of Chicago Press, 2006), 22, 146.
9 "Transcript of Mark Zuckerberg's Senate Hearing," *Washington Post*, April 10, 2018, https://www.washingtonpost.com/news/the-switch/wp/2018/04/10/transcript-of-mark-zuckerbergs-senate-hearing/; Kurt Wagner, "Mark Zuckerberg Says Breaking up Facebook Would Pave the Way for China's Tech Companies to Dominate," *Vox*, July 18, 2018, https://www.vox.com/2018/7/18/17584482/mark-zuckerberg-china-antitrust-breakup-artificial-intelligence.
10 Nicholas Thompson and Ian Bremmer, "The AI Cold War that Threatens Us All," *Wired*, October 23, 2018, https://www.wired.com/story/ai-cold-war-china-could-doom-us-all/.
11 Graham Allison and Eric Schmidt, *Is China Beating America to AI Supremacy?* (Cambridge, MA: Harvard Kennedy School, 2020), 12. See also Amy Webb, *The Big Nine: How the Tech Titans and Their Thinking Machines Could Warp Humanity* (New York: Public Affairs, 2018), 245, 252.

12 *Hearing to Receive Testimony on Emerging Technologies and Their Impact on National Security, Before the Committee on Armed Services, U.S. Senate*, February 23, 2021, https://www.armed-services.senate.gov/imo/media/doc/21-05_02-23-2021.pdf.
13 Tim Wu, "Don't Fall for Facebook's 'China Argument,'" *New York Times*, December 10, 2018, https://www.nytimes.com/2018/12/10/opinion/facebook-china-tech-competition.html.
14 Kiran Stacey and Caitlin Gilbert, "Big Tech Increases Funding to US Foreign Policy Think-Tanks," *Financial Times*, February 1, 2022, https://www.ft.com/content/4e4ca1d2-2d80-4662-86d0-067a10aad50b.
15 Gilad Edelman, "Big Tech Targets DC with a Digital Charm Offensive," *Wired*, March 8, 2021, https://www.wired.com/story/big-tech-targets-dc-with-digital-charm-offensive/.
16 Cade Metz, "Away from Silicon Valley, the Military Is the Ideal Customer," *New York Times*, February 26, 2021, https://www.nytimes.com/2021/02/26/technology/anduril-military-palmer-luckey.html; Jessica Bursztynsky, "Oculus Founder Says Best US Minds Need to Work on A.I. Just Like They Did during the Nuclear Arms Race," *CNBC*, July 19, 2019, https://www.cnbc.com/2019/07/19/palmer-luckey-best-us-minds-need-to-work-on-ai-like-with-nuclear-arms.html.
17 Justin Sherman, "Reframing the U.S.-China AI 'Arms Race.,'" *New America*, March 2019, http://newamerica.org/cybersecurity-initiative/reports/essay-reframing-the-us-china-ai-arms-race/.
18 Sam Biddle, "Why an 'AI Race' between the U.S. and China Is a Terrible, Terrible Idea," *The Intercept*, July 21, 2019, https://theintercept.com/2019/07/21/ai-race-china-artificial-intelligence/; Scarlet Kim and Graham Webster, "The Data Arms Race Is No Excuse for Abandoning Privacy," *Foreign Policy*, August 14, 2018, https://foreignpolicy.com/2018/08/14/the-data-arms-race-is-no-excuse-for-abandoning-privacy/.
19 Tech Transparency Project, "Funding the Fight Against Antitrust: How Facebook's Antiregulatory Attack Dog Spends Its Millions," May 17, 2022, https://www.techtransparencyproject.org/articles/funding-fight-against-antitrust-how-facebooks-antiregulatory-attack-dog-spends-its-millions.
20 Edelman, "Big Tech Targets DC."
21 Patrick Hedger, "TPA Launches New Ad Featuring Former National Security Advisor," Taxpayers Protection Alliance, March 31, 2022, https://www.protectingtaxpayers.org/technology/tpa-launches-new-ad-featuring-former-national-security-advisor-2/.
22 B. J. Bullert, *Public Television: Politics and the Battle over Documentary Film* (New Brunswick, NJ: Rutgers University Press, 1997); Laurie Ouellette, *Viewers Like You: How Public TV Failed the People* (New York: Columbia University Press, 2002).
23 Henry A. Giroux, "Breaking into the Movies: Public Pedagogy and the Politics of Film," *Policy Futures in Education* 9, no. 6 (2011): 689.
24 Sidney Fussell, "This Film Examines the Biases in the Code That Runs Our Lives," *Wired*, November 15, 2020, https://www.wired.com/story/film-examines-biases-code-runs-our-lives/.
25 Fussell, "This Film Examines the Biases"; Ethan Shanfeld and Meredith Woerner, "How Netflix's 'Coded Bias' Breaks Down the Frightening Race and Gener Biased Algorithms That Run the World," n.d., https://variety.com/video/coded-bias-shalini-kantayya-documentary/; "Coded Bias Activist Toolkit" (Coded Bias, 2020), https://static1.squarespace.com/static/5eb23eee707c5356dea97eaa/t

/604a50925387f562551235c0/1615483038648/CODED_Activist_Toolkit_Final_21.pdf; Tristan Harris and Aza Raskin, "Bonus—Coded Bias," Center for Humane Technology Podcast: Your Undivided Attention, n.d., https://www.humanetech.com/podcast/bonus-coded-bias.

26 Toshiya Ueno, "Techno-Orientalism and Media-Tribalism: On Japanese Animation and Rave Culture," *Third Text* 47 (1999): 95; David S. Roh, Betsy Huang, and Greta A. Niu, "Technologizing Orientalism: An Introduction," in *Techno-Orientalism: Imagining Asia in Speculative Fiction, History, and Media*, ed. David S. Roh, Betsy Huang, and Greta A. Niu (New Brunswick, NJ: Rutgers University Press, 2015), 2–3.

27 Roh, Huang, and Niu, "Technologizing Orientalism: An Introduction," 3–4, 98.

28 Allison and Schmidt, *Is China Beating America to AI Supremacy?*, 23, 13.

29 Gabrielle Bruney, "A 'Black Mirror' Episode Is Coming to Life in China," *Esquire*, March 17, 2018, https://www.esquire.com/news-politics/a19467976/black-mirror-social-credit-china/; Clinton Nguyen, "China Might Use Data to Create a Score for Each Citizen Based on How Trustworthy They Are," *Business Insider*, October 26, 2016, https://www.businessinsider.com/china-social-credit-score-like-black-mirror-2016-10.

30 Arthur Kroker and Michael A. Weinstein, *Data Trash: The Theory of Virtual Class* (New York: St. Martin's Press, 1994), 15.

31 Léa Steinacker et al., "Facial Recognition: A Cross-National Survey on Public Acceptance, Privacy, and Discrimination," in *Proceedings of the 37th International Conference on Machine Learning*, 2020, https://arxiv.org/abs/2008.07275; Genia Kostka, "China's Social Credit Systems and Public Opinion: Explaining High Levels of Approval," *New Media & Society* 21, no. 7 (2019): 1565–1593; Genia Kostka, Léa Steinacker, and Miriam Meckel, "Between Security and Convenience: Facial Recognition Technology in the Eyes of Citizens in China, Germany, the United Kingdom, and the United States," *Public Understanding of Science* 30, no. 6 (2021): 671–690.

32 Allison and Schmidt, *Is China Beating America to AI Supremacy?*; Paul Mozur and John Markoff, "Is China Outsmarting America in A.I.?" *New York Times*, May 27, 2017, https://www.nytimes.com/2017/05/27/technology/china-us-ai-artificial-intelligence.html.

33 Ben Goertzel, "Seeking the Sputnik of AGI," *H+ Magazine*, March 30, 2011; Ben Goertzel and Joel Pitt, "Nine Ways to Bias Open-Source AGI Toward Friendliness," *Journal of Evolution and Technology* 22, no. 1 (2012): 5.

34 Paul Mozur, "Google's AlphaGo Defeats Chinese Go Master in Win for A.I.," *New York Times*, May 23, 2017, https://www.nytimes.com/2017/05/23/business/google-deepmind-alphago-go-champion-defeat.html.

35 Neda Atanasoski and Kalindi Vora, *Surrogate Humanity: Race, Robots, and the Politics of Technological Futures* (Durham, NC: Duke University Press, 2019), 4, 44–48.

36 Atanasoski and Vora, *Surrogate Humanity*, 47.

37 Shoshana Zuboff, *The Age of Surveillance Capitalism: The Fight for a Human Future at the New Frontier of Power* (New York: Public Affairs, 2019), 392–393.

38 Jodi Melamed, "Racial Capitalism," *Critical Ethnic Studies* 1, no. 1 (2015): 79.

39 Atanasoski and Vora, *Surrogate Humanity*, 40.

40 Carolijn van Noort, "On the Use of Pride, Hope and Fear in China's International Artificial Intelligence Narratives on CGTN," *AI & Society* 39 (2022): 299–301, https://doi.org/10.1007/s00146-022-01393-3.

41 Webb, *The Big Nine*, 86, 212.

9

Techno-Futurehistory and the Sojourners of Global China

―――――――――――◦―

A Threefold Reading of *The Wandering Earth*

SHANA YE

Adapted from the novella *Liulang Diqiu* 流浪地球 (2000) by Liu Cixin, the most internationally acclaimed science fiction (SF) writer from the People's Republic of China (PRC), *The Wandering Earth* (2019) and its prequel *The Wandering Earth II* (2023), directed by Frant Gwo, have garnered vastly contrasting receptions among national and international audiences. While Western-based media remain largely skeptical of the theme of "the Chinese saving the world" and criticize the pronounced nationalistic undertones of the duology, PRC viewership appreciates its celebration of family values, collectivism, and anti-Messianism.[1]

Among all the differences, one particular area of contention is regarding the films' heavy reliance on the aestheticism of futuristic technology. At first glance, the films seem to complicate techno-Orientalism's premodern-hypermodern dynamics by rendering the futurism of Asian/Chinese technological achievement not through liberal individualism but through Chinese-led collective solidarity (a method different from both Japanese cyberpunk in the 1980s and

Hollywood-style SF productions for instance). Yet, the intended message of social togetherness of the oppressed as key to humanity's survival does not land well for audiences (mainly in the West and Global North) who too often mistakenly associate collectivism with totalitarianism, despotism, or Communist brainwashing.[2] Needless to mention, the materialization of futuristic technologies such as the earth thrusters, antimatter bombs, space elevators, and so on through first-class visual effects also exacerbates prevalent global fear and anxiety over the Eastern behemoth, especially during the time of a Sino-U.S. "trade war," the competition over 5G technology, and post-COVID stagflation. These mismatched agendas have a lot to tell us about the messiness of a renewed postcolonial postsocialist world, where the PRC is simultaneously seen as a threatening "other" to Western liberal modernity, a technoeconomic alternative arising from its Third World peasantry roots, and a different type of global hegemon in its planetary (extraterrestrial) expansion.

Through textual, visual, and discourse analysis, this chapter provides a threefold reading of Liu's novella and its cinematic adaptations to explore the intersections of techno-Orientalism and Sinopostcoloniality, paying particular attention to how gender and queerness play out in disrupting and reinforcing norms about the Orient/China, technology and history, and the future.[3] As will be discussed in more detail, both the print and screen versions of the story in their respective ways challenge techno-Orientalist projections of the PRC and decenter Western modernity, including its discourse of technological advancement, individualist heroism, and settler colonialist expansion. While the novella pushes back against the telos of progress through denaturalizing gender and sexual stereotypes, the films deploy heteronormative technonationalism to recuperate Chinese inferiority and assert national superiority. The slippages, I argue, are in fact two sides of the same coin, exemplifying a particular form of Chinese postsocialist postcolonial relations—"sojourner colonialism"—that I will unpack more throughout the chapter.

Different from the "settler" and the "arrivant,"[4] I use the concept of the sojourners to refer to those in the PRC who are motivated by desire or fear to traverse among different worlds, making temporary stays along the way without the ability or willingness to settle. Their journeys are characterized by a pervasive oscillation between national pride and shame, woundedness and guilt, and mobility and stuckness, informed by the PRC's negotiation with a dual trauma of socialist history and neoliberal present. This sojourner ambivalence could lead to the gratification of nationalistic aggression and replication of colonial technologies, oddly intertwined with decolonial, deimperial, and queer possibilities. It not only muddles the boundaries between the East and the West, the colonizer and the colonized, the oppressor and the oppressed, but also asks us to take seriously what it means to live as transient beings at the interface of mismatched agendas, miscommunication, and unarticulated feelings. Providing multiple readings of Liu's original story and its adaptations and their contexts, this chapter intends to unpack these possibilities and oddities to shed light on the ways in which the messiness of geopolitics and histories shapes how futures are imagined and materialized. In what follows, I will use *Liulang Diqiu* or *Liulang* to refer to the print publications and *Wandering I* or *II* to refer to the films.

From the Print to the Big Screen: "Wandering China" and the Ambiguity of Postsocialist Postcolonial Sojourners

Published in 2000, *Liulang Diqiu* depicts an apocalyptic world of a postnational, postfamilial, and posthuman existence.[5] In this story, humanity faces an impending catastrophe as the Sun is about to explode (an event called a "helium flash"). To save the planet, the United Earth Government (UEG) builds Earth Engines to halt the rotation of Earth and eject it from the solar system, in the hope of relocating humanity to the system of Proxima Centauri through a millennia-long journey. The short story is told by a nameless narrator whose childhood, coming-of-age, and adulthood and aging are concomitant with the Earth's "braking age," its "deserting age," and its "escape age." During the journey, he also experiences the deaths of his family members and witnesses the Earth survivors' revolt against the UEG.

The screen adaptations deviate significantly from the original. The director Gwo and his team transpose the novella into a family melodrama of father and son that features three generations of Chinese men.[6] In *Wandering I*, the nameless protagonist is given the name Liu Peiqing (meaning to "cultivate strength"), one who signals national strength. The father and son of the narrator, who both receive passing mentions in the original, become the main characters who move the plots in the films. In *Wandering II*, characters, relations (father-daughter, Chinese-Russian camaraderie), and technology (space elevators, quantum computers, etc.) that are not present in the original are added, just to name a few.

This type of over-the-top adaptation certainly blunts the critical edge of the original, and a calculated guess could be made that these changes are to cater to the consumer market and to cement the Chinese authority. While agreeing with these observations, I want to point out that both the print and the film versions are teeming with ambivalences, including attitudes toward technology, socialist histories, gender and sexuality, and so on, that go beyond what phrases such as "China's flexing its SF muscles" and "space soft power" could simplify. These layered contradictions are symptomatic of China's complex relationship with the world system of capitalism and colonization that is also reconfigured in the age of the PRC's increasing global presence. When it comes to understanding "global China," conventional terms such as colonialism, empire, and hegemony face definitional, empirical, and historical troubles. There is so far no military occupation by the PRC in other continents, no chartered companies with exclusive or sovereign trading rights, no religious proselytizing—all things that typically accompanied colonialism.[7] Chinese intellectuals, especially those inside the PRC, tend to emphasize China's unique histories of semifeudalism and semicolonialism and the concept of South-South solidarity, and see China's peasant- and worker-led socialism as anticolonial and postcolonial and its advancement onto the global stage as a necessary extension for national restoration. This "Chinese exceptionalism" is contrasted with the framework of Chinese neocolonialism that regards the PRC as a failed/failing replica of Western colonialism, which is evident in China's presence in Africa as well as nationalistic projects such as the Belt and Road Initiative. In my opinion, neither of these accounts gives sufficient expression to the multitude of the Chinese reality, especially given that China's global rise relies significantly on a unique figure, the sojourner who, with a particular set of assets and affects, privilege and baggage, aspiration and trauma, navigates through multiple worlds and locations in negotiating colonialism, racism, and advanced capitalism.

Mainly written from the late 1980s to early in the 2000s, Liu Cixin's speculative works capture a particular moment when the PRC began to transition from a "world factory" to a "venture capital investor," seeking larger acceptance and advantage on a global stage. Proceeding from its market economic reforms beginning in 1978 and "connecting track with the world" initiative since the 1980s, the state supplemented the previous *yin jinlai* ("Bring In" 引进来) policy with the *zou chuqu* ("Go Globally" 走出去) strategy, emphasizing the equal importance of overseas markets and international collaboration. This outbound adventure required China to accelerate its escape from the gravity of past socialist isolation and to create a new cohort of subjects as the "wandering agents" who could pilot the voyage of a global "China on the rise."

Enlarging the previous class of *xiahai* (lit. "dive into the sea" 下海) businesspersons in the 1980s and 1990s, the "wandering agents" could be venture capitalists investing in infrastructure and real estate in the African continent; entrepreneurs managing factories, hotels and restaurants in Ethiopia; migrant workers laboring in Zambia's copper mine and construction sites and small-business owners selling lingerie in Egypt.[8] They could also be state officers and their families stationed in developing countries; personnel of Chinese NGOs (nongovernmental organizations) rotating positions abroad; the increasing numbers of students, intellectuals, and scientific researchers who carry the "dreams of flight" in the United States, United Kingdom, or Australia; and the Chinese bosses who bail out American companies depicted in documentaries such as *American Factory* (2019).[9] Unlike Western colonizers commissioned by the state or unwilling settlers forced by slavery and indentureship, the majority of the PRC's sojourners were not directly sent by the state but garnered state incentives, driven by cosmopolitan desire, potential profits, and the national affect of opening and development, to assist China to "test the water" in seeking "spatial and political fixes to the nation's resource and profit bottleneck."[10] They enjoyed different levels of privilege and exploited both locals and Chinese, as well as being subjugated to racial hierarchies and racism along the way in their journeys.

Similar to the formidable uncertainty constantly faced by the Earth survivors in *The Wandering Earth*, the exploration by China's sojourners navigating in the sea of the national and global overaccumulation crisis was teeming with uncertainty and not preordained to guaranteed outcomes. Even with the goal of settling in a new home just like in the journey to Proxima Centauri in Liu's short story, most of them ended up wandering without settling nor returning to what had been lost. Viewed from a different light, the PRC's promotion and facilitation of mass grassroots sojourning during this period of time could also be seen as a form of venture capitalist investment in and by itself that aimed to capitalize on a high-risk, high-reward future of the speculative economy without providing any guarantee to its participants. Speculation, both in the financial and futuristic senses, could be said to be part of the zeitgeist of the 2000s.

This optimistic uncertainty was reciprocated by the speculating "sojourners within." Although Sinophone speculative writing could be traced back to the late Qing and the Republic periods and SF was used by the Chinese Communist Party to promote science education, the PRC saw a sudden revival of SF in the post-1989 era, pioneered by writers such as Han Song, Wang Jinkang, and Liu Cixin. As a result of the *wenhuare* ("cultural fever" 文化热) and the nationwide embrace of Western Enlightenment thought in the 1980s, the lost popularity of revolutionary literature, the boom of sciences and technologies, and the rise of the Internet, what SF scholar Song Mingwei termed the "new wave" was charactered by its cutting-edge experimentalism with subversive cultural and political significance. Different from the Japanese cyberpunk movement in the 1980s that reinvigorated the idea of an essentialist Japanese superiority by reifying and appropriating American techno-Orientalist tropes of gender and nation, the PRC's "new wave" seemed to interact with the techno-Orientalizing of China in more twisted manners.[11]

On the one hand, due to ideological and geopolitical disparities, what could be obviously techno-Orientalizing in American pop culture did not necessarily translate into the Chinese cultural realm when the Sino-U.S. relationship entered a period of honeymoon in the late 1990s and early 2000s. For example, from the Noah's Ark built in China in *2012* (2009), to the futuristic city scene of Shanghai in *Her* (2013), and to secret facilities in Chinese farmlands in *RoboCop* (2014), fetishizing China in premodern and hypermodern tropes could be considered as a telltale sign of China's acceptance on the global stage, and the American catering to Chinese consumers also boosted national confidence and pride. On the other hand, against the backdrop of the further commodification of arts, literature, and moving pictures in the 2000s, revealing the "raw and unpleasant reality" of China was a shortcut for Chinese cultural products to circulate globally.[12] The hands-on technology of self-Othering offered a sure way for "the wandering avant-garde" to step into the global art market and seek economic and political opportunities. Orientalist and Cold War–induced racism and stereotypes found harmony with revolutionary and radical social critiques in turning Chinese history and reality into cultural artifacts under the powered-up face of Sino-U.S. neoliberalism. While Japanese cyberpunk might utilize female cyborgs to deflate Western domination, the Chinese SF tended to focus on issues of socialist illness, postsocialist inequality, and statism through figures of monstrosity, emotionless sub- and perihumanity, and Orwellian-style dystopianism, simultaneously performing the labor of self-techno-Orientalization and cultural and political subversion.[13]

Taking Liu Cixin's oeuvre for example, one can immediately detect this messiness from his oscillation between the internationalization of Western colonialism and resistance to Western modernity. Many of his works contain strong identifications with the Third World and cosmopolitan utopianism, as well as condemnation of imperialism and challenges to key concepts of modernity, such as progress, development, nationalism, and scientism, while attesting to a grandiose narrative of unified humanity and China's ascendency to center stage in global politics. There are stories such as *China 2185* (1989) reflecting on the collective trauma of Maoism and the commodification of social life and memory, and *The Three-Body Problem* (2006) engaging colonial encounters with the alien but told from the viewpoint of the colonized, as well as colonial "revenge porn" story *The Atlantic* (2002), a uchronia that replaces Hong Kong's handover to the PRC with Liverpool's return to the British.[14] Sometimes these works feature critiques of and nostalgia for the socialist past and collectivity such as in *Liulang*, whereas other times the collective future fails or the history of the world(s) is moved precisely by individual decisions, such as in *Death's End* (2010). These contradictory oscillations are the defining characteristics of the PRC's sojourners.

When the short story was transposed to the films, the PRC's cultural industry and its consumers had transformed greatly. The rise of cybernationalism contrasts sharply with the growing awareness of feminism and gender issues in online space. The political climate also changed from the more open liberal years to tighter surveillance under the name of "harmonious society" and "China's peaceful rise to power." The director and the screenwriters, the "new sojourners" of global China, who were born in the 1980s and accumulated social capital, discursive power, and transnational and upward mobility, and who benefited from the state educational system and were perhaps more susceptible to the Party's sociopolitical vision, have different challenges in navigating through the labyrinth of the domestic and international market, audiences and geopolitics.[15] In the next section, I will further unpack how these contradictories of China's "wandering sojourners" play out and what they enable and disable.

Techno-Futurehistory and the Call for Carrying the "Shit"

When it comes to the method of space migration and expeditions (that is, new waves of colonization, or the age of the "second voyage"), there are usually four common types imagined in Western popular science fiction. The first one, as seen in films such as *Interstellar* (2014), *Prometheus* (2012), and *The Fifth Element* (1997), involves building a light-speed spacecraft that carries colonists in cryogenic sleep pods and the genetic material necessary to repopulate a new territory. This method is often combined with the second one, which is to create a biosystem within the spaceship large enough to sustain the life of the travelers, as seen in the film *The Martian* (2015). The third one could be called a "galactic railway system," in which each spacecraft with its own biosystem is a self-sufficient module that can also connect and form a larger system when needed, such as depicted in the Japanese anime series *Mobile Suit Gundam* (1979–1980) and the American TV series *The Expanse* (2015–2022). And the last one is "snowballing," or "island jumping," in which the travelers, with the help of AI, never settle but utilize the resources and energy of each of the star systems along the way as "recharge stations." This concept is found in the *Battlestar Galactica* series (2004–2009) or *The Mass Effect* (2007) video game.

In comparison to these prevailing setups that unmistakably rest on both heteronormative tropes of repopulating new territories with (white) human genes and the imagination of the dangerous yet alluring (sexualized) alien other, Liu Cixin's design of turning Earth into a vehicle and carrying the planet to travel with is indeed unusual, if not revolutionary. Many scholars and critics have pointed out the anticapitalist and decolonial potentials of the story, especially from its takes on technology. For instance, drawing from a Marxist analysis of fetishization and growth, film scholar and literary critic Amir Khan suggests that the film's engagement with technology and Sinofuturism (in this case *Wandering I*) invites the audiences to divorce the idea of technological advancement from the motive of profit and to rethink value as coming from the social relations between people, rather than the accumulation of short-term surplus.[16] Despite the depiction of the Chinese state, soldiers, and cadres as the heroic agents, in both films, it is not the individualist hero who saves the planet, but rather the collective efforts of people from different parts of the Earth. By putting the alliance of the "darker nations" at the center of human survival, these films can be read as subverting the colonial capitalist logic of an imperial system that fetishizes technology and growth.[17] If the death of the Sun equals something like global warming and therefore represents the endgame of global capitalism, the wandering Earth of the oppressed in solidarity could be seen as signaling a state of no longer being colonized.[18]

A similar suspicion of technological advancement and desire for a reformed social relationality are also prevalent in *Liulang Diqiu*. At the beginning of the story, the narrator recalls a school trip to visit the colossal Earth Engines when he was a child. While marveling at the fluorescent blue plasm beams, proofs of technological achievement, he recites a riddle to his teacher Ms. Stella: "You are walking across a plain when you suddenly encounter a wall. . . . The wall is infinitely tall and extends infinitely deep underground. It stretches infinitely to the left and infinitely to the right. What is it?" Seeing Ms. Stella shaking her head in confusion, the narrator leans in and whispers the answer: "Death."[19]

Setting the pessimistic and dark tone, the equation of the terrifying beauty of technology with death casts doubt on the technological reductionism and fetishism in China's statecraft. Although the Chinese state's emphasis on technology and science seems to express a basic Marxist position essential to the nation's anticolonial independence and socialist modernization, the establishing of technological reductionism as a priori in the 2000s signaled a right-wing turn in both the political scene and cultural interventions that would only intensify later on.[20] Yet Liu Cixin's poignant critique does not come from his abstract philosophical ruminations but rather from reflections grounded in existing histories.

For example, in order to transform the Earth into an interstellar vehicle, humanity has to be dislocated and mobilize billions of people and strictly control its population. In the films, half of the human population is sacrificed through a lottery system, and in the novella, only one-third of newly wedded couples are allowed to procreate.[21] These fictional depictions immediately conjure up memories of the PRC's land reforms, the Great Leap Forward, the One Child Policy, and the household registration system. The history of the Cultural Revolution also finds its reminiscence and resonance in the fictional world's devaluation of the arts, philosophy, and anything deemed nonessential to the goal of survival.[22] During the rebellion against the UEG, the survivors are easily agitated and brutally execute the prisoners, putting the readers in mind of the violence of the Red Guards.[23]

What makes *Liulang Diqiu* stand out, however, is that, unlike many mainstream space expedition and colonization SF stories that depict the exuberating discovery of a "new home," Liu insists that we must "carry the shit" that makes and unmakes us. Instead of celebrating the abstract idea of home and homeland to heal historical wounds, the short story reminds us that there is no grand escape for the "wandering China" to wash its hands off, jump into the future, and to be accepted by the rest of the world. Instead, the "wandering sojourners" must live with the gravity of the past and be impacted by the materiality of the "shit" they carry.

Here, Liu's tapping into China's traumatic history departs from a self-Orientalist desire to fetishize its unpleasant history by making it spectacularly "Chinese" to cater to the international taste, nor claim distance from it, as many of the avant-garde artists would do in the 2000s. Instead, under the motif of a planetary catastrophe, the "shitty" local histories of the PRC are placed in the context of the larger "shitty global history" of land grabbing, accumulation by dispossession, colonialist invasion, and the disasters of late capitalism.

For instance, while the land reform in the 1950s was crucial to socialist modernization and the legitimacy of the Communist Party, the state-regulated land distribution would also lead to land grabs and the dislocation of people in the postsocialist era. Although rooted in the particularities of the Chinese political economy, these activities are also part of a global wave of land dispossession afflicting Africa, Latin America, and Southeast Asia, caused and exacerbated by speculative investment in food and agriculture by global financial capital, or by state-sponsored projects of building global cities.[24] Situating the domestic history within what is happening globally allows a deeper understanding of Chinese modernization, both socialist and postsocialist, and renders any blanket claims of colonialism or anticolonialism false and arbitrary.

Following the Earth Engine visit in *Liulang*, the school children are led to visit the seashore, where the narrator witnesses the scene of seawater gushing through the windows of submerged skyscrapers. This massive destruction was caused by the Earth Engine melting the polar ice caps and engulfing two-thirds of the Northern Hemisphere's major cities, a series of disasters that happened even before Earth's escape. As the story thrusts us into an imagined world of the ruthless destruction of nature, homes, cities, monuments of civilization, arts and culture, and money, the readers are pushed to confront a possible reality of the dispossessed land and the dissolution of established social orders, and to identify with the present-day "wanderers" of settler colonialism, capitalist urbanization, and financial imperialism.

If the films' juxtaposition of idealized social relations against technological advancement bears no real-world correlation, the message sent by the original could be a powerful and timely intervention to the official discourse that romanticizes socialist modernization and Third World solidarity as anticolonial.[25] By staging complexity and connections, Liu's short story does not glorify the Chinese as the future savers of humanity, but rather functions as a cautionary tale for the technoelites and leftist scholars to reflect on the "shit" that is being carried forward. Without replicating the grand narrative of national restoration and China's leading role for humanity, this futurehistory is a rare call for larger responsibility at both the state and individual levels to look inward on China's outward journey of global rise beyond the binary of national humiliation and right-wing nationalism, as "carrying the shit" allows the readers to engage with the dual traumas of history without attempting to heal from them.

This insight into cross-spatial and temporal connections also leads to the author's emphasis on relationality (with other human and nonhuman agents) beyond the narrowly defined family. As the school was scheduled to relocate to an underground city during the "deserting age," the narrator reveals that Earth Engines are not as powerful as people think. Although they create the initial momentum to halt and nudge the planet, ultimately, its acceleration and escape rely on repeated "gravity assists" of other larger objects such as Jupiter and the Sun. Earth needs to orbit the Sun fifteen times before it can gain enough velocity to escape and its orbit will increasingly become a flatter ellipse as it speeds up.

Here, the novella interestingly complicates our reading of the dying Sun as representing the end of capitalism and colonization by highlighting the difficulties and costs of escaping. As the story unfolds, we learn that tremendous anxieties, doubts, and conspiracies rise during the Earth's orbit transfer and eventually lead to the rebellion and coup. This mirrors the PRC's own history of battling with different routes to modernity and the contradictions of replicating colonialism. Indeed, the anticapitalist decolonial reading can be compromised as the novella also reinforces colonial ontology and discourse at a deeper level. For example, although Earth has been terraformed beyond recognition, we are surprised to learn that its surface is still marked by old names such as Asia, North America, Oceania, and so on. What is in between the continents, such as the islands and archipelagoes, has been completely wiped out in this type of futuristic historicization. In other words, the human history of the Ante-solar Era cannot escape from the doomed fate of only being remembered through its colonial, capitalist, and modern state–centered historiography. This repetition is confirmed in the narrator's description of an Olympic Games he has attended as a snowmobile racer, where the competition begins in Shanghai, and then moves across the frozen Pacific to New York City. Revived after a two-century hiatus, the Olympic Games is supposed to signal mankind's resilience, love (the narrator also meets his future wife there), and oneness in harshness, but this idealized optimism rests on the careless handling of the entire Asia-Pacific region's islanders and Indigeneity, revalidating the significance of Sino-U.S. alliance and connectivity in the future worldmaking. The cinema adaptations of the story also share these unfortunate pitfalls. As Khan further articulates, the path to appreciating the film's antitechnofetishism and therefore anticapitalist decolonial insight . . . comes precisely via the onscreen rendering of technofetishism, succumbing to the very Western fantasies it seeks to slay.[26]

Queering Sojourner Colonialism: Technonationalism and the Gendered Cost of the "Greater Good"

For feminist, queer, and gender-sensitive readers, Liu Cixin's writings can be distasteful, as they often internalize misogynist and heteronormative logics. *Liulang Diqiu*, however, can be read as challenging the authority of heteronormative patriarchy and its ultimate embodiment—the protective parent-like state—throughout its narrative unfolding. This is done through a surprising construction of a queer world with "deviant" sexuality and intimacy as daily norms that reveal the plasticity of the heteronormative family and denaturalize the symbolic power of the father figure and military and scientific masculinity, another instance to express the PRC sojourner's ambivalence.

In the story, during a family trip to the seashore to watch the Sun during one of the Earth's orbital accelerations, the narrator's father tells the family out of the blue that he has fallen in love with Ms. Stella and will leave his wife and family to live with her. Despite the suddenness of the announcement, the news does not surprise the family and rocks no boat. Two months later, the father comes back home and pleases the family only because he can help switch up the holographic backgrounds of their home, which they are sick of looking at.[27] This diluted emotional attachment also manifests itself in the narrator's complete apathy to his father's death and only a tinge of pain at the passing of his mother and grandfather.[28] Later, the narrator and his son are abandoned by his Japanese wife, who leaves the family for the rebellious army and is eventually killed in action.

There are multiple readings of these scenes: one could be that these thinned emotions highlight the brutality of human social bonds and show the ephemerality and meaninglessness of established social norms, orders, and common senses in the face of a Darwinist struggle for human survival. To me, these suppressed emotions and feelings testify to the persistence and dailiness of collective trauma and the labor of living through the ordinary trauma, especially in a prolonged "state of emergence."[29] But as Cathy Caruth tells us, unclaimed trauma is the eloquent speaker for erased histories; and these minor feelings do speak loudly about the gendered history of who bears the cost of the unified career of human survival.[30]

There is no secret that the Chinese nation-state has long relied on patrilineal family and Confucian ethics as technologies for its governance, shaping social norms, personal relationships, and political institutions. The Confucian emphasis on respect for authority, hierarchy, and duty is translated into respect for the figure of the father and his role as the head of the family and primary decision-makers. Filial piety is also the foundation of the state-family. With the patriarchal norms, women's bodies and reproductive capacities have been utilized by the state. While the One Child Policy has since been abandoned, the government's regulations take different forms, such as seen in coerced sterilizations or abortions, particularly in ethnic minority regions such as Xinjiang, as well as the encouragement for urban middle-class women to have a third child under the new Three Child Policy.

Women's social reproduction labor and sacrifice for the "greater good" are recalled in *Liulang* in a strangely dark way. During the "deserting age," the Earth Engine disturbs the equilibrium of Earth's core, causing disasters such as volcanic eruptions and magna seepage into the underground cities. This is what caused the death of the narrator's mother since the emergency plan arranges the order of evacuation according to age, from the youngest to the oldest.[31] In this context, Liu reminds us that the real technology for Sinomodernity is its people. The denouncement of the family therefore can be read as a desire to escape from the national ideology of sacrifice and service and as a reminder of the gendered cost of progress and advancement.

This is not to say, however, that Liu Cixin proactively addresses gender issues and nonnormative sexuality in his fictional worlds; rather, the unexpected queerness might come from a larger intellectual and political tradition of engaging with pressing social issues. In other words, this queerness is rooted in intersectional concerns rather than the concept of an automatic liberal subject who comes

FIGURE 9.1 Film still from *The Wandering Earth 2*. (Source: China Film Co.)

out of the closet of repression to form rights-bearing identities. At the same time, this ambivalence and murkiness can also be appropriated and assimilated into the national discourse of gender as technology to cement social norms.

Roughly one-quarter through *Wandering II*, the protagonist Liu Peiqing gets himself involved in a fight with a group of rebels who plan to hijack the Space Elevator. During a series of fast-paced Hollywood-style actions, his love interest, Han Doudou, throws him a robotic arm as a weapon to assist him. When Liu Peiqing hits his opponent with the arm, a ring-shaped metal part comes loose and flies toward the point of view of the audience. The slow-motion close-up allows the audience to see the engraved words "Made in China" on the part; no surprise, this metal part later becomes the engagement ring Liu gives to Han (Figure 9.1). Through this hyperdrama of love (in contrast to the hypoemotional detachment in the book), heteronormativity and national projects such as "Made in China 2025," which aims to elevate China's world domination in high-tech industries, form a perfect unity.

The alloy of Chinese technonationalism with the superficial heteronormative love story might seem naive and "out of time" when the mainstream U.S. cinema has turned to queerbaiting, pinkwashing, and (superficial) featuring of women of color. This is not to suggest that the United States is more "advanced," but to point out the multileveled mismatching when China steps up on the global stage.

Another example of these mismatches in *The Wandering II* is about the male laborer, another overdetermined gendered figure when it comes to nationalism, development, and geopolitics. In U.S. techno-Orientalist discourse, the PRC is often treated as both a human factory for the global economy and a consumer market for Western cultural products, a fashion that can be traced back to the 1980s when the PRC began to be recognized as a newly industrialized country, which solidified the U.S. global economic dominance in the 1990s.[32] These well-worn stereotypes evolve over time and often manifest themselves in an oxymoronic manner when it comes to the representation of the nonhuman: on the one hand, the robot-like Chinese labor testifies to China's embodiment of high technology without regard for human concerns; on the other hand,

FIGURE 9.2 The rapid construction of *Fangcang* medical facilities in 2020, following the outbreak of the COVID-19 pandemic. (Source: *Sohu News*.)

China's technological advancements are no more than those of a backward copycat that relies solely on the cheap labor of its dehumanizing workforce. We can easily trace these contradictions, for example, in the video that went viral online at the very beginning of the COVID-19 pandemic, showing the marvelous speed of the Chinese machines building the *Fangcang* quarantine facilities without displaying any human agency (Figure 9.2); and the revelation of the real human labor holding and maneuvering the 897 pieces of printing template at the 2008 Beijing Olympic Games opening ceremony that made the performance appear operated by automation (Figure 9.3). Both examples can be read as humanizing or dehumanizing depending on the point of view.

Many of the films' details reveal the intention of the production team to engage with the techno-Orientalist ambiguities. For instance, in *Wandering II*, the camera pans over a group of workers sitting on a metal beam overlooking the dark construction site of the Earth Engines (Figure 9.4). In contrast to the non-humanness of the machines and infrastructures, this scene attempts to insert the humanity represented by workers. This shot conjures the well-known 1932 photograph *Lunch atop a Skyscraper* (Figure 9.5), which features eleven ironworkers sitting on a steel beam 850 feet above the ground on the sixty-ninth floor of the RCA Building in Manhattan.

The readings of this reference could be manifold: by connecting the Chinese workers with the workers in the United States, the cultural sojourners (the

FIGURE 9.3 Screenshot of the 2008 Beijing Olympic Opening Ceremony. (Source: Baidu.)

FIGURE 9.4 Film still from *The Wandering Earth 2*. (Source: China Film Co.)

production team) celebrate another group of sojourners (the overseas laborers) and credit them as contributing to a global history of modernization, industrialization, and capitalist development. It could also be read as responding to the allegations of China's labor and human rights violations, showing China's appreciation of workers. A third reading could be that this scene mocks U.S.-led modernity and hints at Chinese alternatives since what the American workers were building was the Rockefeller Center, a symbol of U.S. financial capitalism, while the Chinese workers are building for the Earth's "common good." Interestingly, *Lunch atop a Skyscraper* turns out to have been a publicity stunt, arranged by the company as part of a campaign promoting the skyscraper, and the photograph was acquired by the Visual China Group in 2016. Does this irony also reveal the hypocrisy of utilizing the images of workers and confirming their exploitation as human technology (therefore reinforcing the techno-Orientalist construction of China as labor) by global capital both in China and the United States? Although none of these readings is conclusive, the reference does

FIGURE 9.5 *Lunch atop a Skyscraper*. (Source: Wikimedia.)

accentuate the messiness and mismatched agendas when it comes to representing global China at the interface of the consumer market, the cultural industry, and real-world geopolitical tensions, misunderstandings, and mistrust.

Conclusion

Given their large departures and deviations from Liu's novella, one may wonder how the two film adaptations could be endorsed by the author himself. To conclude, I want to recapitulate that this split is precisely a symptom of China's particular history of semicolonialism and socialist and postsocialist encounters with modernity, experienced by its transspatial, transtemporal sojourners.

Going back to the original story, we know that the set destination of Earth's journey is Proxima Centauri, or *bi linxing* 比邻星, the "neighbor star," a red dwarf in the triple-star system Alpha Centauri. If we read this setting in relation to Liu's later masterpiece *The Three-Body Trilogy*, we could safely speculate that a three-star system is unstable and will not make a suitable new home for the earthlings. Even if humanity could reach and settle at Proxima Centauri, Earth would be tidal locked by the star since Proxima Centauri is too small to become a new Sun, leaving the planet uninhabitable. Reading the story against the grain, we might realize the real tragedy that Liu alludes to: The doomed journey is the only journey that humanity could embark on, resonating with the predestination of

China—the doomed postsocialist postcolonial journey that China embarked on in searching for acceptance and prosperity on a global stage is the only journey that leads to its demise through replicating existing colonial world orders.

We see finally that the story is not about an outward journey of technological triumphalism, Chinese national achievement, and Confucian family values, but rather an inward journey to reflect on the Chinese psyches and desires informed by a trauma-laden history and the inevitable brutality of a neoliberal nationalistic future. The search for healthier humanistic values ultimately entraps itself in the repetition of global coloniality structured by previous imperial forces, colonial history, and the renewed apparatus and technology of Sinopostcoloniality. Although Liu's work does provide the reader with explicitly uplifting queer and feminist energy to glimpse at what the "something different" could be, the odd unison of anticolonial/decolonial energy and the internalization of colonialist and imperialist impulses is what global China cannot escape.

Notes

1 For a more detailed discussion on Western's media skepticism about the film, see Muqiang M. Zhang, "What Western Media Got Wrong about China's Blockbuster 'The Wandering Earth,'" *Vice*, April 2, 2019; and Jodi Byrd, *The Transit of Empire: Indigenous Critiques of Colonialism* (Minneapolis: University of Minnesota Press, 2011). Zhang cites, for example, a critic in the *Washington Post* who wrote that the film is "a prototype for exporting an image of China as the leader of the future," and that, "in this fantasy, only Chinese leaders can be relied on in a crisis. Only Chinese engineers know how to effectively manage the complex systems of the future." Similarly, in *Slate* it was written that "*The Wandering Earth* arguably reflects the Chinese party line that bureaucracy can manage doom via central planning and clever engineering. The urgency of the crisis leaves no room for dissent." And the subtitle of the piece was "Authoritarian? Yes. Propaganda? Maybe." Zhang points out, "When the US releases a propaganda film, it is often applauded as a win for patriotism. But when China releases a blockbuster, the film is dismissed as a ploy to brainwash the West" and then continues, "These writers have projected a message of Chinese authoritarianism despite the film's clear theme of international cooperation, the lack of nearly any mention of the Chinese government, and the denouement showing soldiers from all over the world collectively saving Earth. Interestingly, there is no substantial mention of any of these themes in Western media's coverage of the film. Instead, they have dismissed the film as a rip-off of US films, lambasted it as propaganda to brainwash the West, and projected tropes of Chinese people as a yellow peril horde."
2 Amir Khan, "Technology Fetishism in *The Wandering Earth*," *Inter-Asia Cultural Studies* 21, no. 1 (2020): 22.
3 Peng Cheah, "Introduction: Situations and Limits of Postcolonial Theory," in *Sitting Postcoloniality: Critical Perspectives from the East Asian Sinosphere*, ed. Peng Cheah and Caroline Hau (Durham, NC: Duke University Press, 2022), 1–29.
4 See Byrd, *The Transit of Empire*.
5 Ping Zhu, "From Patricide to Patrilineality: Adapting The Wandering Earth for the Big Screen," *Arts* 9, no. 3 (September 2020): 1.

6 Zhu, "From Patricide to Patrilineality," 2.
7 Ching Kwan Lee, "Ching Kwan Lee: The Specter of Global China," *Made in China Journal* 2, no. 4 (October–December 2017), https://madeinchinajournal.com/2017/12/24/ching-kwan-lee-the-specter-of-global-china/#:~:text=In%20her%20new%20book%20The%20Specter%20of%20Global,fixates%20on%20China%20as%20a%20new%20colonial%20power.
8 Howard French, *China's Second Continent: How a Million Migrants Are Building a New Empire in Africa* (New York City: Knopf, 2014); Ching Kwan Lee, *The Specter of Global China* (Chicago: University of Chicago Press, 2018); Peter Hessler, *The Buried: An Archaeology of the Egyptian Revolution* (New York: Penguin Books, 2020).
9 Fran Martin, *Dreams of Flight: The Lives of Chinese Women Students in the West* (Durham, NC: Duke University Press, 2021).
10 Lee, "The Specter of Global China."
11 Mingwei Song, "After 1989: The New Wave of Chinese Science Fiction," *China Perspectives*, 2015, 7–14; Kumiko Sato, "How Information Technology Has (Not) Changed Feminism and Japanism: Cyberpunk in the Japanese Context," *Comparative Literature Studies* 41, no. 3 (2004): 335–355; Toshiya Ueno, "Japanimation and Techno-Orientalism," in *The Uncanny: Experiments in Cyborg Culture*, ed. Ed Bruce Grenville (Vancouver: Arsenal Pulp Press, 2002), 223–236.
12 Jason McGrath, *Postsocialist Modernity: Chinese Cinema, Literature, and Criticism in the Market Age* (Stanford, CA: Stanford University Press, 2010), 13.
13 David S. Roh, Betsy Huang, and Greta A. Niu, "Technologizing Orientalism: An Introduction," in *Techno-Orientalism: Imagining Asia in Speculative Fiction, History, and Media*, ed. David S. Roh, Betsy Huang, and Greta A. Niu (New Brunswick, NJ: Rutgers University Press, 2015), 8.
14 Mengtian Sun, "Alien Encounters in Liu Cinxin's *The Three-Body Trilogy* and Arthur C. Clarke's *Childhood's End*," *Frontiers of Literary Studies in China* 12, no. 4 (2018): 610–644.
15 Zhu, "From Patricide to Patrilineality," 5.
16 Khan, "Technology Fetishism in *The Wandering Earth*," 26–27.
17 Khan, "Technology Fetishism in *The Wandering Earth*," 31.
18 Khan, "Technology Fetishism in *The Wandering Earth*," 35.
19 Cixin Liu, *Wandering Earth*, trans. Ken Liu (New York: Tor, 2021), 6.
20 Chuang, "The Wandering Earth: A Reflection of the Chinese New Right," *Chuangcn.Org*, August 30, 2019, https://chuangcn.org/2019/08/wandering-earth/.
21 Liu, *Wandering Earth*, 26.
22 Liu, *Wandering Earth*, 16.
23 Liu, *Wandering Earth*, 42.
24 Lee, "The Specter of Global China."
25 Khan, "Technology Fetishism in *The Wandering Earth*," 34.
26 Khan, "Technology Fetishism in *The Wandering Earth*," 34.
27 Liu, *Wandering Earth*, 17–19.
28 Zhu, "From Patricide to Patrilineality," 2.
29 Walter Benjamin, *Illuminations: Essays and Reflections*, ed. Hannah Arendt, trans. Harry Zohn (New York: Schocken, 1968).
30 Cathy Caruth, *Unclaimed Experience: Trauma, Narrative, and History* (Baltimore: Johns Hopkins University Press, 1996).
31 Liu, *Wandering Earth*, 23–24.
32 Roh, Huang, and Niu, "Technologizing Orientalism: An Introduction," 5.

Part IV
Machinic Subjects

10

Sacrificial Clones

The Technologized Korean Woman in *Shiri* and *Cloud Atlas*

JANE CHI HYUN PARK

In *Yellow Future: Oriental Style in Hollywood Cinema* (2010), I drew on early scholarship on techno-Orientalism to examine the increasing visibility of East Asian actors, cultures, and styles on the big screen from the 1980s to the early 2000s.[1] During this period, aesthetic forms such as anime from Japan and martial arts from Hong Kong became popular in the United States, constituting what became known as "Asian Cool." I coined the term "Oriental style" to refer to the complex ways in which these forms were referenced and integrated into dominant visual culture—a process of cross-cultural translation that reproduced and reworked the power dynamics of Orientalism and techno-Orientalism.

A little over twenty years later, with the steady rise of *Hallyu*, or the Korean Wave, South Korea has become the latest East Asian country to signify "Asian Cool." Garnering regional audiences first through South Korean television dramas and K-pop in the 1990s, then international ones through South Korean cinema, the Korean Wave is now a well-known global phenomenon that has expanded to include other aspects of national culture such as food, beauty, and tourism, which are linked synergistically to its exported media culture.

This chapter brings together scholarship on the Korean Wave, techno-Orientalism, and cultural representations of Korean and Asian American women to explore the figure of the technologically modified, Orientalized Korean woman in two films that track the ascent of Hallyu: *Shiri*, an action movie hailed as the first Korean blockbuster, and *Cloud Atlas*, a Hollywood adaptation of the titular multigenre novel by David Mitchell.[2] In both films the body of the inauthentic Korean woman functions as a productive site of discourse for examining the costs of South Korean development and globalization.

As Euny Hong notes in *The Birth of Korean Cool*, the appeal of Korean cultural products does not derive from their originality, but rather from the innovative ways in which the South Korean cultural industries have brought together and adapted ideas, technologies, and aesthetics from other cultures to suit local, and now global, tastes—whether this is the Hollywood film, the boy band, or African American fried chicken.[3] I would add to Hong's observation the temporal and affective intensification of these cultural forms in their Koreanization. In this way, Hallyu unyokes the concept of originality from authenticity: Korean New Cinema, BTS, and Korean fried chicken are considered culturally Korean even as they invoke older, foreign forms. This mode of (re)production, which I call *creative imitation*, draws on neo-Confucianism and postindustrial late capitalism and serves as the engine that has driven and continues to drive South Korean economic development. Hallyu exemplifies this mode at the macrolevel in its successful commodification of South Korean culture for global consumption and at the microlevel in the "happy endings" that characterize feminized narrative genres such as the Korean cosmetic surgery makeover and the Korean romantic television drama.

This happy ending is denied in *Shiri* and *Cloud Atlas*. The lives of the female protagonists end in a bad death, an unhappy ending. Their culturally and ontologically mixed bodies are unable to perform this mode of creative imitation in order to reproduce an "authentic" national identity that is recognized and celebrated as "successful" in neoliberal terms—at least while they are alive. In other words, these women embody the discarded and necessary failures of the capitalist machine, the obscured and invisible costs of visibility. I argue that their inability to faithfully imitate—and in so doing, become the ideal neo-Confucian Korean woman whose purpose is to reproduce the patriarchal family as nation, both biologically and culturally—renders them *bad copies*.[4] Put another way, they are "bad copies" because their difference from this ideal remains visible. Because of their cultural and ontological mixedness, the protagonists are denied the role of proper womanhood even as they reproduce the ideology of the nation. In this sense they evoke and rework within a Korean context, the familiar figure of the mixed-race Tragic Mulatta in nineteenth- and twentieth-century American literature and popular culture. Much like the light-skinned Black female characters who comprise this figure, they attempt to "pass" into dominant culture through imitation (as South Korean in *Shiri* and "Pureblood," or biologically human in *Cloud Atlas*). And like these characters, they are doomed to fail and meet an untimely end.[5]

In *Shiri* a North Korean terrorist group tries to sabotage peace talks between the two Koreas at the end of the twentieth century, a cinematic allusion to South Korean president Kim Dae Jung's 1998 Sunshine Policy. Several thousands of years later, in "Orisun of Somni 451," one of six interconnected stories in *Cloud Atlas,* Korea has unified to become the center of a global civilization mostly destroyed by nuclear war and environmental collapse. Played by transnational actors Kim Yun-jin and Bae Doona, respectively, the Korean women central to these films' narratives illuminate anxieties around cultural authenticity that undergird the Korean Wave as well as the divided Korean Peninsula and its diasporas. Like these cultural and geopolitical expressions of "Korea," they complicate boundaries between East and West, North and South, past and future, the fake and the real.

The bodies of the protagonists in these films are altered through cosmetic surgery and created through cloning to serve totalitarian states but instead are used to betray and destroy the state. In this sense, these characters provide case studies of imitation gone wrong, of copies that take on lives of their own. Both films present speculative scenarios of national and global unification through romantic relationships between male protagonists representing democracy and female protagonists representing its opposite. After falling in love, these techno-Orientalized women refuse to play the subordinate roles they have been assigned by a totalitarian patriarchal state—the North Korean regime in *Shiri*, a hypercapitalist future society in *Cloud Atlas*. Instead they sacrifice themselves for the ideologies of their male partners, who awaken in them a desire for liberal humanist values. This pushes them to disavow their formerly "enslaved" selves, liberating them from oppressive systems that treat them as expendable weapon (*Shiri*) and disposable slave (*Cloud Atlas*). Transgressive romantic love transforms them into *individuals* within the paradigm of liberal humanism. Yet ironically, they cannot survive as such: both characters ultimately kill themselves for their partners and the vision of a democratic society. It is only through this act of self-sacrifice that they are transformed into authentic human(ized) women.

I begin the chapter by considering how the trope of imitation permeates popular Western discourse on the Korean Wave and Korean/Asian women. This provides the theoretical and cultural context for my observations on how this trope plays out in the films.

Hallyu as Techno-Oriental Style

Ongoing anxieties of Asia decentering the West appear in familiar stereotypes of Asian people that threaten to take the place of their Western counterparts—whether they be Asian American "whiz kids," Southeast Asian "mail-order brides," or K-pop "idols." All of these stereotypes draw on the idea that Asian people lack agency and subjectivity, propelled by a collective will to produce and reproduce. In this sense they are seen to resemble robots or machines more than human beings—a formation with a long history that Long Bui tracks in *The Model Machines*.[6] The racialized Asian automatons discussed by Bui—Chinese laborers, Japanese soldiers, and Asian sex workers—exemplify the concept of techno-Orientalism by linking yellow peril fears to cultural anxieties around industrial capitalism, migration, globalization, and emerging technologies. The stereotypes mentioned earlier are contemporary figurations of "model machine minorities" embodying the trope of aspirational imitation that undergirds developmental narratives of modernity within and outside the West.

Hallyu provides a particularly rich site to consider the relationships among imitation, Asia, and globalization. Much scholarship has emerged in the past two decades on the Korean Wave. Discussions of its content in the public sphere highlight its cultural hybridity: K-pop as postmodern pastiche, Korean cosmetic surgery as racial imitation, and Korean cinema as "Copywood."[7] South Korean products and people are both celebrated *and* denigrated for their ability to mimic, combine, and transform existing cultural forms to appeal to multiple audiences. These imitative tendencies are often attributed to South Korea's social, economic, and cultural aspirations, and the appeal of Hallyu narratives and aesthetics to the similar aspirations they evoke in their Asian neighbors (or nostalgia, in the case of Japan). Meanwhile, the same tendencies become a point of fascination, derision, or both in the West, much like the mixed reception of non-Western cultural styles and products from other Asian countries such as Japan, Hong Kong, and India.

What this line of thinking elides is how the trope of imitation—and more precisely, the act of *copying*, through and as technology—might be understood within two contexts that are seldom explored in the Hallyu literature, those of North Korea and Korean diasporas in the Global North. Both formations emerged in the Cold War context and more specifically, the Korean War. The geographical, geopolitical, and ideological division between North and South Korea continues to function as a collective psychic wound, *fragmenting* the national body and separating families. Meanwhile, the postwar migration of Koreans to developed Western nations such as the United States, Canada, Australia, and countries in Europe *multiplied* the national body, with different diasporic communities retaining certain elements of Korean culture and identity while diverging from others. In English-language scholarship, the Korean diaspora usually refers to these formations, particularly the Korean American diaspora, eliding diasporas within Asia and the Global South. Thinking about these displaced "copies" together disrupts and questions the teleological narrative of capitalist progress epitomized by South Korean development and globalization.

The presence of North Korea is a constant reminder of ongoing Cold War politics. North Korea plays the "yellow peril" to the "model minority" that is South Korea; the thirty-eighth parallel ties the traumatic past of the Korean Peninsula to its impossibly cool present and potentially dystopic future. Observing how North Korean refugees appeal to human rights discourse to represent themselves in autobiographical narratives, Christine Kim notes this is a strategic response to the ways in which the West has positioned North Korea as the ultimate Orientalized other: "North Korean defectors exemplify the parameters of contemporary Western 'rights' imagination and demonstrate what Jodi Kim calls 'the protracted afterlife of the Korean war.' Within this context, North Korea functions alternately as a metaphor for the inhuman and as a metonym for Asian incivility."[8]

Meanwhile, as the United States represents the largest Korean diaspora in the West, Korean American scholarship, art, and literature bear witness to the ways in which this trauma continues to play out intergenerationally and across time and space, disrupting the cultural nationalism often celebrated in Korean Wave discourse. At the same time, the ongoing appeal of Hallyu has clearly helped to increase the visibility and power of Korean American stories and people in popular media culture. For instance, the content of two, highly celebrated films, both written and directed by and starring actors of Korean heritage—*Parasite*, the first non-English-language film to win an Oscar for Best Picture in 2019, and *Minari* the first Korean American film to be nominated for an Oscar for Best Picture the following year—demonstrate stark differences in the representation of Koreanness but similarities with respect to their reception in the United States.[9]

Depictions of both North Korea and the Korean diaspora gesture toward the ongoing trauma of the Korean War and the geopolitical orientation that emerged from the Cold War. Both formations are characterized by the geographic, cultural, and psychic division and fragmentation that resulted from displacement. And both are represented as the necessary other—the bad copy or double—to the global rise and visibility of South Korea. These traits also define the female protagonists of *Shiri* and *Cloud Atlas*.

Cosmetic Surgery Clones: The Threat of Korean and Asian Femininities

This should come as no surprise since the cultural traumas of national division and diasporic separation in South Korea often have been projected onto the bodies of women—from sexualized figures of shame such as the *jungshindae* (comfort woman) and the *yanggongju* (Western princess) to those of conspicuous consumption such as the "Modern Girls" of the early twentieth century and their contemporary counterparts, the *doenjang nyeo* (a derogatory term for aspirational women who scrimp on essentials to purchase Western luxury goods).[10] These female figures have been and continue to be marginalized because they trouble the traditional neo-Confucian role of women in Korea as "wise mothers and good wives" who sacrifice their individual selves for the good of the patriarchal family and nation.[11]

Indeed, female aspiration, labor, and sacrifice lie at the very heart of the national origin myth. Koreans trace their ancestor to Dangun Wanggeom, the hybrid child of the sky god Hwanung and a female bear. This bear becomes a human named Ungnyeo (Bear Woman), beating out her competition—a less patient female tiger—by diligently following Hwanung's instructions to eat only garlic and mugwort while staying in a cave for one hundred days. Hwanung pities Ungnyeo because she is childless, takes her for his wife, and impregnates her with the seed of what will become the Korean people.

For Seungsook Moon, this myth "suggest[s] that woman's only contribution to the creation of the Korean nation was the provision of a proto-nationalist womb.... In addition, the transformation of a bear into a woman carries the deep social meaning of womanhood epitomized by patience to endure suffering and ordeal."[12] At the same time, as Sang Un Park notes, the Bear Woman's persistence also gives her the ability, albeit within the terms of patriarchy, to transform herself physically and thus transcend her animal fate.[13] This contradiction plays out in discourses around cosmetic surgery in South Korea.

Initially introduced during the Korean War to enhance the appearance of Korean translators and war brides, cosmetic surgery became popular during this time and has since led to South Korea becoming the "plastic surgery capital" of the world, boasting the highest number of cosmetic surgery procedures per capita.[14] Furthermore, going under the knife in South Korea has become a global phenomenon. Dubbed "Medical Hallyu," cosmetic surgery tourism was launched by the bureau of Korean Tourism in 2007 as a "main area of growth and as one of the country's strategic products." In two years the number of medical tourists jumped from 20,000 to 60,000.[15] By 2018 this number had increased to 464,452 foreigners traveling to South Korea, the majority from China, for procedures such as the double eyelid surgery, nose implants, and jaw reduction.[16]

The aesthetic sought after by these tourists, as well as by clients in South Korea and its diasporas, has been sensationalized in Western media for ostensibly *imitating* the implied global standard of Western white female beauty and as a result, *deracializing* those who choose cosmetic surgery procedures, especially those regarding the face. In our analysis of Australian television news shows on "racial cosmetic surgery," Joanna Elfving-Hwang and I point out that this discourse of deracialization is never applied to white women, for whom cosmetic surgery is almost always framed through the rhetoric of neoliberal choice and agency.[17] In the U.S. context, S. Heijin Lee observes that the prevalence of this perspective "reveal[s] more about a US empire in relative decline than... about Korean women."[18] Discussing earlier public discourse around Korean cosmetic surgery in the 1990s, she elaborates, "Then, as now, US obsessions with Korean cosmetic surgery as a primary avenue through which to contend with Korea's newfound affluence (and influence), came at a time when US economic global dominance appeared most threatened, which bespeaks an anxious Western gaze desiring to see itself in places where its hegemony is on the wane."[19]

Finally, Ruth Holliday and Jo Elvfing-Hwang argue that shifts in aesthetic ideals cannot be reduced to Korea's relationship with the West but must also take into account its "strong sense of nationalism, as well as its national relationship with other regional powers," particularly Japan.[20] Specifically, they note that the Western body became desirable in Korea only after the country was liberated from Japanese colonization, suggesting that this new beauty standard was "mobilized in defiance of Japanese standards of beauty—as anti-colonial discourse . . . constitut[ing] as much a rejection of Japanese colonial influence as an embrace of western beauty norms."[21] As well, they point out that in South Korea "successful surgery . . . should look 'natural,' where natural is importantly defined as enhancing *Korean* features," and that this enhanced "Korean look" becomes linked to upper-middle-class status given the high cost of these procedures.[22]

Building on this research, I would suggest that Asian women who aspire toward the surgically enhanced aesthetic of Korean cosmetic surgery end up resembling the female characters of manga and anime as much as traditionally beautiful white women in Hollywood. In a 2011 episode on race and cosmetic surgery on the Australian SBS (Special Broadcasting Service) news program *Insight,* Chinese Australian Heidi Liow states she got chin implants and blepharoplasty for the "stronger features" associated with manga characters like Sailor Moon as well as Hollywood actress Megan Fox.[23] Indeed it is the hybrid, Eurasian, and uncannily *virtual* look of the former that is promoted in South Korean popular media. In this way, contemporary beauty norms in South Korea—codified through cosmetic surgery and advertised through the bodies of Korean celebrities—gesture toward a new global fantasy of *mixed East Asian* beauty. This fantasy melds ideal "Western" and "Eastern" features with a preference for certain national signifiers of attractiveness. The hybrid aesthetic that results marks South Korean modernity as a complex cultural and aesthetic modification of the East Asian female body for local, regional, and international consumption.

Remaking bodies, in this instance, might be seen as one expression of the mode of creative imitation mentioned earlier, since a certain amount of agency is practiced on and by those whose bodies are modified. It follows then, to echo S. Heijin Lee's observation, that the transformation of these women through cosmetic surgery, and that of the Korean national brand through Hallyu, would elicit anxiety in Western audiences and consumers since these makeovers challenge not only the geopolitical centrality of the West as represented by the United States, but also its continued relevance as a leader of cultural globalization.

With these ideas in mind, I turn now to look at how these anxieties play out around the figure of the technologized Korean woman in *Shiri* and *Cloud Atlas.*

Shiri

Directed by Kang Je-gyu, *Shiri* was the first Korean movie to model and market itself after the Hollywood blockbuster with its emphasis on spectacle and special effects, a memorable theme song, a character-driven storyline, and a big production budget. Yet, according to Chi-Yun Shin and Julian Stringer, "This Korean blockbuster is no *clone*. It is special."[24] It is special because it self-consciously references and performs Hollywood style, localizing it to tell a culturally specific nationalist story about and for South Korean audiences.

Shiri brought the Korean film industry out of a slump, grossing $27.5 million, and was called "the little fish that sank the *Titanic*" for overtaking sales of the then top-grossing film in the world.[25] Purchasing a ticket to see *Shiri*—and subsequent South Korean blockbuster films—was regarded as a patriotic act, a show of monetary and affective support for the then growing national film industry.[26] The film performed strongly in the region, particularly in Hong Kong and Japan; became the first South Korean action film to be released in the United States; and garnered critical recognition at home, winning Best Actor for Choi Min-sik and Best Actress for Kim Yun-jin at the Grand Bell Film Awards, the Korean equivalent of the Academy Awards.[27]

In this action thriller an agent of the South Korean National Intelligence Service (NIS), Yu Jung-won (played by Han Suk-kyu), and a North Korean terrorist, Park Mu-young (played by Choi Min-sik), fight for control of the future of the Korean Peninsula. Trapped in the middle of their conflict and metaphorically that of the two Koreas is the female protagonist, North Korean sniper and agent Lee Bang-hee (hereafter, Hee, played by Kim Yun-jin). Hee is a North Korean spy and sniper in the Eighth Special Forces unit, an elite team of the North Korean army. Led by Park, this group has been tasked with starting another Korean War as a catalyst for reunifying the country under North Korea. Hee is assigned to infiltrate the NIS by getting involved romantically with Yu, who, with his partner, Lee (played by Song Kang-ho) has been trying to track her down for years. To play her new role, Hee gets cosmetic surgery in Japan before returning to South Korea. We learn later that she assumes the face and identity of a chronically ill hospitalized woman, Lee Myung-hyun, whom she had befriended earlier while boarding with her family. In Seoul, she poses as Hyun, the owner of a pet fish store where Yu first meets and woos her with a CD that contains the film's hit theme song.

Shiri opens with a brutal six-minute training montage of the Eighth Special Forces set in "September 1992, somewhere in North Korea." Hee appears as a promising young cadet, the only woman among men learning to fight and kill like efficient machines in the countryside. The camera centers her, providing close-ups of her fierce, dedicated facial expressions as she quickly rises above the pack. Park demonstrates typically stoic Korean fatherly approval of Hee when he drops extra meat into her can as the trainees consume their meager provisions in the bitter winter cold. The montage ends with the end of Hee's training. Against a triumphant military soundtrack, she burns a photo of her family in the rain and receives her dog tags from Park as her comrades salute her and she is driven away in a jeep—presumably to undertake her assignments in South

Korea. Another montage sequence immediately follows, this one entirely virtual. We assume Yu's perspective at his computer as photos and reports of Hee's killings from 1993 to 1996 of multiple government, military, and business officials flash rapidly on the screen to a techno soundtrack. These two montages together effectively create a picture of Hee as a dangerous, unfeeling female killing machine—a North Korean techno-Oriental Other to South Korean audiences.

The film cleverly interweaves the action genre (A-plot: finding Hee and later, averting a second Korean War triggered by the assassination of the South Korean president) with the romance genre (B-plot: development of Yu and Hyun's relationship as it moves toward marriage). Until Hee is revealed as Hyun to the audience in the second half of the film, we only glimpse traces of her from a distance, usually from higher vantage points than the two agents who are chasing her. Her disembodied invisibility, stealth, and shooting skills give her almost supernatural qualities, rendering her a dangerous and powerful ghost.

In contrast Hyun first appears in a cheerful yellow raincoat on her bike, transporting pet fish to her store. She and Yu flirt, pretending to be strangers as they discuss the mating habits of fish. It is only when she holds up a half-knitted sweater against his chest, declaring him to be the same size as her boyfriend and they kiss, that we realize they are a couple. The two are associated with fish throughout the film: the kissing gourami that Hyun gifts Yu, telling him that if one dies the other dies too, as she would if he left her, and the titular *shiri*, a freshwater fish endemic to the rivers that flow between North and South Korea.

It is worth noting that two types of fish are invoked here—the gourami whose "kissing" tendencies are reflected in the physical displays of affection between Hyun and Yu and the shiri, an indigenous fish that symbolizes the ability to move freely between North and South Korea. Both reflect a doubling of South Korea with the United States as well as North Korea. The kissing performed by the lead characters imitates the body language of Hollywood on-screen couples and was still quite novel for Korean audiences at the time. Meanwhile, Hee (the dutiful North Korean spy) is imitating Hyun (the kind South Korean fiancée) when she gifts aquariums containing shiri to the NIS. The shiri are weaponized as bugs for the North Korean terrorist group to spy on South Korean security operations. This doubling characterizes and ultimately undoes Hyun and Yu's relationship. It also sets up the central theme of the film: the longing for and impossibility of national unification, expressed through heterosexual romance and its culmination in marriage.

Aaron Han Joon Magnan-Park asserts that the appeal of the tragic romance between Hyun and Yu for South Korean audiences lay in its invocation of "Korea's traditional vision of the ideal heterosexual union: *namnambungnyeo* (southern man/northern woman)" described as follows: "This Korean cultural-bound romantic ideal held that the quintessential Korean heterosexual romantic pairing involved a man from the South . . . with a woman from the North. . . .

When the two meet and amorous sentiments are consummated in marriage, they become the Confucian idealized Korean husband and wife."[28]

The romance between Yu and Hyun is never consummated in marriage. The conflict between Hyun and Park builds as Hyun tries to protect Yu, endangering the work of her fellow spies. Yu finally figures out and thwarts Park's plan to detonate a liquid bomb called CTX during a goodwill soccer game between North and South Korea, which would kill the South Korean president and thousands of soccer fans. This activates Hyun to attempt to assassinate the president at close range. She rages through the crowd with her machine gun and is caught in a standoff with her lover. As the car carrying the president drives by, she pauses, then turns to shoot at the car and is shot dead by Yu—all in slow motion to heartbreaking music.

Hyun's death clearly demonstrates the impossibility of national unification. Like the Korean Peninsula, she is divided within herself and unable to be two women at once: the inhuman North Korean assassin and the humanized South Korean fiancée and mother-to-be (as we learn, with Yu, after her postmortem). Asked at his interrogation how he was unable to suspect Hyun, Yu responds by invoking the figure of the hydra: "You know Hydra? She's a Greek goddess with six heads. She has multiple personalities with one body. Hyun and Hee—they're totally different. She was the Hydra of our times. The reality of a divided Korea turned her into Hydra."

In contrast to the character she plays, the diasporic doubleness of Kim Yun-jin, the 1.5-generation Korean American actress, has led to her success as a transnational actor. Like Bruce Lee and other contemporary Asian American actors such as Maggie Q and Daniel Henney, Kim could not secure roles in Hollywood and so went to Asia to try her luck. Thanks to her fluent language skills, she was able to land roles in Korean TV dramas, which led to her role in *Shiri*. This, in turn, gave Kim international clout in the eyes of producer J.J. Abrams, who in 2004 created a specifically Korean role for her and, later, for Korean American actor Daniel Dae Kim to play her husband, in the long-running American TV show *Lost* (2004–2010). Since then, Kim has continued to work between the United States and South Korea, with lead roles in the American TV series *Mistresses* (2013–2016) and the South Korean hit film *Ode to My Father*, which, like *Shiri* many years before, promoted an intensely emotional patriotic and patriarchal narrative of South Korean national identity.[29]

Cloud Atlas

Like Kim and other South Korean actors such as Lee Byung Hun, Jeon Ji-Hyun, and Han Ye-ri, Bae Doona has crossed over successfully into Hollywood. She came to prominence in South Korea playing spirited young girls in Bong Joon Ho's *Barking Dogs Never Bite* (2000) and *The Host* (2006) as well as the lyrical coming-of-age film *Take Care of My Cat*.[30] In 2009 Bae had her international acting debut, playing an inflatable sex doll that falls in love with a Japanese video clerk, in *Air Doll*.[31] The film met with some controversy in South Korea due to the history of comfort women in Japan. In her second international role in *Cloud Atlas*, she, like the other actors, plays several roles in different narrative segments. These include supporting white and Mexican female characters, and most prominently, a female clone that references the actor's Korean heritage.

Based on the 2004 novel by David Mitchell, the film consists of six connected narratives that span nineteenth-century New Zealand to an off-planet colony in 2321. The central theme of both the film and novel is that we are all connected through space and time through reincarnation, and that we have the responsibility to imagine a future in which everyone is "free." This concept of freedom is literally performed by actors as they cross racial and gender categories. It is worth noting that the forms of transracial performance in the film assume and privilege a U.S.-centric racial politics: blackface never appears, in stark contrast to brownface (which appears in the New Zealand segment) and yellowface (which dominates the segment on which this chapter focuses).

In 2144 a unified Korea has absorbed surrounding areas into the last civilization on earth. Its name, Nea So Copros, alludes to the New East Asian Sphere of Co-Prosperity—Japan's early twentieth-century pan-Asian imperial dream that became a nightmare for its colonies, including, most prominently, Korea. The narrative segment "An Orison of Sonmi-451 reprises the techno-Orientalism of now-classic cyberpunk films like *Blade Runner* and *The Matrix*, replacing Japan and Hong Kong with Korea. This future is a postcapitalist, overpopulated urban hell that runs on the labor of Asian female clones. Somni, as representative of this slave class, is rescued by a group of "Korean" male rebels—all played by white and Black men in yellowface, including Jim Sturgess, who becomes her lover (Hae-joo Chang) and trains her to become the leader of the Union Rebellion.

This future is not the final utopian future, which is located in the "off-world" colonies and populated by the Black and white mixed-race offspring of Zachry (Tom Hanks), a primitivized white "Valley man," and Meronym (Halle Berry), a technologically sophisticated Black "Prescient." Instead it is a point of transition, the necessary precursor to and dystopic double of the progressive American "happy ending." Neo-Seoul functions as a site of labor extraction and exploitation much like the slave plantation in nineteenth-century New Zealand. The film explicitly draws a parallel between the two narratives through their romantic pairing of Bae Doona and Jim Sturgess, discussed in more detail shortly. It also symbolically pairs Bae as Somni with David Gyasi, who plays Autua, the escaped African slave in Adam Ewing's sea voyage from New Zealand to San Francisco. Somni the clone is "awakened" to the unequal power dynamics of her society by Hae-Joo Chang (Sturgess) while David the slave "awakens" Ewing (Sturgess) to the horrors of slavery, which compels him to become an abolitionist. In both narratives the nonwhite characters are humanized through and for the gaze of the liberal white male and by extension, the liberal white audience.

In the scene that introduces Somni, she is being interviewed by an Archivist (James D'Arcy) before her execution for escaping Papa Song's—the fast food restaurant she has been genetically engineered to work in as a server her entire life—and joining the Union, a rebel group that seeks to build an egalitarian society by overthrowing the dominant group, Unanimity. Somni becomes a tool for actualizing this goal after her friend, fellow Fabricant Yoona-939 (Xun Zhou), is killed for resisting the sexual advances of a male Consumer and trying to escape. Yoona is "awakened" by her manager, Seer Rhee (Hugh Grant), an embedded Union agent. The process of awakening (presumably to human consciousness), for Yoona, and later for Somni, involves having sex with a male human or "Pureblood" and reiterating the words of English-language texts. After she strikes back at the offending Consumer, Yoona addresses the customers who are staring at her in shock with the quote, "I will not be subjected to criminal abuse." This is a line borrowed from a clip from *The Trials of Timothy Cavendish* that Rhee showed her and that she later shows Somni, referencing a comic storyline in the film about an English publisher who is locked up in a senior citizens' home by his brother.

Somni then functions as a kind of double imitation, referencing both Pureblood women and Yoona, the first Fabricant chosen by Union. Like the other servers, she is infantilized and hypersexualized through her appearance (tight cheongsam uniforms, conventionally attractive figures and features) and passive behavior with respect to Pureblood men. Both she and Yoona display Lolita-esque qualities that presume their sexual agency and desire in situations in which they actually have no power. Her romance with Hae-Joo Chang develops as he introduces her to Western philosophical texts and other Union members, and protects her from Unanimity police as they whiz by in flying vehicles around the Neo-Seoul nightscape. The climactic revelation that leads to Somni agreeing to broadcast Union ideology occurs toward the end of the film. General An-Kor Apis (played by an African American actor, Keith David, in yellowface) arranges for her to witness an Xultation—the Fabricants' version of a happy retirement when they have reached the end of their work contracts. Instead of the heaven she has been told awaits her and her sisters, Somni learns the truth—that the Fabricants are killed and turned into the "soap" that other Fabricants are given as food. The image of numerous naked Asian female Fabricant corpses hung from the ceiling, moving in assembly line fashion recalls the horrific scene in *The Matrix* (which itself references *Soylent Green*) where a close-up of a human baby being fed human remains intravenously cuts out into an image of multiple babies, representing the future human race enslaved by AI.

"An Orison of Somni-451" seems to break from earlier depictions of Oriental style in Hollywood films in that the protagonist of the narrative is a Korean female character played by a Korean actor. Yet strangely, even as the protagonist, Somni functions primarily as a decorative backdrop and mouthpiece for the liberal white American progressive message of the film as well as the non-Korean actors who ironically play "Pureblood" Koreans. Somni's body is closely associated with the Neo-Seoul cityscape and the technologized texts that she consumes and later quotes. Throughout the film, and especially in the far future storyline, her presence is referenced through recordings, spiritual texts, images, sculptures, and architecture.

Ultimately, as LeiLani Nishime points out, this Korean female service clone sacrifices herself for a liberal humanist future created by progressive white men—a future in which she can participate only as a disembodied recorded voice in the far future, and the wife (in whiteface) of one of its progenitors. Nishime locates this within a larger trend in contemporary Hollywood science fiction films that seek to remedy historical yellowface practices by whitewashing female Asian characters: "Asian American scholars have convincingly rooted yellowface performances . . . in the erotics of Asian difference. The undertheorized practice of whitewashing . . . is indebted to a more ambivalent story of racial progress enacted through the transformation of Asian bodies into white ones, echoing shifts in both the rhetorics and practices of global labor migration. The prevalence of whitewashing enables science fiction films to imagine future technologies while disavowing the unequal transpacific labor practices that underwrite the production of those future technologies."[32] In this way, the relationship between yellowface and whitewashing echoes the one between Orientalism and techno-Orientalism. The former is seen as a product of a racist past that is "washed" clean by depicting a fantasy in which technologies enable the total assimilation of Asians into whiteness, the absolute erasure of their racial difference as signified by their "yellow" faces.

This scenario eerily echoes the discourse of racial imitation and assimilation in debates around South Korean cosmetic surgery and its purported whitening of Korean and other East and Southeast Asian ethnic faces. The face of Bae Doona as Somni-451 provides a striking example of the synergy among bodies, aesthetics, and media within the globalization of Hallyu. Bae promotes Korean visibility on the world stage by presenting a palatable, pan-Asian female face and body that are recognizable through the lens of cinematic techno-Orientalism. Meanwhile, the fact that her biggest role in the film is that of a messianic clone reflects ongoing Western anxieties about Korea—and perhaps "Asia" more generally—as an ultimately radical Other mimicking the values of liberal democracy.

Conclusion

Working between the cultures of the United States and South Korea, Bae Doona and Kim Yun-jin humanize the Oriental and techno-Oriental stereotypes that have characterized the representation of Asian peoples and cultures in the West. At the same time, they are limited in what they can do as transnational actors by the cultural literacies and expectations of the target audiences for the films and television shows in which they appear. It is telling that, for South Korean audiences in 1999 and global ones (read Western, Anglophone) more than a decade later, the figure of the Korean woman in these films is distinctly techno-Oriental—both in its construction (through cosmetic surgery and cloning) and representation (as imitative "Others" of South Korean and human women). At the same time, as unfaithful "bad copies" of "authentic Korean/Asian women," these techno-Orientalized Korean women also render visible cultural and ontological forms of mixedness that question and arguably queer the myth of authenticity underlying the patriarchal formations of South Korean cultural nationalism and U.S.-centered globalization. Perhaps in this sense, they illuminate the complex and ambivalent ways in which the rising global visibility of South Korean culture reproduces earlier forms of Oriental style even as it gestures toward something new, in which the West and its gaze is decentered as something increasingly obsolete and marginal.

Notes

1 Jane Chi Hyun Park, *Yellow Future: Oriental Style in Hollywood Cinema*. (Minneapolis: University of Minnesota Press, 2010).
2 *Shiri*, directed by Je-kyu Kang (Samsung Entertainment, 1999); *Cloud Atlas*, directed by Tom Tykwer, Lana Wachowski, and Lilly Wachowski (Warner Bros. Pictures, 2012). For studies of the Korean Wave, see Beng Huat Chua and Koichi Iwabuchi, eds., *East Asian Pop Culture: Analysing the Korean Wave* (Hong Kong: Hong Kong University Press, 2008); Youna Kim, ed., *The Korean Wave: Korean Media Go Global* (London: Routledge, 2013); and Soyoung Kim, *Korean Cinema in Global Contexts: Post-Colonial Phantom, Blockbuster, and Trans-Cinema* (Amsterdam: Amsterdam University Press, 2022). For studies of techno-Orientalism, see David Morley and Kevin Robins, *Spaces of Identity: Global Media, Electronic Landscapes, and Cultural Boundaries* (New York: Routledge, 1995); Lisa Nakamura, *Cybertypes: Race, Ethnicity, and Identity on the Internet* (New York: Routledge, 2002); and David S. Roh, Betsy Huang, and Greta A. Niu, eds., *Techno-Orientalism: Imagining Asia in Speculative Fiction, History, and Media* (New Brunswick, NJ: Rutgers University Press, 2015). And for studies of cultural representations, see Elaine H. Kim and Chungmoo Choi, eds., *Dangerous Women: Gender and Korean Nationalism* (New York: Routledge, 1998); Grace Cho, *Haunting the Korean Diaspora: Shame, Secrecy and the Forgotten War* (Minneapolis: University of Minnesota Press, 2008); and LeiLani Nishime, "Whitewashing Yellow Futures in *Ex Machina*, *Cloud Atlas*, and *Advantageous*," *Journal of Asian American Studies* 20, no. 1 (2017): 29–49.
3 Euny Hong, *The Birth of Korean Cool: How One Nation Is Conquering the World Through Pop Culture* (London: Simon & Schuster, 2014).
4 Seungsook Moon, "Begetting the Nation: The Androcentric Discourse of National History and Tradition in South Korea," in *Dangerous Women: Gender and Korean Nationalism*, ed. Chungmoo Choi and Elaine Kim (New York: Routledge, 1998), 33–66; Hyaeweol Choi, "'Wise Mother, Good Wife': A Transcultural Discursive Construct in Modern Korea," *Journal of Korean Studies* 14, no. 1 (2009): 1–34.

5 Werner Sollors, *Neither Black Nor White Yet Both: Thematic Explorations of Interracial Literature* (Oxford: Oxford University Press, 1997); Eva Allegra Raimon, *The "Tragic Mulatta" Revisited: Race and Nationalism in Nineteenth-Century Antislavery Fiction* (New Brunswick, NJ: Rutgers University Press, 2004).
6 Long T. Bui, *Model Machines: A History of the Asian as Automaton* (Philadelphia: Temple University Press, 2022).
7 Christina Klein, "'Copywood' No Longer: The South Korean Film Industry Shows It Can Hold Its Own By Combining Local Themes with Hollywood Style," *YaleGlobal Online*, October 11, 2004, https://archive-yaleglobal.yale.edu/content/copywood-no-longer.
8 Christine Kim, "Figuring North Korean Lives and Human Rights," in *The Subject(s) of Human Rights: Crises, Violations, and Asian/American Critique*, ed. Cathy J. Schlund-Vials, Guy Beauregard, and Hsiu-Chuan Lee (Philadelphia: Temple University Press, 2020), 221.
9 *Parasite*, directed by Bong-joon Ho (CJ Entertainment, 2020); *Minari*, directed by Lee Isaac Chung (New York: A24 Films, 2021).
10 Kim and Choi, *Dangerous Women*; Cho, *Haunting the Korean Diaspora*; Lauren Kendall, *Under Construction: The Gendering of Modernity, Class and Consumption in the Republic of Korea* (Honolulu: University of Hawai'i Press, 2001); Seungsook Moon, *Militarized Modernity and Gendered Citizenship in South Korea* (Durham, NC: Duke University Press, 2005).
11 Choi, "'Wise Mother, Good Wife.'"
12 Moon, "Begetting the Nation," 41.
13 Sang Un Park, "'Beauty Will Save You': The Myth and Ritual of Dieting in Korean Society," *Korea Journal* 47, no. 2 (2007): 46.
14 Sharon Heijin Lee, "Gender, Beauty, and Plastic Surgery: Towards a Transpacific Korean/American Studies," in *A Companion to Korean American Studies*, ed. Rachael Joo and Shelley Sang-Hee Lee (Leiden: Brill, 2018), 485–489.
15 Sharon Heijin Lee, "Beauty between Empires: Global Feminism, Plastic Surgery, and the Trouble with Self-Esteem," *Frontiers* 37, no. 1 (2016): 21.
16 Sung-sun Kwak, "Foreign Patients Receiving Plastic Surgery Hit Record High in 2018," *Korea Biomedical Review* October 16, 2019, https://www.koreabiomed.com/news/articleView.html?idxno=6611.
17 Joanna Elfving-Hwang and Jane Chi Hyun Park, "Deracializing Asian Australia? Cosmetic Surgery and the Question of Race in Australian Television," *Continuum* 30, no. 4 (February 2016): 1–2, https://doi.org/10.1080/10304312.2016.1141864.
18 Lee, "Beauty Between Empires," 8.
19 Lee, "Beauty Between Empires," 5.
20 Ruth Holliday and Joanna Elfving-Hwang, "Gender, Globalization and Aesthetic Surgery in South Korea," *Body & Society* 18, no. 2 (2012): 66.
21 Holliday and Elfving-Hwang, "Gender, Globalization and Aesthetic Surgery," 69, 74.
22 Holliday and Elfving-Hwang, "Gender, Globalization and Aesthetic Surgery," 62.
23 Elfving-Hwang and Park, "Deracializing Asian Australia?," 3.
24 Chi-Yun Shin and Julian Stringer, "Storming the Big Screen: The Shiri Syndrome," in *Seoul Searching: Culture and Identity in Contemporary Korean Cinema*, ed. Frances Gateward (Albany: SUNY Press, 2007), 64. My italics.
25 Shin and Stringer, "Storming the Big Screen," 57.
26 Aaron Han Joon Magnan-Park, "*Shiri* (1999) and the Reunifying Korean Romantic Fantasy of Namnambungnyo," *Quarterly Review of Film and Video* 37, no. 4 (2019): 364.

27 Magnan-Park, "*Shiri* (1999)," 364.
28 Magnan-Park, "*Shiri* (1999)," 368.
29 *Ode to My Father*, directed by J. K. Youn (CJ Entertainment, 2014).
30 *Take Care of My Cat*, directed by Jae-eun Jeong (Cinema Service, 2001).
31 *Air Doll*, directed by Hirokazu Koreeda (Asmik Ace Entertaiment, 2009).
32 Nishime, "Whitewashing Yellow Futures," 30.

11

Assembling Mitski

The Aesthetics and Circuits of Techno-Ornamentalism

RACHEL TAY AND JAEYEON YOO

"Your mother wouldn't approve of how my mother raised me / But I do, I think I do." trailed off Mitski Miyawaki, the Japanese American musician who goes by her first name. In a packed concert venue in Raleigh, North Carolina, Rachel and I held our breaths for the ending of "Your Best American Girl," the song that catapulted Mitski to indie music fame in 2016. The musician was on tour with her 2022 album, *Laurel Hell*, and I remembered the countless times I had listened to this song. I felt connected to Mitski. I felt like I was listening to her, and—in an embarrassing cliché—I felt heard.

"Shoutout to women of color!" A woman yelled loudly, before the last guitar note had died out. I whipped around to look at her. Blonde, white-presenting, American accent. My surprise at the interruption melted into an angry possessiveness I felt ashamed to admit: What do *you* know about this song? Or Mitski? I thought.

While I was unnerved by the shoutout at the concert, it was only later that I had this sense of discomfort confirmed. The call, which had made conspicuous Mitski's racial and gender background, had felt to both of us—of East Asian descent—more unsettling than expected. It affirmed presumptions that we did not so willingly want to admit: that Mitski, like us, is bound by the racial configurations of the United States, and that our looks render us interchangeable for some. Further unsettling was how we had both immediately jumped to categorize the person shouting with the same rubric of visual-coded racial politics.

A week later, I watched as the singer-songwriter took to Twitter in a rare expression of her opinion. She appealed to her fans to put down their phones during her shows: "I'm not against taking photos at shows, but . . . when I'm on stage and look to you but you are gazing into a screen, it makes me feel as though those of us on stage are being taken from and consumed as content."[1] These tweets sparked an online debate about not just whether phones should be permitted at live music events, but also whether her request would have drawn such a backlash were she not a woman, Asian American, or markedly a minority in any other way—given that many performers have asked the same in an age of smartphones, social media, and real-time content creation. Speculations on why Mitski's ask had provoked such outsized controversy then began to be drawn along the lines of her racial identity—a topic well beyond the scope of her tweets. Propelled through the accelerated circuits of the Internet, what had once been a simple matter of decorum soon became fraught with the friction of identity politics.

This chapter proceeds from these two episodes of racialization to examine the perplexing dynamics at play between performer and audience, musician and fan, commodifiable content and the hand that holds up a phone to capture it, particularly when Asian femininity is made the object of consumption. We stress the conspicuous and contradictory role of race *as* value production in Mitski's case; for as much as Mitski—and the music industry, at large—has profited from her creative output, she has long refrained from centering her personal identity in it. Contrary to the profitable glorification of minority representation in popular culture, the musician has sought to abstain from any emphatic identification with her background.[2] "I talk about being Asian and then that becomes the article," she explains early in her career, "It stops there and everyone goes back to their day."[3] Despite this refusal, Mitski has been celebrated by fans and music writers as a "Japanese-born American singer-songwriter."[4] Journalists declare her the "non-white savior" of indie rock, or "More Than Another Sad Asian American Girl," making her into more *and* less than her identity categories.[5] Because her celebrity ensures that she stays in our online and offline imaginaries—to be variously received, retrieved, and refashioned by her audiences—her retreat from, and reconstitution by, the public eye thus never seems complete.

Given this, Mitski's lack of response against the public's often-racialized perceptions of her may appear contradictory: it may be read as both a gesture of withdrawal *and* a deliberate aesthetic choice, a stereotype of acquiescent Asiatic femininity *and* an opposite attempt at the artist's own de-essentialization. Further complicating Mitski's racialized reception is her biracial heritage (white/East Asian) and her international upbringing—facts that are often glossed over when she is claimed as an Asian American "icon." As LeiLani Nishime argues, "rather than [being] exceptions to racial rules, multiracial bodies help us decipher the playbook" of racialization.[6] What, then, might her audience's insistent misrecognition of her background illuminate of our extant "playbook" of "racial rules"?

Drawing on Anne Anlin Cheng's concept of "ornamentalism," our chapter examines Mitski's ambiguous identity and her (de)racialized representations across media platforms. Tracking Mitski's public persona through its technological circuits of production, we find a peculiar mode of being we call "techno-*ornamentalism*": an aesthetics of ornamental ambiguity, which Mitski wields as a self-objectifying "technology of personhood" *and* a prosthesis for survival.[7] Using Cheng's theorization and a techno-Orientalist framework that views "the Asian body as a form of expendable technology," we conceptualize the technologized ornament as a means of identity formation propelled by the engines of racial capital.[8] We analyze, on one hand, the sedimentation of life, value, and significance that results from Mitski's objecthood, and on the other, the labor of objectification performed by fans and critics on her behalf. Through this lens, we probe at the simultaneous process of cocreation and co-constitution that unfolds between artist/object *and* fan/consumer.

Throughout, we observe the ways that the *operationalization* of media has transformed Mitski's racial identity into a generalizable and thereby infinitely productive manner of value accumulation.[9] As the sociotechnical ensembles of multimedia decompose into their audio, graphic, and textual parts, processes of racialization are extended from representational to nonrepresentational fields: shifting from signifiers of identity to a signifying signals, the calculus of subject formation draws into its circuits increasingly imperceptible *relations* between entities.[10] Thus, our analysis centers the most conspicuous instances wherein race appears and disappears in Mitski's body of work—that is, how Mitski's (de)raciality circulates as a site of both economic production *and* social cooperation, where commodifiable "value is produced as inseparable from . . . the invention and diffusion of common desires, beliefs and affects."[11]

With Mitski as an exemplary, but not exclusive case study—which, we admit, is another way of using her as ornament and object—this chapter traces how her aesthetics of ambiguity and technology have isolated her racial identity as a category to be deployed, obscured, or perceived circumstantially.[12] Ultimately, in observing how she harnesses the potency of ambivalence to stir her audiences' affective and creative engagements, we contend that our collective capacity to invoke polymorphous reimaginations of her—as racialized or otherwise—is precisely how she is reconstituted, manifested, and endures as a techno-ornamental object.

Techno-Ornamentalism

In *Ornamentalism*, Cheng troubles the yellow woman's recurrence in Orientalist images and their critiques, contending specifically that the motif's materialization of ornamental aesthetics blurs the line between object and vital life. In this, her work recognizes how the ornament is preserved—and comes to preserve itself—through Orientalism's objectifying discourse. Likewise, no contemporary cultural object—aesthetic commodities themselves—can ever exist in a foreclosed form. Just as a tweet may be mined for its data *and* discursive content, the screens through which one views a musician on stage ripple out into recordings, remixes, or new codifications of concert etiquette. Such an operationalization of media transforms Mitski's (or any celebrity's) digital footprints into an open field of perception, gesturing toward "endless permutations [of] data" beyond what is intelligible to us—a dynamic "space of pattern recognition" for audiences to facilitate assemblages of her persona.[13] In this context, to construe race as a technology is to operationalize its internal logics; to conceptualize the ornament as a supplemental "technolog[y] of personhood" is to disperse the process of subject formation throughout interlocking orders of abstraction, semiosis, and signification. So, while Orientalism describes the appropriation of the "Orient" as a mechanism of racial capitalism, our formulation of techno-ornamentalism attends to the innumerable diffractions of life encapsulated within: it observes how the self-effacing labor of racial performance has become increasingly incomplete and insidious with its digital mediation but, in this nonfinality, can in fact safeguard one's life.

If, as Cheng suggests, the "flesh[y]" life of "Asian American female personhood" subsists within the transformation of the yellow woman into an ornament, we are led to ask, where does her objectified being reside when she is rerouted through increasingly complex circuits of mediation?[14] From where does her (de)racialization occur, and from where does it derive its capacity to endure? If race is no longer so stable a property encrusted on the techno-ornament, but a "flickering signifier," how is it seen only wherever it is convenient, when an audience hopes to see it, and when the nonresponse of an object can easily be transmogrified into an attestation of one's race?[15] These questions lead us to Nishime's suggestion: "If we take multiracial Asian representations seriously, they can give us a place to explore the *why* rather than the *what* of race."[16] Hence, we focus on the diverging audience receptions of Mitski's variable appearance; her shifting racial perceptibility among each group of consumers and across platforms of consumption speaks to the strange malleability and tenaciousness of racial identity—or the endurance of race in its structural malleability.

Be(coming) The Cowboy: Asianness as Ornamental Performance and Fantasy

Following her breakout success in 2016, Mitski's lyrics and discography have veered away from the autobiographical: her 2018 album, *Be the Cowboy*, contains explicitly fictional narratives, while her music for film (e.g., *Everything Everywhere All At Once*, *After Yang*) and graphic novel soundtracks introduce fantastic worlds. These works demonstrate her effort to supplant her own subjective voice with those of other characters. For instance, when discussing the concept of *Be the Cowboy*, she notes that the "exaggerated myth of the western cowboy" invokes a charisma that she seeks to imitate, for this quintessential Americana trope introduces an archetype so distinct from audiences' expectations of her that she does not have to "[act] too much like [her] identity."[17]

We read these choices as a reworking of narrative, one that detaches her creative output from biographical authority. Despite this, audiences continue to view Mitski as a representative of Asian America. Reviews of *Be The Cowboy*, for instance, praise her "introspective, stomach-pummeling lyrics" for her "honesty," "candid" negotiations with fame, and "story of self-discovery, resilience, and shifting identity."[18] Glossing over Mitski's musical craftsmanship, fans and critics regard the artist's shape-shifting selves as a way for her to safeguard her inscrutable Asian American interiority from her "mainstream" fame.[19] Summer Kim Lee has suggested that such a phenomenon may be due to Mitski's genre: "Mitski performs and works within a music genre [the indie singer-songwriter] constituted by a perceived authentic transparency, through intimate, confessional narratives of the self."[20] Anchored by the genealogical baggage of its genre—and heightened by the gendered expectation that female writing is "confessional"— Mitski's music orients our attitudes toward it: it anticipates its subsequent reception as an honest representation of the artist's life.

While Lee explains our affective expectations of singer-songwriters, it does not account for why Mitski's audiences stick with this imaginary of Asian America; what underlies this tenacious association is a more insidious codification of race. As Wendy Hui Kyong Chun and Lynne Joyrich write, race, as it has been formalized in the West, "maintains an uneasy relation to the visual."[21] Wedded by music videos, photographs, and stage and recorded appearances to an Orientalized body, Mitski's music then comes to give voice to not just her individual being, but her racialized experiences. Granted, identitarian categories do not always have to entail a pernicious burden of representation. Yet, they are often mobilized and charged with sociopolitical import. As Rey Chow asserts, discourses of (anti-)normativity preassign "the non-Western X ... a politicized status ... as the witting or unwitting harbinger of repair and purification to Western thought."[22] Hence, individuals are reduced into not just their social identities, but reified categories (mis)appropriable for the commodification of diversity in an era of neoliberal multiculturalism.[23] Against this instrumentalization of identity, Mitski's "cowboy" can only remain a fantasy. Her musical inventions, once attached to her name and appearance, remain confined by an expectation that has been predefined by the visual and discursive logics of race. Subsequently, her attempts at distancing herself from the contentions of identity politics only consign her to others' raced definitions of her.

One may observe such a gap between her ornamental (synthetic, produced, objectified) performances of self and the neoliberal subjecthood prescribed to her, for instance, in the cover art for *Be the Cowboy*, which simultaneously punctuates and punctures the sense of play-pretend that the album's title invokes. In a photograph that portrays the musician having her makeup done, her white hairnet and typically feminine makeup—dark lids, red lips, and heavy mascara—depict not an unembellished truth, but a tabula rasa brimming with potential. As an anonymous hand moves into the frame to touch up her eyeliner, her steely gaze at the camera implicates her audience as those who will determine her character. Relinquishing control over her own image, Mitski submits herself to the interface of artistic consumption that has historically "chopp[ed] up ... the star's image-body into identifiable ethnic and racialized modes."[24] Whether the artist wants to "be the cowboy" here is no longer significant; it is up to her audiences to make her up as such, or anything that we want for her to be. Perhaps this is why Mitski's name and that of the album are figured on the cover art by a mere white outline, an unobtrusive font that blends into the photo's bright background. The lore and person behind the album have melded into the commodity itself.

Presenting herself, in her music and cover art, as an object for her audiences' consumption, Mitski transforms into an evocative visual signifier that commands our participation "in a friction-free transaction that ... generates capital in the form of racialized images and performances."[25] This, however, does not entail the artist's total self-erasure: she is both the "cowboy" and the "Asian woman" we see, the commodity and the force that entices her own commodification. While her fictions render her beguiling to us—precisely because they do not trope a straightforward biography of Asian American subjectivity—we cannot forget that she has curated the incongruous hermeneutic frameworks that fascinate us. She produces herself as an alluring enigma, for it is as such an ornamental object that she can be reproduced and sustained—in her audiences' admittedly profitable imaginations of her.

Stay Soft: The Sonic Illusion of Intimacy

While it is largely in her visual mediation that we observe the artist's ornamental practices, Mitski's sound remains critical to her body of work. Her voice—seemingly unembellished and exposed—may seem a stark contrast to her curated image of cool reserve, but this is a productive disjunction: such a *sound* of vulnerability serves to activate her audiences' imaginaries of her lost subjecthood, drawing us to confront our various (de)racialized projections of her. Mitski's visual techniques of (dis)identification may, in this sense, lay the terrain for processes of racial formation (and/or our negotiations and negations of these processes). Yet, it is the crafted vulnerability of her music that ostensibly incites audiences' fantasy of its "authenticity." Such an illusion of intimacy and relatability is what established Mitski's earlier fan base after all: listeners corralled around the "raw" pain of Mitski screaming into her guitar (as she does in her

NPR Tiny Desk concert), for instance, or the "real," self-deprecating tweets that Mitski used to write from her personal account.

Even as her performances have shifted from a "grungier," do-it-yourself style to a sleek and highly produced sound accompanied by choreographed movements, her oeuvre continues to be marked by an undertone of intimacy. This is no coincidence: for example, her audio engineers deliberately chose "[not to] layer [her] vocals with doubles or harmonies, to achieve that campy 'person singing alone on stage' atmosphere" for *Be the Cowboy*.[26] Applying a slapback delay—a sound effect commonly used to make voices sound less "dry"—to her music, they give her single-tracked voice a touch of resonance to generate a semblance of bare vulnerability, warmth, and intimacy.[27] The aural insinuation of physical and emotional proximity conjures a sense of Mitski's immediate presence. Consequently, called by an audio technique to relate to a seemingly accessible and all-accommodating object, fans are moved to establish the grounds of relatability to Mitski's music—to at once author and authorize its truth.

Consider, for instance, the illusions of proximity stirred by the sonic effects of "Stay Soft," a single from *Laurel Hell* (2022): after a loud instrumental, Mitski croons, "I am face down on my bed / Still not quite awake yet, thinking of you—/ I tuck my hand under my weight." Here, the slapback delay clarifies the vocalist's enunciation, rendering her consonants particularly crisp in a line layered with *t*'s and *k*'s. Mitski's vocals cut through the distorted guitar, pulsating bass, synths, and drum lines below, transporting her listeners from a busy soundscape to a solitary one. The result is one of extreme nearness, as if we have trailed Mitski into her bedroom and shut the louder world outside. Combined with the lyrics' suggestive content, such artful sound production takes us alongside Mitski in her performance of masturbation. It attests to the singularity of the singer and her words, offering us an impression that we are bearing witness to her experiences. More importantly, it affords us the promise of an object's authentic life—a glimpse, perhaps, at the vocalist who enlivens her music.

Compelling us to breathe life, narratives, and imagined subjectivities into her material, the almost-campy loneliness of Mitski's voice urges us to reach for an artist—even if all that lies behind it is only a sonic "intimacy" as real as the track's synthesized drones and beat drops. Still, audiences need only such an intimation of life to hope that the "real" Mitski lies in her music. As evident in her unwaning popularity, fans and journalists continue to invest their interpretive labor into Mitski: whereas her queer, Asian, and women of color fans discern signs of Otherness in her music, others may articulate a connection with her that circumvents her minoritarian identity.[28] Mitski's assembled persona impels our *activities* of speculation, association, and desire—their own animating force. We stress the aesthetic quality of what induces movement here, for these traits are not borne intrinsically but by design. The ornament's "racial personhood is assembled not through organic flesh but instead through synthetic designs and inventions."[29] No symbolic discrepancy, narrative incoherence, or even a vocal crack in

Mitski's self-presentation is in this sense unplanned, for it is precisely such shadows of self-exposure that enable ornamentalism's "transformative magic"—the enchantment of commodity fetishism—to cohere.

Working for the Knife: Ornamental Personhood as Systematic and Interactive

While our analysis thus far has regarded the race-neutral, unmarked, and unaccented *sound* of Mitski's music as a complementary technique of her ornamental image—the inkling of a universal humanity that sparks her fans' animation of the ornament—online, this semblance of generality brings her new consumers with no investment in her identity at all. Specifically, through their rapid recirculation on social media platforms, her music and words have reached the most unexpected of markets, but only as racially and ontologically ambiguous entities. Although Mitski's fans usually hear her music alongside an imaginary of her racialized subjectivity, current media technologies have made it possible for her sound to be anonymized. In this way, her melodies and lyrics may be disseminated apart from her image and name—even if only briefly.[30] Such a digitization of Mitski's oeuvre renders its constituent parts isolable from each other; with this, her visually signaled racial background is now spectralized. Mitski's ornamental aesthetics thus comes to take on new significance, producing the *phantasm* of an unauthored text completely unburdened by the constraints of subjectivity.

Consider the broadcast of Mitski's lyrics on Twitter, a uniquely text-based platform, where @mitskilyricsbot holds court to over 160,000 followers. The bot's concept is simple: random excerpts from Mitski's lyrics are tweeted out by the hour. Despite the arbitrariness of the account's selections, its tweets have been shared so widely—given the ordinariness of its subject matter—that the songwriter is now recognized on the platform for her lyrical tropes. Accordingly, because her lyrics tend to underscore universal experiences of failure and loneliness, the bot's Twitter virality effectively abbreviates her work into a homogeneous note of sadness, whereby its followers enshrine her as their grand priestess of misery.

The same material, however, has found another life on TikTok, a short-form video hosting and creation platform, where users can edit text, audio, and video to produce multimedia clips. Such a form of collage has seen Mitski's composite body of work being mined for its separable parts: "The hashtag #Mitskiistherapy, referring to her emotionally evocative lyrics, has close to 50 million views," whereas her song "Strawberry Blonde" has been remade into a backing track for videos invoking cottagecore trends.[31] Notably, even though "Strawberry Blonde" speaks to unattainable love—it juxtaposes its chipper instrumentals against heartrending lyrics about one's aspiration toward blondeness, producing an ironic and poignant effect—the clips that sample it bear little resemblance to the song's narrative. Featuring frolicking girls, strawberry dresses, and floral arrangements—"romanticized [iconographies] of western agricultural life"—what these remixes derive from Mitski's music is only its breezy tune.[32] Evidently, as the discretized fragments of Mitski's media persona meet the recombinatory practices of our platform technologies, their spans are thus reanimated and spread throughout far-flung spheres of social media. What results is the proliferation of her variegated or even contradictory identities: the "sad" Mitski, the "therapist" Mitski, the carefree "Strawberry Blonde," and so on.[33]

Curiously, as these facets of her being are extrapolated from data patterns cleaved from their producer's image-body, they are no longer so clearly raced or gendered. As Mitski's music is subordinated to computational media's "content indifferent . . . matrix of abstraction," their digitized trails seem to lose the particularities of their context of production.[34] Still, we cannot neglect the enduring potential of derivative data to summon their sources, even when these parcels of Mitski's oeuvre have been anonymized, decontextualized, and made unrecognizable: "Strawberry Blonde" might have gained a new audience as the soundtrack of "cottagecore," but its (mis)appropriation online has also angered Mitski's Asian American fans, sparking a recursive feedback loop of (anti)identitarian discourses.

The artist's nonappearance in her work is thus only ephemeral—the labor that undergirds it is present even as it is invisible. But what is most remarkable is that it is Mitski's decisive abnegation of her production that enables it to gain such currency across contexts of consumption: making her output *sound* as generalizable as possible, she offers us a makeshift veneer of an unghettoizable universality upon which we can project ourselves—or our creative labor. In this way, while fans and music industry insiders are led to dream of the human in the ornamental object, the casual consumer is not freed from her enticements to give body to her nonvisual creations either. It is precisely in our disparate and mutually exclusive claims to the artist's truth that a productive friction is harbored. For the specter of Mitski's identity will always loom over its absence; the prospect of its recovery remains tempting. Her deracialized afterlives on social media can be said to turn on the very same aesthetics of ambiguity that activates her racialized imaginaries among canny fans and journalists. Mitski's conscious abstention from all self-reference in her work hence comes to serve another function here: it stages a guise of ambivalence that may be most seamlessly integrated with global networks of media consumption.

The artist's constant and active dissociation of her public personae from her racial background culminates in either of two partial ways of survival: in her audiences' belated ascriptions of racialized objecthood to her, or in a body of work estranged from her. Both are irreconcilable to each other; both preclude the unification of a self-sufficient subject. Perhaps this is why Mitski sings of the costs of survival in the 2021 single "Working For The Knife." Expressing a newfound resignation to her status as a popular musician in the track, she describes a "li[fe] in a world that sucks the humanity out of you at every turn."[35] Because the hype-machine that sustains her career is predicated on an "affinity for being thinglike . . . indebted to states of objecthood"—hence her diligent acts of self-erasure—Mitski must perpetually distill herself into the ornamental novelties that she offers us.[36] She must submit her sound, words, and image to an all-consuming machine of cultural production, to drive along her own ultimate endurance/extraction.

Accordingly, regardless of how she has been taken up—in her music, interviews, or even as a TikTok soundtrack—Mitski seldom reacts to her public reception. Only on occasion has she demurred, as with the aforementioned Twitter request, against her "being . . . consumed as content."[37] For just as ambiguity yields the greatest potential for information, controversies about Mitski's biracial identity call forth more publicity for her. It is imperative for us to note that Mitski's appeal has served little purpose (phone photography and recordings remain a mainstay at her shows) except to augment the networked spheres of social media wherein she circulates and is consumed.

Noticing Nobody: The Negated Subjecthood of Techno-Ornamentalism

In our account of a technocapitalistic media ecology that subsumes all relations under its expropriative designs, we have described the self-sustaining synchrony of labors between Mitski and her fans, critics, and other audiences that realizes Mitski as a consumer commodity. Yet, the interpretive relation between an artist and her audience has always been a cooperative practice in the first place. Cheng's concept of ornamentalism reminds us that race is an open process of (un)doing personhood, a way of surviving an enervating system without proscribing other modes of social being.

Despite this, we must also acknowledge that the ornamental object's generative potential is almost entirely compatible with capitalism's productive drive—especially with "the *production* of racial meaning" in now-valuable terms of multiracial visibility.[38] As Mitski notes, "I am a musician, but the reason they really pay me the big bucks is to be . . . the black hole where people can dump all their shit."[39] Providing a material basis that instigates all our interactive imaginaries of her, she transforms her audiences' devotion into "the big bucks." The evocative capacity of her aesthetics of ambiguity can be problematized as an apparatus of music industry capitalism. Interrogating the extractive means by which Mitski's racialized image and work are staged for us, we must question the operations of racial discourses. Similarly, we cannot deny our investments in Mitski and imaginaries of Asian America.

But the questions and problems of "Asian America" surrounding Mitski do not have to be contained so neatly within neoliberalism's foreclosed circuits. As Kandice Chuh stresses, the concept of "Asian America" already denotes an insoluble difference; it is and has always been a "subjectless discourse" that, in its negations of the white, Western subject, offers us a "space to prioritize [the discursive constructedness of] difference."[40] Like the chimerical forms of Mitski's public persona, "Asian America" cannot be reencapsulated by fixed notions of subjectivity. Something always escapes to oblige our interrogations into social difference, and through such inquiries return us to neoliberal, racial capital—or not. To "imagine [Mitski] otherwise" is therefore to not so readily accept the instrumentality of her ornamental aesthetics to capitalist production.[41] Instead, it is to refuse to settle into any one mode of relating with her: it is to probe the effects of our ambivalence, to admit the possibility of our complicity in our cruel attachments to the mechanisms of capital, and to endeavor to countervail their effects.

So, if our critiques have been pessimistic about the mediation of intimacy as a form of capital, Asianness as a detachable, commodified technology, and the impossibility of subjecthood for life as a techno-ornament, it is because we cannot afford to be careless. The impetus to go on imagining—and even imagining the worst—is only a prophylaxis for our survival. It is what allows us to hold out for the possibility that techno-ornamentalism might still gesture toward a mode of not just being, but vibrantly creating alongside each other—as the very art/artifice of Asianness.

Notes

1 Mitski Miyawaki (@mitskileaks), "When I'm on Stage and Look to You but You're Gazing into a Screen, It Makes Me Feel as Though Those of Us on Stage Are Being Taken from and Consumed as Content, Instead of Getting to Share a Moment with You," Twitter, February 24, 2022, 11:43pm, https://x.com/mitskileaks/status/1496979142981660672.

2 Examples include Netflix series like *Beef*, *The Chair*, and *To All The Boys I've Loved Before*; blockbuster hits such as *Crazy Rich Asians* and *Everything Everywhere All At Once*; or the success of Asian American musicians such as Japanese Breakfast and The Linda Lindas.

3 Kristin Yoonsoo Kim, "Mitski Is Not Here to Save Indie Rock (But She Might Save You)," *Complex*, June 17, 2016, https://www.complex.com/music/2016/06/mitski-puberty-2-profile.

4 "Mitski," Wikipedia, https://en.wikipedia.org/wiki/Mitski, last modified January 21, 2025; "Mitski," Last.fm, December 9, 2023, https://www.last.fm/music/Mitski/+wiki; Kayeleigh Hughes, "Album Review: Mitski Outdoes Herself on the Stunning Be the Cowboy," *Consequence*, August 14, 2018, https://consequence.net/2018/08/album-review-mitski-outdoes-herself-on-the-stunning-be-the-cowboy/.

5 Kim, "Mitski Is Not Here to Save Indie Rock (But She Might Save You)"; Zoe Hu, "Mitski Is Much More than Another Sad Asian American Girl," *Buzzfeed News*, August 24, 2018, https://www.buzzfeednews.com/article/zoehu/mitski-bad-cowboy-asian-american-singer-songwriter.

6 One could extend this thought to argue that Mitski's white/Asian heritage is the suitably ambiguous, white-leaning common ground to advertise a mass media–palatable Asian femininity. See LeiLani Nishime, *Undercover Asian: Multiracial*

Asian Americans in Visual Culture (Champaign: University of Illinois Press, 2013), xvii.
7   Anne Anlin Cheng, *Ornamentalism* (New York: Oxford University Press, 2019), 14.
8   David S. Roh, Betsy Huang, and Greta A. Niu, "Technologizing Orientalism: An Introduction," in *Techno-Orientalism: Imagining Asia in Speculative Fiction, History, and Media*, ed. David S. Roh, Betsy Huang, and Greta A. Niu (New Brunswick, NJ: Rutgers University Press, 2015), 11.
9   By "operationalization," we follow Harun Farocki's—and later Jussi Parikka's—deployment of the term to describe media that are themselves instruments that function to synthesize our perceptions of the world. Jussi Parikka, *Operational Images: From the Visual to the Invisual* (Minneapolis: University of Minnesota Press, 2023), vii.
10 Parikka, *Operational Images*, 42.
11 Tiziana Terranova, "Attention, Economy, and the Brain," *Culture Machine* 13 (2012): 13.
12 Other artists of Asian descent have likewise refused to center their identities in their work, such as Esther Yi (author of *Y/N*).
13 Many thanks to Betsy Huang and David Roh for highlighting this connection. Also see Parikka, *Operational Images*, 68.
14 Cheng, *Ornamentalism*, 22.
15 N. Katherine Hayles, "Virtual Bodies and Flickering Signifiers," *October* 66 (1993): 69.
16 Nishime, *Undercover Asian*, 159.
17 Ann-Derrick Gaillot, "Mitski Wants You to Be Your Own Cowboy," *The Outline*, August 14, 2018, https://theoutline.com/post/5810/mitski-be-the-cowboy-interview.
18 Hughes, "Album Review: Mitski Outdoes Herself on the Stunning Be the Cowboy"; Quinn Moreland, "Album Review: Be the Cowboy," *Pitchfork*, August 17, 2018, https://pitchfork.com/reviews/albums/mitski-be-the-cowboy/; Cheri Amour, "Mitski—Be the Cowboy: Album Review," *The Skinny*, August 14, 2018, https://www.theskinny.co.uk/music/reviews/albums/mitski-be-the-cowboy.
19 Mitski's popularity may, indeed, be heightened by her alluring inscrutability. See also Sunny Xiang, *Tonal Intelligence: The Aesthetics of Asian Inscrutability during the Long Cold War* (New York: Columbia University Press, 2020); and Vivian L. Huang, *Surface Relations: Queer Forms of Asian American Inscrutability* (Durham, NC: Duke University Press, 2022).
20 Summer Kim Lee, "Staying In: Mitski, Ocean Vuong, and Asian American Asociality," *Social Text* 37, no. 1 (March 2019): 38.
21 Wendy Hui Kyong Chun and Lynne Joyrich, "Race and/as Technology," *Camera Obscura: Feminism, Culture, and Media Studies* 70 (2009): 14.
22 Rey Chow, *A Face Drawn in Sand: Humanistic Inquiry and Foucault in the Present* (New York: Columbia University Press, 2021), 22–23.
23 See also Grace Kyungwon Hong, *The Ruptures of American Capital: Women of Color Feminism and the Culture of Immigrant Labor*. (Minneapolis: University of Minnesota Press, 2006).
24 Lisa Nakamura, *Digitizing Race: Visual Cultures of the Internet* (Minneapolis: University of Minnesota Press, 2007), 29.
25 Nakamura, *Digitizing Race*, 17.
26 Derrick Rossignol, "Mitski Announces an Intimate New Album with The Epic Single 'Geyser,'" *Uproxx*, May 14, 2018, https://uproxx.com/music/mitski-be-the-cowboy-new-album-geyser-song/.

27 This production technique is not unique to Mitski. For example, Ella Fitzgerald is sometimes so closely mic-ed that we can hear her every breath and traces of spit: Brooks Frederickson, interview by Jaeyeon Yoo, February 7, 2022. It is a recording technique that allows listeners to be closer to the performer than physically possible in performance.
28 Her 2022 album made "No. 5 on the Billboard 200 albums chart (dated Feb. 19[, 2023]), and was] by far the best debut of Mitski's career," "Five Burning Questions: Mitski Debuts in the Billboard 200's Top 5 With 'Laurel Hell,'" *Billboard*, February 15, 2022, https://www.billboard.com/music/chart-beat/mitski-laurel-hell-five-burning-questions-1235032017/.
29 Cheng, *Ornamentalism*, 19.
30 While we recognize that the sampling of creative content on social media has posed great challenges to attributability and copyright, a fuller discussion of intellectual property lies beyond the scope of this paper. Nonetheless, we would like to stress that our contemporary computational media ecology entails that no information is truly lost from an object: however unrecognizably Mitski's work is circulated online, she can always be and has been retroactively reidentified as its source.
31 Lindsay Zoladz, "Mitski Is More than TikTok," *New York Times*, March 11, 2022, https://www.nytimes.com/interactive/2022/03/11/magazine/mitski.html.
32 "Cottagecore," Aesthetics Wiki, last accessed April 22, 2024, https://aesthetics.fandom.com/wiki/Cottagecore.
33 While the immaterial labor of social media users cannot be sufficiently explicated here, their role in the production of Mitski's cultural value must be problematized, as the Internet's participatory culture depends on exploiting nonpaid labor. As Tiziana Terranova asserts, such a pervasive mode of generating extractable value challenges what we might view as production and consumption: Terranova, "Free Labor: Producing Culture for the Digital Economy," *Social Text* 18, no. 2 (2000): 35.
34 Jonathan Beller, *The World Computer: Derivative Conditions of Racial Capitalism* (Durham, NC: Duke University Press, 2021), 19.
35 Angie Martoccio, "Mitski Cuts Sharp and Sure With 'Working for the Knife,'" *Rolling Stone*, October 5, 2021, https://www.rollingstone.com/music/music-features/mitski-working-for-the-knife-new-music-1236649/.
36 Cheng, *Ornamentalism*, 85.
37 Miyawaki, "When I'm on Stage and Look to You."
38 Nishime, *Undercover Asian*, xv.
39 Ben Beaumont-Thomas, "Mitski, the US's Best Young Songwriter: 'I'm a Black Hole Where People Dump Their Feelings,'" *Guardian*, February 4, 2022, https://www.theguardian.com/music/2022/feb/04/mitski-us-best-young-songwriter-im-a-black-hole-where-people-dump-feelings.
40 Kandice Chuh, *Imagine Otherwise: On Asian Americanist Critique* (Durham, NC: Duke University Press, 2003), 9.
41 Chuh, *Imagine Otherwise*, 149.

Part V
Extensions

12

Asian Solarpunk

Between Utopia, Collective
Futures, and Remedies for
Climate Panic

AGNIESZKA KIEJZIEWICZ AND
JUSTIN MICHAEL BATTIN

Solarpunk is an emerging genre form without a clear definition, yet within distinct disciplines can be found consistent characteristics, each of which highlights the process of searching for remedies to dystopian futures. One of them is urban planning, which, through the ideal fusion of architecture and nature, is a visual manifestation of the philosophical postulates of sustainable futures. Solarpunk, corresponding with the exhaustion of the cyberpunk and postcyberpunk genres, can be traced to the gardencentric Ecotopia, a utopian project described by Ernest Callenbach.[1] Taking a position of a futuristic flâneur, Callenbach juxtaposes the design of an ideal country with 1970s America, building his conception on sociopolitical criticism. In his visionary project, the indicators of a sustainable life are not limited to renewable resources but feature, as well, a woman-dominated political system, which might be the key to the peaceful existence of a nation.[2] The collective planning and the return to nature-based philosophical framing of societies have also been featured by science fiction writers like Ursula K. Le Guin and numerous authors of solarpunk short story anthologies.[3] The tropes of Callenbach's ideas or references to American science fiction literature can be listed among many foundational elements incorporated into Asian solarpunk from American culture.

As an emerging aesthetic and cultural phenomenon, one can observe that solarpunk could be considered a subgenre of science fiction. Like cyberpunk, it rejects all improbable elements unrelated to possible technological development, such as extraterrestrial beings. Solarpunk encompasses the already-existing ideas for a utopian, sustainable future and proposes remedies for climate panic, built from the traces of ecofuturistic discourses and surprising semiotic connections. It has not progressed beyond its initial phase, as the solarpunk iconography remains fragmented and still morphing, depending on the cultural background of the authors repeating Callenbach's approach through a critical analysis of their closest surroundings. The lack of a consistent genre theory indicates that solarpunk remains a fan-driven phenomenon constructed from film elements or animation narratives and contemporary social philosophy. Although mainstream examples can be found, particularly in cinema, this subgenre primarily develops through fans drawing from existing visual media and literature to produce their own fiction, such as short stories, photographs, graphics, and speculative discussions on social media channels, forums, or blogs. All mentioned platforms and digital tools support different spectrums of fan bases' creative and informational exchange. Recently, artificial intelligence (AI) image-generation tools have opened new ways for the development of solarpunk-themed graphics, thus improving the possibilities of visualization of green futures. It is also worth adding that the fans' aesthetic approach to solarpunk often comes from discussions about sustainability. Consequently, solarpunk fashion and decor must be made out of natural, recyclable components and solarpunk visual art (i.e., sculptures) should mix with the natural environment, considering nature as a medium in following the conceptual principles of the land art movement.[4]

Cinematic Asian solarpunk, which has existed longer than its Western counterpart, features Studio Ghibli's ecological animations, such as *Nausicaä of the Valley of the Wind* (*Kaze no tani no Naushika*, 1984), *Princess Mononoke* (*Mononoke-hime*, 1997), and *Howl's Moving Castle* (*Hauru no ugoku shiro*, 2004). Those were created long before the emergence of the "solarpunk" term and were classified as part of the subgenre only later by fans. Films with solarpunk elements consist of such titles as the Korean American coproduction *Okja* (2017), directed by Joon-ho Bong, and the Asian American *After Yang* (2021) by Kogonada. Among Western solarpunk examples, Marvel's *Black Panther* series (2018 and 2022), the animated *Zootopia* (2016), the visions of Panem Capital City presented in the *Hunger Games* films (2012–2015), and the use of real Singaporean locations in the third season of the *Westworld* series (2016–2022) are the most visible.

Drawing on urban planning, in this chapter we investigate the connections between cinematic Asian solarpunk and building the collective imagination of the concept of futuristic, sustainable cities. By focusing on Asian solarpunk and its aesthetics inspired by the shape of Asian urbanscapes, our aim is to contribute to the broader understanding of the forms of presentation of the coexistence of technology and nature and the place of philosophical posthumanism, or a deviation from anthropocentric perspectives and concerns, inside this vision. Pointing out the visual inspirations of the filmmakers, whose imagination interweaves with the already-existing urban solutions in Asia, we want to show that the worldwide imaginary of a sustainable city is built on Asian-infused aesthetics and philosophy. This includes, for instance, a Buddhist contemplation of nature and the nostalgic Japanese concept *furusato*, which can be translated as a "home," "cradle," and "native place," symbolizing the nonexistent imaginary place of childhood nostalgia, as well as the national urge to return to premodern, mythologized nature.[5] Modeling solarpunk narratives on the existing Asian locations transfers the message of the ecological potential of a specific modern Asian lifestyle.

Solarpunk, as an emerging genre and a way of framing the future, challenges techno-Orientalist thinking by disrupting the aestheticization of Asian technological development. Contemporary urban planning, both real and as depicted in fictional narratives, allows us to see beyond not only the dominant understanding of Asians as ecopolluters but also comprehend the underpinning desire to reconnect with nature. As solarpunk in its main objective focuses on building a new identity compatible with societal sustainable goals, Asia is given back a voice and considered a leader in green practices. Techno-Orientalist reductionist tendencies are therefore subverted by the possibilities (and urgent need) of global application of the Asian ideas and implementations of solar living.

Eco Discourses around Asian Urban Planning

In the mid-twentieth century, the German philosopher Martin Heidegger famously remarked that humanity had become *enframed* by technological thinking (he used the term *Das Gestell*). The things of the world, which include everything from the natural environment to even ourselves, manifest into one's experiential view as resources (or *bestand* [standing-reserve], to use his term) to be optimized. In one specific example, he remarks that a human being's orientation to the Rhine River reveals neither a symbol of culture, the nation, or a source for poetic inquiry, but rather electrical power and energy to facilitate modern industrial societies.[6] Today, too many solutions addressing climate change and global warming are informed by such an understanding of the world; nature is still perceived through an adversarial lens, to be challenged, managed, and controlled.

Examples of such solutions can be seen through contemporary sustainable developmental projects across Asia. Singapore's Gardens by the Bay, Tokyo's Roppongi Hills, Ho Chi Minh City's in-progress Thu Duc City, and numerous private residential developments are, to varying degrees, reflective of the established solarpunk aesthetics and narratives further discussed in the second part of this chapter, with Singapore offering, perhaps, the most visible and striking example. The Gardens by the Bay receives much global attention for its status as an internationally renowned tourist attraction. Yet, in various social media channels, from Instagram to Reddit, the "gardens" are equally praised for their solarpunk style. From the cloud forest to the supertree grove, solarpunk aficionados are drawn to the futuristic weaving of humanmade structures with the natural environment. While these visual feasts are viewed as enduring symbols of the solarpunk aesthetic, one should be attentive to the fact that the Gardens by the Bay plays a relatively minute role in the city's sustainability ambitions. As a low-lying island state, Singapore is especially vulnerable to the myriad impacts of climate change, and in the previous fifty years it has been an exemplar to other countries for how to systematically confront this existential challenge. This is evident not only in the progress thus far made, but also in the future policies to be enacted, which include living in harmony with nature, achieving cleaner and better uses of energy, implementing a circular economy, enabling green finance and research, and building national resilience.[7] Singapore, thus, claims to strive for the utopian ideal of a highly developed and sustainable human settlement.

In addition to Singapore, Tokyo's Roppongi Hills integrates public garden spaces within a dense urban landscape, such as those within the Mori Arts Center, TV Asahi headquarters, and the Keyakizaka Complex, some of which are used for artistic purposes, and others for the cultivation of fruits and vegetables. The initiative, which includes networks of greenways and gardens, was initially conceived to alter the identity of the Roppongi neighborhood, which historically is associated with yakuza and controversial nightlife activities, as well as provide a solution to the heat island effect.[8] Following the destruction of ponds and green spaces that accompanied Tokyo's urbanization, the city witnessed an increase in temperature of three degrees Celsius over the previous one hundred years.[9] Contemporary heat maps, however, show that the surface temperature of greened spaces like Roppongi Hills is five to fifteen degrees cooler than the asphalt of the surrounding areas.[10]

In Ho Chi Minh City, developments like Thu Duc City, an area intended to spur economic growth in the city's technology sector, are planned with significant attention to sustainable living, such as the development of parks and ecologically friendly construction.[11] Moreover, considering the climate of Vietnam's southern metropolis, significant space for tropical plants is frequently included in the architectural designs. Across Ho Chi Minh City, and Vietnam in general, green buildings are gaining momentum as vehicles for investment, in part due to the emergence of a market for sustainable building materials and partnerships with foreign entities like the United Nations Development Programme and the World Bank.[12]

At first glance, these urban plans are reflective of the immediate concerns wrought by climate change and a need to develop a less anthropocentric approach to the natural environment. Closer scrutiny, however, reveals that the primary motivation for these developments is the lucrative financial gains to be made by implementing sustainable solutions. The Vietnam Urban Green Growth Development Plan 2030 openly positions economic opportunity as the key motivating factor for the adoption of sustainable policies and practices. The Plan suggests that "green" economic development will improve the country's competitiveness and facilitate job growth, which will eventually result in alleviating poverty, improving the well-being of the people, reducing greenhouse gas emissions, and, broadly, improving the country's resilience to climate change.[13]

The economic prioritization evident in these plans is reflective of an oscillation between two approaches, *eco-efficiency* and *eco-effectiveness*. The latter is a concept that prioritizes closed-loop supply chains: "The philosophy is one of bio-mimicry in which processes consume natural capital at sustainable rates and do not produce wastes that cannot be absorbed by the natural environment at a sustainable rate."[14] While the radicalism of this concept is often praised, given that it requires a complete overhaul of dominant supply chain models, Michael and Joyce Huesemann note that its negative impact on the economic bottom line renders it as a long-term goal rather than a feasible approach in our current context.[15] Thus, eco-efficiency is often the preferred approach. Eco-efficiency, in contrast to eco-effectiveness, affords equal consideration to ecological and economic concerns but in practice prioritizes economic benefits.

The enduring persistence of eco-efficiency and eco-effectiveness in societal developments toward addressing climate change (and anthropogenic solutions more broadly) can be partially explained by the long-established utilitarian and instrumental relationship between humankind and nature. While these three developments are geographically and culturally rooted in Asia, the ideological and ontological foundations are predominantly Western in origin, yet have become more pervasive in the Asian region, particularly following the continent's embrace of global economic systems and the resultant lifestyles. Historically, particularly in a medieval context, nature has been distinguished as a binary opposite to civilization. Lynn White Jr. notes that the widespread return to agricultural labor in the early medieval period, coupled with the rise of Christian anthropocentrism, fostered a dualistic rift between humanity and nature.[16] Nature, therefore, became interpreted as an exploitable object for the whims of humankind. Yet, to restrict such an interpretation to purely the developmental (such as for food cultivation or building materials) misses a key use; nature also metamorphosed into a source for healing and rejuvenation (such as for monks and royalty) and a sanctuary or refuge for others (such as criminals and other societal outcasts). In both instances, nature is comprehended for what it provides, placing it entirely under the realm of human dominion.[17]

This conception of nature was furthered during the European colonial conquest of the Americas and subsequent western expansion of the United States, through which the natural environment became an antagonistic force. Indigenous inhabitants and wildlife were considered dangerous obstacles to overcome in the name of territorial expansion, societal progress, and divine providence. Similar to the European medieval age, nature retained a sense of renewal, yet a clear distinction can be made from the earlier epoch: a sense of achievement. The American transcendentalist Henry David Thoreau, for instance, abandoned city life "to live deliberately in the woods" to rediscover "the essential facts of life."[18] Thoreau's quest, while inspired by a desire to become more attuned to nature, to appreciate it more deeply, is equally indicative of a challenge; stripping himself of everything except his resolve, Thoreau aims to see if he can withstand nature's most arduous trials. In the end, Thoreau's *Walden* depicts his endurance as a humbling experience, yet it is also illustrative of *furusato*.[19] By abandoning "civilization" and embarking on an *ethical topos* through nature, Thoreau returns to a more primal, almost premodern form of existence, one where he is able to reflect upon corrupting social and cultural conditions of progress.[20]

It is unsurprising that the abovementioned urban planning developments have taken this approach, as each was created broadly for commercial purposes (such as tourism, in the case of Singapore, or economic development in Vietnam). The Japanese idea of a *furusato*-like settlement can be witnessed and explored, however, in a variety of private housing across the three countries identified above. In Singapore, a development dubbed "forest town," formerly the site of military training and brickmaking factories, was unveiled in 2016, having been designed to reduce carbon emissions, as the country's per capita emissions exceed those of several other postindustrial countries.[21] The site's use of innovative cooling technologies and solar power will result in a reduction in carbon dioxide emissions equivalent to the removal of 4,500 automobiles from the road.[22] Additional strategies include smart lighting systems, central trash storage, and mobile applications that allow residents to monitor their energy output. The ultimate outcome, in addition to the reduction of emissions, is to encourage a reconnection with nature and appreciation for biodiversity, which can be a catalyst for broader behavioral changes.

In Japan, the Nishizawa Garden & House is a multiusable space for domesticity and work (the site houses two writers) that epitomizes the phrase "to do more with less." The building has been developed vertically because of spatial restrictions; the site is a mere four meters wide. The architect for the building, Ryue Nishizawa, wished to ensure the building could be lighted naturally, a difficult undertaking given the cramped conditions; the site is nestled between two large buildings each roughly thirty meters in height. The building therefore is designed around two slabs and steel beams for support, with a single vertical staircase being used to access the building's five stories. The walls are mostly made of glass, which can be opened to access the structure's external areas. These features allow for not only natural light to enter each area, but also air when opened, an intentional design choice to encourage the growth of gardens.[23] The site also employs a floating illusion emblematic of minimalist Japanese houses (*jutaku*), and the absence of opaque walls, customary in contemporary housing, deconstructs the binary of interior and exterior, thus encouraging a closer relationship with the natural elements.

In Vietnam, some private homeowners are following a similar trajectory. In Thu Duc City, the aforementioned area within the Ho Chi Minh City metropolis, one single-house project eschews concrete structures, instead employing wooden sliding doors (and glass) to create an open-space concept, one that allows for the owners to more easily access and feel connected to the tropical setting that surrounds the house. The use of locally sourced wood helps to integrate the house more seamlessly with those surroundings. For cooling, the sliding doors allow for a breeze effect, which helps to mitigate the need for artificially produced and carbon dioxide–emitting air conditioning.[24] Outside of Thu Duc City, another private house in District 11 has used twenty-five thousand alternating bricks to create a series of ventilation holes to improve air circulation. The front of the house includes dense vegetation and a pond, which facilitates the creation of a microclimate, and the roof has been fitted with tropical plants, a vegetable garden, and an automatic watering system. The house's architect suggests that the goal of the roof design is to reduce the heat absorbed by the roof, a common problem in tropical countries, and provide the homeowner with intimate access to nature, despite being in the middle of a crowded urban area. The house has also been fitted with skylights to allow for natural light, thus reducing the need for electric lighting and allowing indoor plants to thrive, furthering the integration between the built environment and greenery.[25]

Based on the above examples, which are a mere few among many, we can see that sustainable, green futures are more visible in private housing than in commercially driven urban development plans. These private developments, particularly the examples from Vietnam and Japan, are reflective of the *furusato* concept, returning to the cradle as provided by nature. They are also, however, emblematic of a Buddhist framework to architecture. Although Buddhist-inspired architecture in Asia varies across the continent's geographical climates, the designs tend to integrate local building materials, particularly those that allow for integration within the natural surroundings, and include elevated floors, staircases, and rectilinear forms, all of which are visible in the abovementioned private structures, which are largely unaffordable to the general public, also because of their cost.[26] It could therefore be said that the values of solarpunk, made most visible by way of the peaceful union between the natural environment and human beings, are better processed and enacted by individual citizens rather than by policymakers or commercial urban developers, as their reliance on the eco-effectiveness and/or eco-efficiency approaches still positions nature as an optimizable resource.

Moving forward, any successful push for sustainability requires the revival of *furusato*, which is a key area of emphasis within a range of Japanese solarpunk animations produced by Studio Ghibli. Humanity, after wrecking its home, returns to the primordial state, acknowledging that only nature allows for one to get back to a "true" self, one without *Das Gestell* or with any focus on efficiency. According to this postulation, it can be observed through these films that solarpunk aesthetics and corresponding lifestyles cannot be driven by eco-efficiency or eco-effectiveness, but rather by ecocentrism. Solarpunk-related ecocentrism raises skepticism about the position that the environment is a manageable object, something to overcome, inviting us to question our egocentric view of the world and find intrinsic value in all of nature on ethical, evolutionary, spiritual, and ecological terms.[27]

Asian Ecotopia (through *Furusato*) in Film and Animation

The most significant feature of solarpunk is the hope for prosperous and joyful futures.[28] This subgenre adds a new approach to the existing speculative fiction genres, such as cyberpunk, steampunk, the lesser-known dieselpunk, or stonepunk. However, while cyberpunk and other -punk subgenres manifest an extrapolative character, solarpunk's interpolative estimation of future scenarios is based on the technological solutions already applied by humanity.[29] The applicable nature of this subgenre results from its framing of observable problems such as water pollution or global warming. Bringing bright visions of the future, solarpunk films and animations do not fall into the category of postapocalyptic, "pre-trauma cinema." Forming his ecological film theory, Robert Geal observed that "the act of surviving the environmental apocalypse is generally depicted as a triumphant transcendence over nature's destructive powers... films set after such apocalypses have the potential to subvert this transcendence, and depict the consequences of survival as unpleasurable and continuing traumatic struggles against the powers of nature."[30] In solarpunk, even if the disaster is part of a narrative, its purpose is to build ecological consciousness within the audience and shift the paradigm of anthropogenic actions.[31] In this context, solarpunk offers an antidote to the fears other speculative genres create, of a dystopian, gloomy future filled with constant struggles and dialectic chaos.

There are two narrative solutions reappearing in both Western and Asian solarpunk film, animation, and literature. Firstly, it can be observed that the reoccurring iconography of this subgenre is the futuristic landscapes, which feature highly developed human settlements or biology-related technological solutions in progress. Secondly, these texts discuss the foundations of the ultimate happiness of a sustainable society and the moral or sociopolitical price of the development. The plots of solarpunk narratives are often based on sudden twists, in which unexpected events, such as political upheaval, destroy an initial tranquility. All in all, the struggle often leads to the verification of the initial policies and necessary improvements.

The first traces of solarpunk-related cinematic narratives on Asian ground can be observed back to the revolutionary Japanese animation film *Nausicaä of the Valley of the Wind* (1984) by Hayao Miyazaki.[32] Generations of film scholars have analyzed *Nausicaä* in an ecological context and Miyazaki's authorial style.[33] However, *Nausicaä* has been rediscovered by the fans of the emerging solarpunk subgenre, which has brought exciting reinterpretations in the spirit of sustainable futures. For instance, fans on various internet forums praise Miyazaki's visual style, considering it as the foundation of solarpunk aesthetics.[34] His models of a possible future approach to urbanized nature have influenced further depictions of solarpunk settlements.[35] In *Nausicaä*, the director depicts a postapocalyptic world filled with overgrown living creatures and a toxic jungle. The newly emerged habitats are dangerous and inaccessible to humans, who have gathered in the last untouched areas, creating sustainable rural settlements. The titular Valley of the Wind is a wind-powered village made of tiny stone houses blending with nature. The community is portrayed as a traditional society living in happiness thanks to the well-established rules and orally transferred culture. The feeling of belonging is underlined as the crucial element of the success of green communities, which are based on the exchange of abilities and cultivation of communal causes. It is a complete opposition to cyberpunk dystopia, which alienates individuals and encourages living in highly polarized societal structures. Miyazaki's didactic message, which will resonate in further solarpunk narratives, suggests that the future is collective and not driven by instrumental reasoning. However, referring to the previous part of this chapter, nature in *Nausicaä* needs to be tamed and researched, approached with fear of failure that would bring destruction.

It is significant that in many solarpunk narratives, even the earliest ones, complete coexistence with nature is depicted as an impossible project. For example, similar reflections can be found in both the Cold War–influenced *Laputa: Castle in the Sky* (1986), another Miyazaki animation film, and in Ursula K. Le Guin's work.[36] Le Guin's *The Word for World is Forest* (2010/1976), the fifth book of the Hainish Cycle, revolves around the bitter reflection that the connection between technological civilization and nature remains an interrupted attempt, a project based on inequalities and exploitation from the very beginning.[37] This protosolarpunk book presents the culture clash between two societies—the little humanoids coexisting with nature for ages, and the newcomers, applying their postcolonial approach brought from the (dying) Earth. The first mistake of humans is that they do not try to understand the rules of the new planet and its inhabitants, considering nature and pacifist humanoids as objects for use and economic profit. Similarly to *The Word*, Miyazaki's *Laputa* investigates the problems with the switch of perspectives from the anthropocentric mentality of conquerors to that of a sustainable nation. Even if technologically perfect, the project of the self-sufficient flying castle in *Laputa* was abandoned because its creators could not accept coexistence with nature. In this story, Miyazaki shows a ray of hope for posthuman self-reflection. As the inhabitants of the Laputa castle understood their arrogance and lack of appreciation toward nature, they abandoned their civilization. They came back to Earth, giving humanity more time to mature. However, Akimoto observes that they did not destroy their technological development, leaving the flying castle as the symbol of the critique of excessive modernization.[38] As Pamela Gossin underlines, Miyazaki's films are built on contradictions, ecological thinking, teaching moments, and fascination with posthuman technology.[39] Those elements can also be found in *Princess Mononoke* (1997) and *Howl's Moving Castle* (2004), which could be further analyzed in the context of solarpunk tropes.

In the solarpunk narratives mentioned above, sustainable futures are depicted as providing the opportunity to recreate *furusato*. Relatedly, there are "healing anime" (*iyashikei*), which emerged in the early 1990s as part of a wave of soothing cultural artifacts during the economic crisis in Japan.[40] Healing anime features slices of life composed into a slow narrative without sudden plot twists, violence, or dangerous quests. The intangible element of healing anime is the closeness to nature, which is perceived as the ultimate escape, such as in the animated series *Mushishi*.[41]

Conclusion

The inconsistencies of the solarpunk semiotics and discrepancies in the original aims of the cinematic visions and their possible readings through philosophical lenses reveal the limitations of the collective futures. First, solarpunk can be perceived as an impossible project, focused on building demonstrative, sustainable surroundings in the chosen, representative areas—as in the context of Singapore, Ho Chi Minh City, and Tokyo. In the films, the solarpunk project questions the human capacity to embrace the intrinsic value of nature and reject egocentric perceptions in favor of absolute collectivism. Finally, a mythical *furusato*—a perfect natural escape—seems beyond our reach, as the live-action and animation films depict.

Even though Japan still plays the main role in solarpunk development, two vivid examples from recent years express solarpunk ideals through instances of negativity, highlighting the consequences of excessive consumption and labor technologies. *Okja* (2017) and *After Yang* (2021) immerse the viewer in the not-so-distant future worlds, which are built on nature/civilization oppositions. In *Okja*, young Mi-ja and her grandfather live deep in the Korean mountains with a prototype genetically modified giant pig-like creature, the titular Okja, considered one of a kind. However, when a meat-producing American corporation steals the pig, Mi-ja pairs with the Animal Liberation Front to bring her animal friend back home. This is a presolarpunk story of the formation of the green future, showing humanity's radicalization before reaching the ideal symbiosis with nature. In this narrative, in radical futures, solarpunk becomes an ideology based on opposition against capitalism and authoritarianism, acknowledging that unrestrained consumption has reached its capacity. Resembling Callenbach's postulates, ecological thinking is connected to feminism and the laws of minorities and the oppressed, including animals. The film's most active and aware person is a young girl who becomes a symbol of future generations raising ecological awareness and challenging anthropocentrism. In *Okja*, the opposition between the mythologized Asian green way of living and a sterile, consumption-oriented Western civilization is clearly visible.

A different approach is presented in *After Yang*, a quiet drama featuring the grief of young Mika and her family after their android's expiration date (i.e., death). The main discursive point of this serene, "healing anime"–style film is that everything is degradable—especially the androids, which serve their years as nannies, helpers, or ersatz siblings. After their time comes and their internal parts, similarly to biological organs, start disintegrating, they should be brought to the dismantling point for repurposing. In this optics, Yang, an android, even though mechanical, joins the sustainability chain, in which his once-capable mechanical parts will be reused after the expiration date. The main question of the film is about the emotional burden connected to a highly sustainable future in which everything is degradable. The elements of the environment—both biological and synthetic—have clearly stated functions that exclude the possibility of expanding their usability further.

Summarizing our observations, it can be concluded that the scarcity of actual solarpunk films and animations leads to this phenomenon developing dynamically through fans and fandoms. Envisioning a better future, the fans focus primarily on creating video essays explaining their objectives and discussions, as well as graphic design or wallpapers. It is worth mentioning that the entries created by fans do not necessarily contain deeper analysis or manifest an awareness of the philosophical and socioeconomic aspects related to the visions of solar futures. Instead, solarpunk in mainstream perception functions as a fashion trend highlighting the connections of Indigenous nations' patterns, such as Māori symbols, and high-quality breathable fabrics.

The second understanding of solarpunk, most visible in the aforementioned films, animations, and literature, presents this subgenre as a visual imaginary focused on green cities in opposition to cyberpunk dystopia. Finally, solarpunk on the Internet platforms functions as the new, ecology-oriented anticapitalist and anticonsumerist ideology.

As solarpunk is in the formative phase as a pop-cultural phenomenon, the following years will certainly bring new, exciting research perspectives. For instance, the first solarpunk digital games, such as sandbox Solarpunk (developer: Cyberwave), were scheduled to be released in 2024. Moreover, solarpunk films could be further researched in the context of transnationality or with the methodologies provided by production studies—focusing on international cooperation and permeation of influences.

Notes

1. See Dougal McNeill, "Future City: Tokyo After Cyberpunk," *Review of Asian and Pacific Studies* 44 (2019): 89–108; Ernest Callenbach, *Ecotopia* (Berkeley, CA: Banyan Tree Books, 2004).
2. Callenbach, *Ecotopia*.
3. See Liz Grzyb and Cat Sparks, eds., *Ecopunk! Speculative Tales of Radical Futures* (Greenwood, Western Australia: Ticonderoga Publications, 2017); Gerson Lodi-Ribeiro, *Solarpunk: Ecological and Fantastical Stories in a Sustainable World*, trans. Fábio Fernandez (Albuquerque, NM: World Weaver Press, 2018).
4. See Robert Smithson, "A Sedimentation of the Mind: Earth Projects," *Artforum* 7, no. 1 (1968): 44–50.
5. See Jennifer Robertson, "Furusato Japan: The Culture and Politics of Nostalgia," *International Journal of Politics, Culture, and Society* 1, no. 4 (1988): 494.
6. Martin Heidegger, "The Question Concerning Technology," in *Martin Heidegger: Basic Writings*, ed. David Farrell Krell (San Francisco: Harper, 1977), 297.
7. "A City of Possibilities," Singapore Green Plan, last accessed December 30, 2022, https://www.greenplan.gov.sg.
8. Roman A. Cybriwsky, *Roppongi Crossing: The Demise of a Tokyo Nightclub District and the Reshaping of a Global City* (Athens: University of Georgia Press, 2011).
9. Annelise Giseburt, "How a Scientific Approach to Urban Greening Could Cool Japan's Concrete Jungles," *Japan Times*, April 30, 2023, https://www.japantimes.co.jp/news/2023/04/30/national/green-infrastructure-heat-island-effect/.
10. "Countermeasure to Urban Heat Island Phenomenon," Mori Building Company, 2023, https://www.mori.co.jp/en/sustainability/environment/biodiversity.html#about_contents07
11. "Master Plan of Thu Duc City in Ho Chi Minh City Until 2040," Asem Connect, January 27, 2022, http://asemconnectvietnam.gov.vn/default.aspx?ZID1=14&ID1=2&ID8=114619.
12. Thu Nguyen and Siddartha Bhatla, "Green Buildings in Vietnam: How Sustainable Are They?" *Vietnam Briefing*, March 28, 2022, https://www.vietnam-briefing.com/news/green-buildings-in-vietnam-how-sustainable-are-they.html.
13. "Conference on Implementation of Viet Nam Urban Green Growth Development Plan to 2030 and Circular of Urban Green Growth Indicators," Global Green Growth Institute, October 4, 2018, https://gggi.org/conference-on-implementation-of-viet-nam-urban-green-growth-development-plan-to-2030-and-circular-of-urban-green-growth-indicators/.
14. Paul Nieuwenhuis, A. Touboulic, and Lee Matthews, "Is Sustainable Supply Chain Management Sustainable?" in *Sustainable Development Goals and Sustainable Supply Chains in the Post-Global Economy*, ed. Natalia Yakovleva, Regina Frei, and Sudhir Rama Murthy (New York: Springer, 2019), 20; See also William McDonough and Michael Braungart, *Cradle to Cradle: Remaking the Way We Make Things* (New York: North Point Press, 2002).
15. Mette Mo Jakobson, "The Relation of Eco-Effectiveness and Eco-Efficiency—an Important Goal in Design for Environment," in *DFX 1999: Proceedings of the 10th Symposium on Design for Manufacturing, Schnaittach/Erlangen* (Germany, 1999), 101–104; Michael Huesemann and Joyce Huesemann, *Techno-Fix: Why*

Technology Won't Save Us or the Environment (Gabriola Island, BC: New Society Publishers, 2011).

16 Lynn White Jr, "The Historical Roots of Our Ecologic Crisis," *Science* 155, no. 3767 (1967): 1203–7, https://doi.org/10.1126/science.155.3767.1203.

17 See also Roderick Frazier Nash, *The Rights of Nature: A History of Environmental Ethics* (Madison: University of Wisconsin Press, 1989).

18 Henry David Thoreau, "Walden," in *The Portable Thoreau*, ed. Carl Bode (New York: The Viking Press, 1964), 343.

19 Thoreau, "Walden."

20 Viriato Soromenho-Marques, "'Walden': A Tale on the 'Art of Living,'" *Configurações* 25 (2020): 25–35.

21 Oscar Holland, "Singapore Is Building a 42,000-Home Eco 'Smart' City," CNN, February 1, 2021, https://www.cnn.com/style/article/singapore-tengah-eco-town/index.html.

22 "Empowering My Tengah," SP Group, May 27, 2023, https://www.mytengah.sg.

23 "Garden & House, Japan," Arquitectura Viva, accessed June 24, 2024, https://arquitecturaviva.com/works/jardin-y-casa-10.

24 Ngoc Diem, "Saigon House Uses Brick Walls, Plants to Fight Heat," *VNExpress International*, May 23, 2023, https://e.vnexpress.net/photo/style/saigon-house-uses-brick-walls-plants-to-fight-heat-4599030.html.

25 Diem, "Saigon House."

26 Sonam Chuki, Raju Sarkar, and Ritesh Kurar, "A Review on Traditional Architecture Houses in Buddhist Culture," *American Journal of Civil Engineering and Architecture* 5, no. 3 (2017): 113–123, https://doi.org/10.12691/ajcea-5-3-6.

27 Haydn Washington et al., "Why Ecocentrism Is the Key Pathway to Sustainability," *Ecological Citizen* 1 (2017): 35–41.

28 For example, the authors of the first Western solarpunk anthology featuring mostly Asian authors, *Multispecies Cities. Solarpunk Urban Futures*, build their narratives around the question "What If Stories Could Plant the Seeds of Hopeful Futures?," considering the act of reading as a powerful, formative experience. Christoph Rupprecht et al., *Multispecies Cities: Solarpunk Urban Futures* (Albuquerque, NM: World Weaver Press, 2021).

29 As it was defined by John W. Campbell Jr., Isaac Asimov, or Judith Merril, science fiction and its subgenres (before the emergence of solarpunk) create possible future scenarios by extrapolating from current events, asking about the conditions in different space-time continua. See Brooks Landon, "Extrapolation and Speculation," in *The Oxford Handbook of Science Fiction*, ed. Rob Latham (New York: Oxford University Press, 2014), 26–27. The interpolative character of solarpunk means that this subgenre expands the imagined scenarios by inserting the results of speculation into real-life contexts, such as green architecture or fashion. However, the emergence of new visions of possible futures does not mean that cyberpunk has disappeared from the world's cultural landscape. Instead, cyberpunk evolved into postcyberpunk, exploring more contemporary motifs, rethinking the archetypal protagonist, and referring to up-to-date technologies. This evolution, which was conceptually introduced by narrative solutions in *The Matrix* (1999), which depicts a community-oriented main character, can be considered as the exhaustion of the formula but not the cease of the existence of cyberpunk-related aesthetics. In the first postcyberpunk novel anthology, James Kelly and John Kessel indicated that the postcyberpunk adjusted cyberpunk's clichés according to twenty-first-century politics, social-thinking paradigms, and new popular culture archetypes.

James P. Kelly and John Kessel, eds., *Rewired: The Post-Cyberpunk Anthology* (San Francisco: Tachyon Publications, 2007).

30 Geal points out that postapocalyptic and cyberpunk narratives, showing the viewers the fall of the well-known order and humanity in general, can be considered as the anticipation of the yet-to-be trauma. Robert Geal, *Ecological Film Theory and Psychoanalysis: Surviving the Environmental Apocalypse in Cinema* (New York: Routledge, 2021), 187–188.

31 Geal, *Ecological Film Theory and Psychoanalysis*, 187.

32 It is worth indicating that Japanese solarpunk films and solar urban developments, such as the aforementioned Roppongi Hills, have common origins associated with historical, societal, and economic issues. Japan is the leader in animated solarpunk, as the narrative solutions mirror the real conditions of Japan and provide possible future visions. These discussions concerning sustainable futures come from the acknowledgment of limited natural resources in Japan and the significance of the relationship with nature visible in the Japanese national religion—Shintō. See Elaine Haglund, "Japan: Cultural Considerations," *International Journal of Intercultural Relations* 8, no. 1 (1984): 61–76, https://doi.org/10.1016/0147-1767(84)90008-7

33 See Colin Odell and Michelle Le Blanc, *Studio Ghibli: The Films of Hayao Miyazaki and Isao Takahata* (Harpenden, UK: Kamera Books, 2010); and Susan J. Napier, *Anime from Akira to Princess Mononoke: Experiencing Contemporary Japanese Animation* (New York: Palgrave, 2000).

34 This and other interpretations mentioned in this article in the context of Reddit can be found on the open Reddit group named "Solarpunk—hope for the future" (r/solarpunk). The group can be found at https://www.reddit.com/r/solarpunk/.

35 Here should be mentioned the short animated film *Dear Alice* (2021), which is also an advertisement for Chobani dairy products. The 80-second video features the letter of a grandma to her granddaughter. The woman shows her daily life on a solarpunk farm. See *Dear Alice*, directed by Bjørn-Erik Aschim (London: The Line, 2021), https://www.youtube.com/watch?v=z-Ng5ZvrDm4.

36 As Akimoto observes, there are several semiotic traces allowing us to observe Miyazaki's intentions to point out the Cold War era when *Laputa* was produced. One of them is the shape of the thunder, which resembles a nuclear mushroom cloud, a reference to Cold War atomic weapon tests. See Daisuke Akimoto, "Laputa: Castle in the Sky in the Cold War as a Symbol of Nuclear Technology of the Lost Civilisation," *Electronic Journal of Contemporary Japanese Studies* 14, no. 2 (2014), http://www.japanesestudies.org.uk/ejcjs/vol14/iss2/akimoto1.html. It is possible that Miyazaki was familiar with Le Guin's works, as in 2006, his son Gorō Miyazaki directed for Ghibli *Tales from Earthsea* (*Gedo senki*), based on Le Guin's book cycle under the same title.

37 *The Word for World Is Forest* features the events happening before the stories presented in the first four books of the series, so that it can be considered as a separate narrative. It introduces and describes the foundation of the sociopolitical order of the world described in the series.

38 Akimoto, "Laputa: Castle in the Sky."

39 Pamela Gossin, "Animated Nature: Aesthetics, Ethics, and Empathy in Miyazaki Hayao's Ecophilosophy," *Mechademia: Second Arc* 10 (2015): 217–222, https://doi.org/10.5749/mech.10.2015.0209.

40 See Paul Roquet, "Ambient Literature and the Aesthetics of Calm: Mood Regulation in Contemporary Japanese Fiction," *Journal of Japanese Studies* 35, no. 1 (2009): 111–187.

41 *Mushishi*, directed by Hiroshi Nagahama (Showgate, 2006).

13

Animated Bodies

◦

Project Itoh and the Afterlives
of Techno-Orientalism

BARYON TENSOR POSADAS

Techno-Orientalism, which may be understood as a kind of recoding of Japan through the iconography of the science fiction (SF) genre and vice versa, speaks to the point that "SF" names not only a genre of fiction or film or some other cultural object. Rather, it also functions as a mode of critical discourse—or, in the words of Istvan Csicsery-Ronay, "a kind of awareness we might call *science-fictionality,* a mode of response that frames and tests experiences as if they were aspects of a work of science fiction."[1] Elsewhere, I have argued that one of the implications of the prevalence of a techno-Orientalizing gaze is that it effectively aligns the fields of Japanese studies and science fiction with one another as interlinked critical practices.[2] Such an alignment takes on even greater significance in the contemporary conjuncture. Much has already been written about how the SF genre now faces an impasse as the future is increasingly rendered predictable, made knowable in advance.[3] But just as SF now must operate in an environment after the so-called end of the future, an argument can be made that Japanese studies today continues in an environment after "Japan" has ceased to be all that meaningful as an organizing or disciplinary category of analysis, what with all the conceptual baggage of the nation form, the Cold War backdrop of the intellectual genealogy of area studies, and the logic of Orientalism that structures the field. Against this backdrop, it would therefore appear that the question at hand is what happens to the phenomenon of techno-Orientalism that links these fields of study together when the historical conditions that gave rise to it—most of all, as David Morley and Kevin Robins highlight when they first coined the term, the economic challenge Japan posed to the United States during its postwar growth culminating in the 1980s economic bubble—are no longer in operation.[4]

Relevant here are some recent theoretical work and cultural criticism for articulating a politics of posthumanism, the most noteworthy of which would be Sarah Juliet Lauro and Karen Embry's "Zombie Manifesto," which advances the claim that the zombie is the successor to Donna Haraway's cyborg, and that it is a more properly posthuman figure appropriate to the contemporary conjuncture.[5] Such a shift from the cyborg to the zombie as the central critical object of inquiry brings out some potential consequences to the critical examination of Japanese SF in the context of the techno-Orientalist gaze. After all, the cyborg has so often served as a key conceptual framework for understanding the cultural politics of Japanese SF, often used as a device for apprehending such concepts as techno-Orientalism, postwar hybridity, and the intersections of gender politics and technology. As such, if there is indeed a turn from the cyborg to the zombie, it seems only proper to pose the question of what perspectives might become visible through an examination of Japanese SF oriented around the concept of the zombie.

It is with this question in mind then that I examine the work of the late Japanese SF author Project Itoh (Itō Keikaku, 1974–2009) alongside recent animated adaptations of his work in this essay.[6] While Project Itoh, the literary pseudonym of Itō Satoshi, may not be all that well known outside of the Japanese SF community, it would be difficult to overlook his impact within the SF genre in Japan, despite the fact that he completed only two original novels during his short writing career—*Genocidal Organ* (*Gyakusatsu kikan*, 2007) and *Harmony* (*Haamonii*, 2008)—with a third—*Empire of Corpses* (*Shisha no teikoku*, 2012)—published posthumously. Although his career was cut short by his untimely passing in 2009 after a long struggle with cancer, numerous works inspired by Itoh have since appeared, including the multivolume *Project Itoh Tribute* anthologies.[7] Beyond these tributes, though, perhaps the more telling sign is the emergence of the term "post–Project Itoh" as a periodizing marker of the contemporary conjuncture of Japanese SF.[8] For my part, I want to link this positioning of Project Itoh with a feature of his body of work that strikes me as noteworthy, namely, the employment of the figure of the zombie across his writings. This has taken all manner of forms, everything from the mass-murdering hordes, to philosophical zombies that appear as beings without interiority, to more traditional resurrected corpses serving as soldiers and industrial laborers. Reading the work of Project Itoh through the lens of the zombie opens up a space to properly historicize his writing within larger shifts in the cultural imaginary, to unearth the broader structure of feeling underpinning the popularization of his work, and, in so doing, to raise the issue of whether what has been called post–Project Itoh SF might be more productively understood as postcyborgian SF.

This question becomes only more relevant in the case of the animated adaptations of Project Itoh's novels, which were released as a trilogy of animated films from 2015 to 2019 from three different studios and three different directors, given how much of the writing on Japanese animation all too often emphasizes its role as a bearer of cultural allegory, often manifesting in the form of a cultural hermeneutics organized around a techno-Orientalizing gaze focalized through the figure of the cyborg. Relevant here is Thomas LaMarre's criticism that while much of the existing writing on anime often deals with images of technology—cyborgs, giant robots, virtual reality, and so on—these discussions need to be further extended to provide an account of the "animetic": how animation itself is a media apparatus that conditions forms of visual movement, practices of spectatorship and consumption, and by implication ways of thinking technology at a more fundamental level distinct from other media technologies (such as the cinematic).[9] To this point, I would add that animation itself is arguably generative of literal zombies, of animated images that can simulate human affect and intelligence. As such, it seems to me then only apt for the discussion of the conceptual passage from the cyborg to the zombie to articulate its theoretical implications not only in terms of narrative, but also at the level of the interface with the animetic apparatus itself. In other words, it becomes also necessary to consider the question of how to apprehend the figure of the zombie not only as a representational object of inquiry within the texts under examination, but as a critical prism for articulating the stakes of visualizing the nonhuman Other by raising the issue of our own practices of spectatorship, our own visual pleasure in consuming mass-produced animated bodies in the aftermath of the techno-Orientalist habits of looking.

Project Itoh's Philosophical Zombies

Within Project Itoh's body of work, it is the posthumously published *Empire of Corpses* that most visibly employs the zombie as a motif and therefore makes for a productive starting point for this discussion. *Empire of Corpses* explores an alternate nineteenth-century world populated by mass-produced animated bodies without human subjectivity or interiority, brought about by the invention and mass adoption of a form of resurrection technology. This version of history diverges from our own primarily through the extrapolation of the material and social changes that would ensue when historical SF texts—in this case, primarily (but not limited to) Mary Shelley's *Frankenstein* (1818)—are treated as actual historical events. In *Empire of Corpses*, Dr. Frankenstein's successful resurrection of a corpse brings about massive social upheavals as the subsequent development of his discovery enables the mass production of animated corpses to serve as soldiers for the British Empire as well as a cheap and docile industrial labor force. From this overarching premise, the story then follows its narrator, a young John Watson (of Sherlock Holmes fame), as he searches for the original notes of Dr. Frankenstein, which are rumored to contain the secret to the proper re-ensoulment of a corpse. In his adventures, Watson is joined by a motley crew from all manner of sources. Friday, from Daniel Defoe's *Robinson Crusoe* (1719), has been reanimated from the dead and serves as the group's scribe. Frederick Burnaby (a historical military adventurer from the British Empire) is assigned to Watson as his bodyguard. Other characters they encounter at various points include the robot Hadaly from Auguste Villers de l'Isle-Adam's *L'Eve Future* (1886) and Nikolai Krasotkin and Alexei Karamazov from Fyodor Dostoyevsky's *The Brothers Karamazov* (1879–1880), among others.

Evident from this brief description of the world of *Empire of Corpses* is how much the idea of "reanimation" permeates all aspects of the work At the most basic level, there are, of course, the depictions of corpse reanimation technology as an increasingly normalized part of everyday life at the level of the story, a detail hinted at as early as the novel's prologue, which opens with some observations about the rise of incidents of thefts of cadavers and grave robbing stemming from a shortage of available corpses. This highlights how much the world of *Empire of Corpses* runs on corpse labor, how much the "nation's free market economy is underpinned by corpses."[10] Later in the scene, when the technology of corpse reanimation is discussed, it is revealed that while all corpses are reanimated via the installation of punch card–based basic software, on top of this, more specialized programming modules encoded with job-specific skills can also be added, suggesting that the use of corpse labor is sufficiently widespread across a range of industries.[11] In this regard, the titular corpses of the novel are less akin to versions of the zombie that are supernatural undead beings or even more modern iterations that give them a disease-based origin, but more like programmable organic automatons.

However, this notion of reanimation is not something that features only as a narrative device at the level of the story of *Empire of Corpses*. In certain respects, it is the logic that organizes the novel as a whole, rendering it into a zombie text at multiple levels. To begin with, the premise of the novel is the reanimation of various public domain literary characters to be repurposed for its own alternate history narrative. Beyond this, the novel in itself could be understood as a kind of reanimation of Project Itoh's writing itself, in a manner comparable to how Watson tasks his deceased and reanimated friend Friday with the role of his group's scribe, documenting their adventures as if serving as the text's implied author. After all, at the time of his passing, only some thirty or so pages, mostly from the novel's prologue, had been written. It was left to his friend, SF author and 2012 Akutagawa Prize winner Enjō Tō (1972–), to complete the novel, giving some credence to the argument that *Empire of Corpses* is really more a work of Enjō's than of Project Itoh's.[12] Indeed, such a sentiment is only reinforced by the world of the novel itself, which features a plethora of intertextual references to historical figures and characters from classic SF works alongside a metafictive tendency, both of which are known hallmarks of Enjō's style. Nonetheless, I believe that there is merit in reading *Empire of Corpses* as a Project Itoh work, as doing so allows for the highlighting of the threads of continuity running through his major works of fiction. Indeed, I would go so far as to say that in this act of reanimation of Itoh, *Empire of Corpses* has the effect of retroactively calling attention to the centrality of the zombie to his explorations of the posthuman across his body of work. While this interest may only appear in a disguised or sublimated form in Itoh's earlier work, its importance becomes retroactively more visible when it is reconsidered in light of the explicit employment of the zombie in *Empire of Corpses*.

Consider, for example, the case of Itoh's second novel, *Harmony,* which shares a similar interest in questions of posthuman consciousness while also subtly gesturing toward the figure of the zombie. Written at a time when Itoh was in and out of hospitals for cancer treatment, the novel initially appears to be a satirical dystopian narrative about a world whose functions of governance have been taken over by the World Health Organization (WHO), which operates on an ideology of Lifeism, defined as a "politically enacted policy or tendency to view the preservation of health to be an admedistration's [*sic*] highest responsibility," in a move that reads like a literalization of Michel Foucault's concept of "biopolitics."[13] At an ideological level, this involves shifts in social attitudes that have made the consumption of products like alcohol, caffeine, or tobacco essentially cultural taboos. But beyond these changes in lifestyle, at a more brute technological level is the effective elimination of all disease through the provision of a real-time health monitoring network, through which the health and well-being of every citizen is subject to constant monitoring at the molecular level. This technology is known as "WatchMe," which consists of swarms of networked nanomachines implanted into the bloodstream of every citizen upon reaching adulthood. In tandem with the household medcare units capable of manufacturing whatever cocktail of medicine is required, WatchMe regulates a person's health and reports it to a medical server, allowing for not only the maintenance of everyone's bodies in a disease-free state but also rapid interventions in the event of an emergency.

As the plot of the novel unfolds, however, Itoh's interest in programmable posthuman consciousness manifests itself. The story follows the protagonist Tuan Kirie as the WHO tasks her with investigating a recent wave of simultaneous mass suicides with victims numbering in the thousands. During her investigations, Tuan discovers that the WatchMe network is far more pervasive than is publicly known. It not only regulates physiological functions; more importantly it includes a hidden function for regulating human behavior as well. As a consequence, the system can also be hacked so as to generate and manipulate affective responses (e.g., producing a desire to die). At the heart of the plot is her old friend Miach, who seeks to create a crisis of a wide enough scale to compel the WHO to activate the latent fail-safe program called "Harmony," so named because it would deploy the WatchMe network to literally harmonize competing human desires, with the consequence of ultimately removing the very need for a human consciousness that evolved merely as a mechanism for mediating the feedback from these desires. But by the end of the story, this is exactly what happens. The outcome is a society populated by zombies, not of the kind often seen in horror films, but what philosopher Robert Kirk conceptualized as philosophical zombies (or p-zombies), hypothetical beings with no consciousness or interiority but who are otherwise indistinguishable from human beings in their external behavior.[14]

If there is a common thread running through both texts that can be identified here, it is their repeated meditations around concepts of human consciousness and subjectivity. And it is primarily through this figure of the so-called p-zombie that these ideas are explored. Of course, strictly speaking, the concept of the p-zombie has little to do with the zombie of popular cinema and culture. On the contrary, p-zombies—individuals who are capable of expressing signs of language and intelligence but are otherwise devoid of any interior sentience or consciousness—are largely meant to be intellectual constructs, purely hypothetical beings postulated as thought experiments in the fields of cognitive science and philosophy of the mind.[15] That said, I believe there is merit in drawing such a conceptual connection in the case of Project Itoh's work. In part, this is because it allows us to extend Itoh's meditations on posthuman conceptions of consciousness and place it into conversation with a substantial body of cultural criticism that has taken a keen interest in theorizing the figure of the zombie in recent years.

In their essay "A Zombie Manifesto," Sarah Juliet Lauro and Karen Embry argue that the zombie is a conceptually rich figure for articulating the fundamental irreconcilability of capitalism and humanism. Borrowing their title and building upon the ideas from Donna Haraway's influential essay "A Cyborg Manifesto" (1985), Lauro and Embry employ the figure of the zombie to articulate what they see as the limits of the concept of the cyborg as a model for posthumanist thought. What distinguishes the zombie from the cyborg conceptually is its paradoxical quality that allows for the possibility of utter negation. In other words, if the cyborg introduces a hybridity through an additive process (in other words, human *and* machine), the zombie's hybridity is structured around a negative dialectics, that is, *neither* alive *nor* dead. In their words, "Haraway's 'Cyborg Manifesto' sought to resolve the antagonism between subject and object binary by reimagining the chasm between the two through the hybrid. . . . We contend that the only way to accurately model a *post* human state is the 'neither/nor' of the zombie, which rejects both subject and object categories and is irreducible, anticathartic, antiresolution, and working in the mode of negative dialectics."[16] So, for Lauro and Embry then, it is this emphasis on negation that serves as the locus of politics for the figure of the zombie. Rather than offering it as an alternative liberatory figure to the cyborg, they suggest that its significance lies in how it marks the impasse of the cyborg as one that still centers the human subject as its baseline, in contrast to the zombie's radical negation of it.

Along similar lines, albeit with a more specific historical claim, Eric Cazdyn, in his book *The Already Dead* (2012), notes the seemingly increasing inadequacy of the cyborg as a critical prism in recent years, writing that "there was a crucial limit in these cyborg narratives from the 1980s and 1990s. For example, although the cyborgs necessarily exceeded the human subject in terms of the technologies of the body, their ideological investments (in terms of gender identification and sexual desire, in particular) were still deeply rooted in the human."[17] In place of the cyborg then, Cazdyn suggests that what marks the present moment are profound transformations in the experience of temporality, which he characterizes as an emergence of a "new chronic." In alignment with the emerging interest in the temporality of disability, Cazdyn calls attention to the changes taking place in medical practice wherein the interest in finding a singular cure has been superseded by a greater attention placed on stabilization, on controlling symptoms without in fact removing the disease.[18] For Cazdyn, this marks a move away from the logic of the terminal (whether in the form of a cure or death) to that of the chronic, a shift that extends beyond just medical practice to encompass broader changes in the lived experience of time that implicates other domains such as culture and politics, most significantly in how the logic of the chronic forecloses forms of political action that radically transform existing historical conditions, such that it "effectively colonizes the future by naturalizing and eternalizing the brutal logic of the present." Within such a regime, political consciousness moves away from the cyborg's embodiment of hybridity and new identity formations to what he terms "the already dead," the "state when one has died but has yet to be killed."[19]

Zombie Animetics

It is with this critical context in mind that attending to the animated adaptations of Project Itoh's work becomes most productive. These adaptations, all produced and publicized as a trilogy by different animation studios, were released in reversed order from their source novels, with Wit Studio's *Empire of Corpses* appearing first in 2015, followed by Studio 4°C's *Harmony*, and concluding with Geno Studio's *Genocidal Organ* in 2017.[20] Given how the cyborg has so often been understood as functioning in the service of a metacommentary on the artificial bodies of animated characters, accounting for this shift from the cyborg to the zombie makes for an effective critical prism for articulating the broader implications for thinking of the animated body as itself a kind of literal zombie. Beyond this, what is interesting specifically about the animated adaptations of

Project Itoh in this regard is how the visualizations of their worlds and characters seemingly stage precisely this passage from the cyborg to the zombie that Lauro and Embry and Cazdyn each discuss in their respective writings.

For instance, the adaptation of *Harmony*, including the various approaches it adopts to visualize the WatchMe system at the heart of its narrative, is instructive here. In the source novel, a visible trace of the constant presence of the WatchMe system manifests in the language of text of itself, which makes use of a hypertext-like markup called ETML: "emotion-in-text markup language." Not unlike any other markup system (such as HTML), ETML incorporates annotations that tag metafunctions like descriptions, operations, or textual structures. With ETML, though, these tags do not just include those marking recollections, lists, or external links to dictionary definitions and background music, but also extend to markers for activating affective responses like anger, panic, sentiment, and even sarcasm.

In the animated adaptation, however, this textual style is interpreted in two ways. For much of the animation, this code is treated as a representation of the constant health surveillance of the swarms of nanomachines that constitute the WatchMe system. This takes the visual form of a kind of augmented reality system that appears almost like an experience of information overload, as seen, for example, in a sequence from the protagonist Tuan's point of view when she first arrives in Japan near the beginning of the story. Here, we find a typical example of a machine-mediated mode of visuality, taking on a set of characteristics that seemingly enacts the kind of networked, hypermediated quality closely associated with the figure of the cyborg (Figure 13.1). Yet in the scenes of the prologue

FIGURE 13.1 Point-of-view shot of Tuan Kirie as she arrives in Japan, showing the augmented reality interface of the WatchMe system. From *Harmony* (*Haamonii*, dir. Michael Arias and Nakamura Takashi, 2015). (Source: *Harmony*, FUNimation Productions, 2016, DVD.)

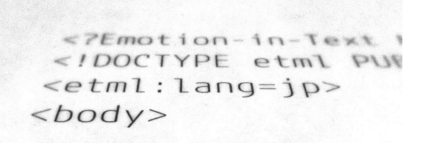

FIGURE 13.2 Prologue from *Harmony* (*Haamonii*, dir. Michael Arias and Nakamura Takashi, 2015), showing raw etml code. (Source: *Harmony*, FUNimation Productions, 2016, DVD.)

and epilogue of the animation, both of which take place following the activation of the Harmony program and the erasure of human consciousness altogether, we get a different picture of the ETML code. Here, we see none of the previous visual strategies of hypermediation. In its place, we see only pure unprocessed ETML code straight from the text of the novel appear directly on screen, displayed on a plain white obelisk, as if staging the negation of the visuality of the adaptation (Figure 13.2).

In contrast, *Empire of Corpses* takes a different, perhaps more conventional approach in giving visual expression to the passage from the cyborg to the zombie as the dominant cultural metaphor for the contemporary conjuncture. Take, for instance, the way the adaptation visually establishes its steampunk setting, which largely makes use of a retrofuturistic visual aesthetic in its mechanical designs, ranging from the punch cards that serve as the storage medium for the corpses' necroware to the oversized gears and mechanical typewriter keyboards of various pieces of machinery. Yet an interesting facet of these designs is the extent to which they are indebted to the iconography of the cyborg. For example, necroware is installed into corpses via cables to plug into the backs of their necks in a manner reminiscent of the classic cyberpunk imagery of jacking into a computer (Figure 13.3). The notes of Victor Frankenstein at the center of the story turn out to be not just some paper notebook, but a complex punch card–based data storage medium in the animation. One could almost be forgiven for thinking that the zombies of *Empire of Corpses* are little more than reskinned, retrofuturistic cyborgs.

By linking the iconography of the cyborg to the zombie via this retrofuturistic aesthetic and setting, *Empire of Corpses* seemingly suggests that rather than simply coming after the cyborg as a posthuman figuration, the zombie historically precedes it, and perhaps even serves as its condition of possibility. In doing so, it also arguably stages a metacommentary about the problem of animation itself as a media technology generative of literal zombies, of animated images

FIGURE 13.3 Corpse laborers linked via cybernetic control cables. From *Empire of Corpses* (*Shisha no teikoku*, dir. Makihara Ryōtarō, 2015). (Source: *Empire of Corpses*, FUNimation Productions, 2016, DVD.)

without interiority that can nonetheless simulate the appearance of human affect and intelligence. Relevant here is Alan Cholodenko's argument, from his essay "First Principles of Animation," that what characterizes the foundations of the technology of animation is an "inextricable, deconstructive commingling of animism and mechanism," that is, the comingling of movement and life.[21] For Cholodenko, in other words, it is in the interface between technology and life itself—in effect, between what is embodied by the cyborg and the zombie—that the animatic apparatus—or what he also terms the "animatic automaton"—resides. In his words, "Animation—the simultaneous bringing of death to life and life to death—[is] not only a mode of film (and film a mode of it) but the very medium within which all, including film, 'comes to be.'"[22]

I believe that this fundamental property of the technology of animation is precisely what the figure of the zombie-as-cyborg (or cyborg-as-zombie) in the animated adaptation of *Empire of Corpses* brings into focus. Consider, for example, the opening sequence of the animation, which differs from the novel in that instead of using the scene of a medical school lecture to introduce the corpse reanimation technology, it instead offers a scene of the protagonist John Watson in his private laboratory employing a technological apparatus to give life—to literally reanimate—his friend Friday. He plugs a cable into Friday's neck to install his necroware, the software that powers the corpses of this world (Figure 13.4). Notably, what marks the success of the procedure is movement: Friday transforms from an inanimate corpse to an animated body.

Immediately, this presents a problem, however, as while ostensibly, within the story-world of *Empire of Corpses,* Watson is alive and Friday is dead, they are still

FIGURE 13.4 Watson stands before Friday as he tests the newly installed necroware. From *Empire of Corpses* (*Shisha no teikoku*, dir. Makihara Ryōtarō, 2015). (Source: *Empire of Corpses*, FUNimation Productions, 2016, DVD.)

both, in practice, animated beings. As Thomas LaMarre has put it in his discussion of this work, "There are two human beings, largely indistinguishable in terms of physical appearance, yet one is supposed to be living, while the other is dead. The scene aims to convey a distinction between life and death, yet, since this is animation, the distinction doesn't readily take hold."[23] LaMarre further notes that the primary technique for distinguishing the living from the dead in the animation is through character movement. While Watson is animated conventionally through limited animation techniques, Friday is given jerky, uncoordinated movements. Paradoxically, then, it is excessive movement, excessive liveliness that characterizes the visualization of the zombie characters. This has the ultimate effect of calling attention to the flatness, to the artifice of the character animations in general, metafictively foregrounding their ontological status as literal animated bodies, as literal zombies. As LaMarre puts it, "Both Watson and Friday are, in fact, animated beings: depending on one's viewpoint, they may be considered alive, dead, or both at once, or something else altogether."[24] Indeed, as if to punctuate the point, *Empire of Corpses* concludes with Watson, having retrieved Victor's notes, repeating the procedure from the opening scene, albeit with himself as the subject this time, effectively switching places with Friday, in order to seal the information of the notes within his own mind while also wiping out his previous memories and consciousness.

Necro-Orientalism

The question at hand then is, where might a shift in emphasis from a model of media theory organized around the cyborg to an analysis of zombie animetics lead as far as its implications for techno-Orientalizing discourses in the contemporary conjuncture? Relevant here is Steven Shaviro's discussion of the of the classic zombie films of George Romero, drawn from his conceptualization of affective spectatorship and the cinematic body. For Shaviro, what is distinct about zombies is their visceral and excessive corporeality, that they are "in a sense all body: they have brains but not minds, ... lumps of flesh that still experience the cravings of the flesh, but without the organic articulation and teleological focus that we are prone to attribute to ourselves and to all living things."[25] This characteristic of the figure generates a tendency toward the short-circuiting of cinema's mechanisms of identification with the images on the screen, replacing its more cognitive, psychoanalytic dimensions with a purely affective, purely corporeal form of spectatorship. The spectators are reduced to watching in horror, themselves effectively becoming a corporeal cinematic body who can respond affectively but not cognitively. In other words, it engenders the realization that, in the words of Slavoj Zizek, "at the most elementary level of our human identity, we are all zombies."[26]

In certain respects, the zombie film in Shaviro's discussion only reveals what is already present as a potentiality in all animation. As Thomas LaMarre reminds us, animetic spectatorship (in contrast to cinematic spectatorship) is organized around a distributive field of visuality: whereas cinema operates on the principle of valorizing the illusion of depth through the construction of perspective, and the placement of the spectator in a transcendental subject position, with Japanese animation, the flat aesthetics and compositing generate a practice of viewing that operates on the basis of a dehierarchized visual field.[27] The effect of this superflatness is to constitute a spectator whose practice is organized not around visual focal points but instead around a kind of information scanning or, in the term of Azuma Hiroki, a form of database consumption. For Azuma, the characteristic feature of database consumption is a process of desubjectification, a process of animalization wherein the spectatorial response is reduced to something purely affective, purely behavioral.[28] Animetic spectatorship, in other words, has always, at some level, exhibited the qualities Shaviro associates with zombie films in its obsessive corporeality and affective animality. Indeed, it is perhaps in this experience of animalization, of becoming little more than an animated body stripped of all subjectivity, where the image of the zombie finds its resonance. Insofar as the fascination with the zombie as a cultural figure can be located in its allegorization of the dehumanizing character of the labor process, this points toward a recognition that this capitalist logic of exploitation also operates at the spectatorial level.[29]

Yet despite this argument that animetic spectatorship reveals the foundation for cinematic spectatorship, it must be pointed out how such a recognition tends to be occluded via the racialization of animation, allowing for it to be projected as something other. As Sianne Ngai has pointed out, animation or animatedness is implicitly racialized, especially around notions of an excess or lack of animation.³⁰ Indeed, this is a point that only becomes more significant in light of how the existence of the term "anime" itself in the English language to designate a specifically Japanese animation suggests that there is something so fundamentally distinct about it, whether in aesthetic or cultural terms, to warrant its own separate name.³¹ As such, if the animated bodies of Japanese popular culture are to be read as instances of the zombie, then just as the logic of capital produces vast swaths of unevenness, it is still imperative to keep in mind that such processes of spectatorial zombification also do not take place on an even terrain, that it potentially engenders a necro-Orientalism analogous to the techno-Orientalism of the cyborg. Parallel to how Donna Haraway suggests that the cyborg is intrinsically linked to the third world woman (as a consequence of industrial production migrating to East Asia), if we are to take the figure of the zombie seriously, then it also necessary to recognize that the zombie is an already racialized entity.³²

If techno-Orientalism remade Japan into the image of the future, we might say that this passage into a new epoch does not so much negate this image as transform it. No longer is the future that Japan represents visible in the neon-lit skyscrapers and street cool of bubble economy–era cyberpunk. No longer does it represent the economic threat it once did since its entry into its seemingly never-ending lost decades. Instead, it premediates a paradoxical future end of futurity, a future marked by stagnation and decline, by that sense of an endless everyday, to use the terms of Miyadai Shinji.³³ In this regard, while the coincidence of the timing of Itoh's passing with the aftermath of the 2008 global financial crisis is certainly just that, a coincidence, it is nonetheless an interesting one. The so-called post–Project Itoh era of Japanese SF might be understood as marking the passage from the cyborg to the zombie as the figuration of the posthuman. It is perhaps no accident that this passage should coincide with an economic crisis that should have properly marked the end of the neoliberal model of globalization, yet nevertheless did not. Instead of ending, it has somehow continued to trudge on for another decade in a kind of zombie neoliberalism. The challenge at hand, therefore, is to map out how the zombie's politics of negation might impact any new imaginaries of Asian bodies under these new historical circumstances.

Crucially, what distinguishes the treatment of the zombie from that of the cyborg in so many of its fictional representations is how it most often functions as the object of mass violence. Gerry Canavan put it succinctly when he observes that "the zombie's mutilation is not one that we easily imagine for 'ourselves,' however that 'we' is ultimately constituted; the zombie is rather the toxic infection that must always be kept at arm's length. Because zombies mark the demarcation between life (that is worth living) and unlife (that needs killing), the evocation of the zombie conjures not solidarity but racial panic."³⁴ Viewed in these terms, the shift from the cyborg to the zombie as the racialized figuration for posthumanism carries troubling implications, for it seems to signal a shift back from what Etienne Balibar and Immanuel Wallerstein have termed "inclusive" racism to "exclusive" racism.[35] Against the backdrop of an emerging new Cold War in Asia and the spike in anti-Asian violence since the beginning of the COVID-19 pandemic, perhaps it is no accident that contemporary iterations of the zombie in film and television often mark the figure as a diseased body whose origins are often located elsewhere, with East Asia being typically named as the source of contagion and, by implication, a potential site of violence.

In the end, it is their identification with the zombie, their refusal to treat the zombie as an absolute other to be exterminated (as so many zombie films do), that is most important about the works of Project Itoh. The conclusion of *Harmony* provides an instructive illustration of this gesture. Following the activation of the Harmony program, the novel concludes with an epilogue that reveals that the preceding story is itself nothing more than a fictive reconstruction of the events prior to the activation of the Harmony program. The significance of this move can be clarified by reading it in conjunction with the explanations for the presence of the ETML code interspersed throughout the text that the epilogue offers. They are intended to function as markup signaling the simulation of affective responses in the presumed post-Harmony reader. The immediate effect of this is to align the reader with the zombie, that is, the subject position of a post-Harmony human who no longer has any interior consciousness that is not already coded in advance. Through this move, the novel suggests that it is not sufficient to struggle against zombification, to simply refuse to become a zombie (as portrayed by so many zombie narratives) at the expense of those who are no longer able to make such a choice. Rather, the more radical strategy of refusal in the political imaginaries of SF must embrace, must be accountable to—and in solidarity with—those who are already reduced to animated bodies, those who are already mediated as zombies.

Notes

1 Istvan Csicsery-Ronay, *The Seven Beauties of Science Fiction* (Middletown, CT: Wesleyan University Press, 2008), 2.
2 Baryon Tensor Posadas, "Beyond Techno-Orientalism: Virtual Worlds and Identity Tourism in Japanese Cyberpunk," in *Dis-Orienting Planets*, ed. Isiah Lavender (Jackson: University Press of Mississippi, 2017), 157.
3 Steven Shaviro, "Unpredicting the Future," *Alienocene*, stratus 1 (2018), https://alienocene.files.wordpress.com/2018/04/unpredicting-to-print.pdf.
4 David Morley and Kevin Robins, *Spaces of Identity: Global Media, Electronic Landscapes, and Cultural Boundaries* (New York: Routledge, 1995), 141.
5 Sarah Juliet Lauro and Karen Embry, "A Zombie Manifesto: The Nonhuman Condition in the Era of Advanced Capitalism," in *Zombie Theory: A Reader*, ed. Sarah Juliet Lauro and Karen Embry (Minneapolis: University of Minnesota Press, 2017), 398.
6 Seeing as how it has been officially used as the English translation of his nom de plume, I have opted to use "Project Itoh" here instead of his Japanese pseudonym Itō Keikaku.

7 *Itō Keikaku toribyūto* [Project Itoh tribute], vol. 2 (Tokyo: Hayakawa shobō, 2015).
8 See, for example, Akira Okawada, *Sekai naisen' to wazuka na kibō* ["Global civil war" and a glimmer of hope] (Tokyo: Shoen shinsha, 2013), 8. Another example is the edited volume *Posuto-hyūmanitiizu: Itō Keikaku ikō no SF* [Post-humanities: SF after Project Itoh], ed. Genkaiken collective (Tokyo: Nan'undō, 2013), which carries the subtitle "SF after Project Itoh."
9 Thomas LaMarre, *The Anime Machine: A Media Theory of Animation* (Minneapolis: University of Minnesota Press, 2009), xxxi.
10 Itō Keikaku and Enjō Tō, *Shisha No Teikoku* [Empire of corpses] (Tokyo: Kawade shobo, 2012), 10.
11 Itō and Enjō, *Shisha No Teikoku*, 20.
12 Itō and Enjō, *Shisha No Teikoku*, 526.
13 Project Itoh, *Harmony* (San Francisco: VIZ Media, 2010), 42.
14 Robert Kirk, "Sentience and Behaviour," *Mind* 83, no. 329 (1974): 43–60.
15 David J. Chalmers, *The Conscious Mind: In Search of a Fundamental Theory* (New York: Oxford University Press, 1997).
16 Lauro and Embry, "A Zombie Manifesto," 400.
17 Eric Cazdyn, *The Already Dead: The New Time of Politics, Culture, and Illness* (Durham, NC: Duke University Press, 2012), 76.
18 See, for example, Sarah Lochlann Jain, "Living in Prognosis: Toward an Elegiac Politics," *Representations* 98, no. 1 (2007): 77–92; and Ellen Samuels, "Six Ways of Looking at Crip Time," *Disability Studies Quarterly* 37, no. 3 (August 31, 2017), https://doi.org/10.18061/dsq.v37i3
19 Cazdyn, *The Already Dead*, 6.
20 *Empire of Corpses*, directed by Makihara Ryōtarō (Wit Studio, 2015); *Harmony*, directed by Michael Arias and Nakamura Takashi (Studio 4°C, 2015); *Genocidal Organ*, directed by Murase Shūkō (Geno Studio, 2017).
21 Alan Cholodenko, "'First Principles' of Animation," in *Animating Film Theory*, ed. Karen Beckman (Durham, NC: Duke University Press, 2014), 101.
22 Cholodenko, "'First Principles' of Animation," 104.
23 Thomas LaMarre, "Animation and Animism," in *Animals, Animality, and Literature*, ed. Brian Massumi, Bruce Boehrer, and Molly Hand (New York: Cambridge University Press, 2018), 294.
24 LaMarre, "Animation and Animism," 294–295.
25 Steven Shaviro, *Cinematic Body* (Minneapolis: University of Minnesota Press, 1993), 86.
26 Slavoj Zizek, "Discipline between the Two Freedoms: Madness and Habit in German Idealism," in *Mythology, Madness, and Laughter: Subjectivity in German Idealism*, ed. Markus Gabriel and Slavoj Zizek (London: A&C Black, 2009), 100.
27 LaMarre, *The Anime Machine*, 110.
28 Hiroki Azuma, *Otaku: Japan's Database Animals* (Minneapolis: University of Minnesota Press, 2009).
29 See Jonathan Beller, *The Cinematic Mode of Production: Attention Economy and the Society of the Spectacle* (Dover, NH: Dartmouth College Press, 2006), 1.
30 Sianne Ngai, *Ugly Feelings* (Cambridge, MA: Harvard University Press, 2007), 95.
31 Patrick W. Galbraith and Thomas Lamarre, "Otakuology: A Dialogue," *Mechademia* 5, no. 1 (2010): 362.
32 Donna Haraway, *Simians, Cyborgs, and Women: The Reinvention of Nature* (New York: Routledge, 1990), 166.

33 Miyadai Shinji, *Owaranaki Nichijô o Ikirô* [Live the endless everyday] (Tokyo: Chikuma Shobô, 1995).
34 Gerry Canavan, "'We Are the Walking Dead': Race, Time, and Survival in Zombie Narrative," *Extrapolation* 51, no. 3 (January 2010): 433.
35 Etienne Balibar and Immanuel Wallerstein, *Race, Nation, Class: Ambiguous Identities* (New York: Verso, 1991), 39–40.

14

Settler Orientalism, Asian American Techno-Environmentalism

―――――――――――◆◇▶―

The Network Novel under Japanese and U.S. Empires

ADHY KIM

Techno-Orientalism, a Western-centric technology of racial and chronological differentiation,[1] reveals itself to be a politics of land occupation. If North American settler Orientalism, in the words of Juliana Hu Pegues, "holds the promise of a non-Native future, a future that cannot include either Asian or Native bodies," techno-Orientalism perpetuates settler colonial futures in the Americas by using white supremacy as the grounding premise and demographic filter by which Asian presence becomes alien, artificial, and threatening.[2] Techno-Orientalism is a colonial strategy of representation that, like settler Orientalism, casts Asian presence and futures as uniquely undesirable for American ecosystems, environments, and geographies. The anti-Asian animus of Orientalism and techno-Orientalism arises from the naturalization of white nativism—an outgrowth of settler colonial capitalism that misrecognizes Asian alien labor as an embodiment of the "abstract evils of capitalism" and the "environmental ruination of the frontier."[3]

Techno-Orientalism's embrace of the white proprietary possession of Indigenous lands extends its logic to U.S. imperialism overseas. In identifying the

Orientalist naturalist tendency to configure Asianness through the "undecidable relationship between ... human and animal or animal and machine," Colleen Lye points to the early twentieth-century context of the United States's great-power competition and strategic diplomacy with China and Japan: namely, the U.S. search for an extended capitalist frontier throughout the Pacific and Asia forced the continental and transpacific empire to reckon with the "nightmare of [Asian] nonidentity" and the prospect of reversible white-Asian hierarchies.[4]

Throughout the U.S.-Japan interimperial competition for dominance over Asia and the Pacific, the Japanese/American body occupied the interstices of colliding imperial expansionisms, colonial occupations, and "frontiers." As Jodi Byrd writes, the figure of the Japanese American undergoing World War II removal and incarceration "[was] made to bear [the identity of] both cowboys and Indians" within the racial biopolitics determining exclusion from and/or inclusion into the U.S. national body, while Indigenous peoples on the continent continued to be rendered as aliens on their own lands.[5] Japanese Americans' wartime dispossession, expulsion, and containment laid bare the "condition of being disposable" that structurally enjambed their alienation with the colonial drive to eliminate Native peoples.[6] At the same time, though, Japanese Americans themselves participated in the "distortive parallactic effects" of Japanese and U.S. imperial policies enacted to deny Indigenous sovereignty.[7]

Historical memories of interimperial conflict between Japan and the United States raise the ongoing question of how East Asian bodies are racially signified in the Western Hemisphere in terms of their claimed proximity to or distance from settler and Indigenous positionalities. Techno-Orientalism traffics in these racial significations by casting the Asian American as antithetical to the white settler and therefore open to dehumanization. Indeed, the approximation of dispossessed Japanese Americans to "Indianness" reveals the dehumanization embedded in the ecological Indian trope while it furthers the erasure of Native sovereignties by turning Asian and Indigenous peoples into interchangeable racial minority categories.[8]

In the face of techno-Orientalism's dehumanizing effects, Michelle Huang proposes "revers[ing] the metaphor" of inhuman Asians by plumbing the posthumanist possibilities found in "the divisions between and intermixing of the human and nonhuman"; in Huang's words, "narrative features [like] nonhuman characters, environments, [and] technology can illuminate different elements of Asian American racialization."[9] Following Huang's insight, I use the term "techno-environmentalism" to identify Asian American attempts to transcend techno-Orientalism through speculative fiction. Techno-environmentalism casts Asian Americans not as dehumanized embodiments of modernization's ecological destruction but as more-than-human bodies that coextend with ecosystems, hemispheres, and planets. This nonrealist mode of Asian American worldmaking seeks to revise the "space-time dimensions" of settler colonialism whereby

Asian migrants were considered modern laboring subjects yet could not properly occupy settler space.[10]

But such environmental revisions of techno-Orientalism pose potential problems when the "space-time dimensions" of settler colonialism are incompletely addressed. Insofar as techno-Orientalism is about the "project of modernity," techno-environmentalism accepts the Asian American subject as constitutive of modernity while confronting both the fantasy and failure of capitalism to realize universal human "freedom" through the technological development of productive forces.[11] Indeed, Asian American techno-environmentalism may show how speculative fiction gets stuck in time loops of recursive discovery, conquest, and settlement.[12] As I argue, Asian American techno-environmentalism is laced with tensions in that its critique of racism coincides with the tendency to reproduce the capitalist and settler environmentalisms that authorize racial violence. If American space-time colonialism dictates that "Asians can never be 'here' and Native peoples can never be 'now,'" Asian American techno-environmentalism risks recuperating Asians within the time and place of a settler modernity that still hinges on the disappearance of Native peoples.[13]

The peculiar significance of the World War II–era Japanese American—made to emblematize "both cowboys and Indians" in the crosswinds of two empires—bears out with a particular intensity the racial unevenness of capitalist imperialism lasting into the present across Asia and the Americas. This chapter addresses the political problem that endures in the aftermath of the twentieth-century U.S.-Japan imperial conjuncture, formed out of the multiple interactions between U.S. and Japanese settler colonialisms, imperialist wars, and capitalist extractive economies. As Japanese American techno-environmentalism grapples with the pasts, presents, and futures of U.S. Orientalism, it also finds itself at a fraught ideological crossroads between capitalist modernization, anticapitalist critique, and settler colonial apologia.

Highlighting novels by Karen Tei Yamashita and Sequoia Nagamatsu, I ask how Japanese American speculative fiction conceptualizes the meaning of Asian American modernity in the most literal terms: as a geographic signifier of uneven power relations with both Asian and American lands in the ruins of Japanese and U.S. empires. Yamashita and Nagamatsu seem to propose that once-excluded Asian Americans have a role to play in large-scale spatiotemporal sagas contextualizing their human characters within vast nonhuman worlds. In doing so, both authors engage with the literary form of the network novel, which connects different characters across far-flung locales and/or time periods and simulates the simultaneous compression and dilation of space-time afforded by long-distance communication technologies. The network novel is an apt vehicle for techno-environmentalism in that it experiments with how border-crossing technical infrastructures connect disparate characters and locales not just to each other but to the ungraspable totality of the planet; it involves a "participatory process of mapping" that exceeds total comprehension but nevertheless perceives or channels a particular historical consciousness.[14] As Min Hyoung Song claims, the transhemispheric Asian American network novel enacts "a belonging in transitory and deterritorialized spaces" in its imagination of a "planetary becoming."[15] American Orientalism's so-called nightmare of Asian nonidentity is turned on its head such that the modern Asian American subject, historically buffeted between the contested frontiers of imperial borders, might now surrender one's humanism and "find oneself" in a "total belonging" with the place and time of the Earth.[16]

However, as long as the territorial and geopolitical coordinates of settler empires remain unchallenged, Asian American techno-environmentalist network novels fail to correct for the foundational premises of techno-Orientalism. They may either disrupt or perpetuate the ecomodernist myth that capitalist development via technological innovation can be socially and ecologically innocent.[17] But in both its procapitalist or anticapitalist iterations, Asian American techno-environmentalism structurally tends to subscribe to a multiracial version of the myth of the ecological settler who awakens into a sense of belonging by romanticizing "wild" nature and appropriating or erasing Indigeneity in order to possess the land.[18] The Asian American network novel can—to quote Wendy Chun—"resurrect the frontier" in its attempt to map out and navigate an Asian American sense of interconnected historical and environmental reality.[19]

Beyond its literary instantiations, Asian American techno-environmentalism may refer to an ideology spanning the liberal and leftist political spectrum that construes the Asian-led development of productive forces as environmentally and politically redeemable within the colonial landscape, even while one's complicity in colonial projects may be acknowledged. This chapter identifies both the utopian aspirations and colonial reproductions of Asian American techno-environmentalism by considering how Japanese and U.S. colonialisms have shaped the ideological trajectory of the Japanese American network novel. Yamashita's and Nagamatsu's deterritorial speculative fictions dramatize the limits and possibilities of the Japanese American search for alternative spatiotemporal philosophies from within the crucibles of capitalist imperialism and colonial violence.

Through the Arc of the Rainforest and Karen Tei Yamashita's Latin America

Karen Tei Yamashita's 1990 novel *Through the Arc of the Rainforest* satirizes U.S.-led capitalist developmentalism while accounting for a then-burgeoning climate consciousness.[20] Set in Brazil and centering on a mysterious plastic substrate in the Amazon rainforest called the Matacão, the novel links characters of various backgrounds, including a Japanese immigrant, mixed-race Brazilians, an Indigenous subsistence farmer in the Amazon, and a white American corporate executive. The Japanese character Kazumasa Ishimaru emigrates from Japan to Brazil and is accompanied by a strange ball that has orbited his head since childhood—debris from the impact of a supposedly extraterrestrial object that we later learn is made of the same material as the Matacão. As the novel progresses, the Matacão becomes profitable raw material for commodity production due to its miraculous shape-shifting qualities. J. B. Tweep, a New York speculator whose acquisitiveness is symbolized by his three arms, semi-forcibly recruits Kazumasa to use the magnetism of his ball to detect underlying deposits of Matacão along the Amazon basin, which Tweep's corporation subsequently mines. We eventually learn that the Matacão is the accumulated by-product of nonbiodegradable trash that has undergone geologic pressure, pushed as far into the earth as the lower mantle and materializing on the surface in "virgin tropical forest" and other landmasses like Greenland, Australia, and Antarctica.[21] In other words, the Matacão has entered the stratigraphic record.

Kazumasa, as the character most closely tethered to the materiality of the Matacão, coextends with the Matacão's local, global, and planetary characteristics. His companion ball, for one, gives him extraordinary "technical gifts," and the ball is also the novel's omniscient narrator, functioning like a television antenna that tunes into the motley lives of the novel's decentralized cast.[22] The extraterritoriality of the Matacão carries environmental and sociopolitical implications as well. Since it is closely likened to plastic (a synthetic product of reconstituted petroleum), the Matacão's artificiality suggests that Kazumasa's synthetic appearance might also evoke the history of Japanese "alien" status across the Americas: as Aimee Bahng asks, "If the most alien-looking entity in this work of speculative fiction turns out to be of the earth, then can one extrapolate that migrant subjects are not so easily defined as alien or native either?"[23] Kazumasa's role in the novel as the human host for the novel's networked, decentralized narrator speaks to a Japanese American techno-environmentalist premise where the Japanese immigrant becomes both a synthetic and planetary entity: an "alien-looking," deterritorialized communication device that coextends with geologic time. Techno-environmentalism in this case affirms the industrial-grade Japanese body as a hybrid product of fossil fuels—both "alien" and "native" nature.[24]

That the Matacão surfaces in the Amazon's "virgin areas" alerts us, however, to who in fact inhabits these areas, like the novel's Indigenous character Mané, and how the causes and effects of capitalist development turn these inhabited lands into both landfills and frontiers for extraction. Kazumasa factors into these layered histories of land appropriation, exploitation, and degradation as part of the reconstituted plasticity of empire across the historical transitions of the twentieth century. As Quynh Nhu Le incisively points out, Yamashita's novel neutralizes the specter of alien Asianness through settler vocabularies of multiraciality: upon Mané's death Kazumasa settles in the forest with his mixed-race Brazilian wife and establishes plantations of "pineapple and sugarcane, sweet corn and coffee" on the "rich red soil of their land."[25]

Kazumasa's settlement echoes Japanese settler patterns during and after Japan's formal imperial project, in which Japanese Americans played an integral role as Japan territorially navigated their competition with white empires.[26] Japanese migration to Brazil in the 1920s and 1930s was inextricably tied to Japan's state-sponsored and private settler colonialisms in East Asia, as well as to Japanese Americans' secondary migrations to and investments in the frontiers of a "racially friendly" Latin America.[27] The Amazon rain forest occupied a central place in these imaginaries, regarded as the last open frontier on Earth with supposedly uninhabited forests that promised the possibility of plantation-style agriculture. And unlike North America with its exclusion laws, the Amazon looked like a racial paradise where Japanese settlers, encouraged by Japanese state agencies to intermingle with local Brazilians in an effort to counteract yellow-peril discourse, could propound an affinity with Indigenous forest-dwellers ostensibly descended from "the ancient Asiatic."[28]

These migrant-colonial projections carried over into the postwar period, rehearsing older settler colonial discourses by way of organizational structures like the Ministry of Forestry and Agriculture, which had recruited, trained, and settled migrants in Manchuria in the 1930s. These same organizations peddled postwar agricultural migration, land exploration, and land acquisition in Latin America as a way for Japan to again prove itself as a "model nation of owner-farmers."[29] The southern part of the Western Hemisphere represented a frontier eminently suited for the Japanese, whose mode of intensive farming supposedly distinguished them from the exploitative practices of Europeans.[30] Under the Cold War American hegemony, this settler ideology revised the history of Japanese colonialism into a story about the hardworking Japanese people as paragons of democratic capitalist modernization.[31]

Notably, Karen Tei Yamashita's second novel, *Brazil Maru*, published in 1992 as somewhat of a companion piece to *Through the Arc of the Rainforest*, similarly closes with a Japanese Brazilian's declaration of settler belonging: "The Japanese have origins in the land. And having settled virgin soil, our responsibility to the land is great; farming is our contribution."[32] The claim to "virgin soil" via the Japanese migrant farming legacy reinforces the novel's conspicuous allusions to the idea that Japanese share genetic kinship with Indigenous Americans.[33] As Eiichiro Azuma points out, Japanese settler colonialism's historical manifestation as large-scale undertakings in agrarian technological modernity—involving the institutionalization and transplantation of Japanese Americans' expertise in "U.S.-style continental farming" to Manchuria and Brazil—combines technocratic agricultural planning with a settler environmentalist ideology purporting a Japanese "native" intimacy with colonized land.[34]

Kazumasa's newfound agrarian lifestyle in the rain forest revives the old Japanese settler memory of racial and environmental paradise on Indigenous lands—an always-already techno-environmentalist fantasy. *Through the Arc*'s settler denouement holds a mirror up to Japan's complex postwar reality: it was a subordinated client state of the United States but also, as H.L.T. Quan outlines, a resurrected regional hegemon that strategically collaborated with the Brazilian national security state through a kind of "savage developmentalism," where Japanese technology and finance significantly bankrolled and profited from the Brazilian government's pursuit of capitalist growth at all costs.[35] Indeed, Kazumasa is a conduit for transnational Japanese capitalism, accumulating large amounts of wealth throughout the novel with the help of his brother, "an entrepreneur and investor par excellence."[36] Kazumasa's settler position is thus qualified by both the fractures and continuities of empire afforded by Japan's model-minority status in the transimperial Cold War system: he is complicit in Tweep's extractivist ventures but is ultimately absolved by the novel's rubric of multiculturalism. By granting Kazumasa belated success as a Japanese large-scale farmer, the novel modifies the historical aura around the alienated Japanese diasporic subject in the American Hemisphere. The Japanese settler, Yamashita implies, can be economically productive but also, unlike the whites, in sync with his social and natural environment.

At the same time, Yamashita knowingly engages with the realities of Indigenous dispossession and genocide in narrating the disasters spawned by capitalist extraction, which Kazumasa crudely facilitates based on his technological and environmental connection to the Matacão. The Indigenous character Mané and his family are repeatedly displaced by the Brazilian state and by Tweep's corporation, GGG (a subtle acronym for the Spanish colonial idiom "Gold, God, and Glory"); GGG then commodifies the bird feathers Mané had formerly used for personal therapeutic purposes, setting off a frantic bird hunt for profits that culminates in a catastrophic pandemic with a 90 percent human death rate—another allusion to Indigenous death. In this light, Kazumasa's pastoral happy ending is tempered by the fact that he lives in an environmental hellscape: the race for the Matacão has unleashed toxic chemical waste and biological mutations, while the Brazilian Air Force responds to the pandemic by carpet bombing the Amazon with DDT.

The spectacle of airplanes dropping poisonous bombs on the Amazon is based on historical fact: the Brazilian government had relied on planes to explore and settle the Western frontier since the 1940s, and by the 1960s colonization programs were dropping napalm to clear the forests and secure strategic areas, accelerating their contact with Indigenous peoples.[37] Japanese agrarian histories in the Amazon cannot be separated from the techno-environmental forces of colonialism and militarism. *Through the Arc* illustrates how a Japanese immigrant like Kazumasa may be pulled into the broader orbits of multiply interacting empires as a figure for technological expertise in the making of capitalist biomes and new stratigraphic layers; he rearticulates the figure of the imperial and postwar Japanese settler whose migration parallels and converges with the militarized currents of colonial land exploration and capital accumulation. But from the perspective of Asian American techno-environmentalism, Japanese diasporic settlers complicit in such environmental destruction can have it both ways: they can embody the modern place and time of settler space while being granted a liminal and innocent position between settler and native through the trope of the vanishing ecological Indian and through redemptive narratives of *mestizaje*, multiracialism, and synthetic nature.

How High We Go in the Dark, Alien Indigeneity, and the Colonial Techno-Environmentalism of Sequoia Nagamatsu

Sequoia Nagamatsu's 2022 novel *How High We Go in the Dark,* received by critics as a timely artistic intervention into the climate crisis and the COVID-19 pandemic, rehearses Japanese American techno-environmentalism as a speculative natural history: namely, Japanese American science and technology underwrites the Japanese diasporic identification with large-scale, *longue-durée* settler environments on Earth and abroad. Nagamatsu's stories-in-a-novel feature loosely connected Japanese or Japanese American characters from the year 2030 to six thousand years in the future, after melting permafrost in Siberia releases a thirty-thousand-year-old virus that shuts down normal human organ functions. Written in the years immediately preceding the COVID outbreak, Nagamatsu's novel was hailed as "unnervingly prescient" and "in sympathy with our lived present," instilling a sense of wonder for an "evolving reality" made possible by feats in genetic modification, space travel, and artificial intelligence.[38] The novel also pursues the astrobiological hypothesis that nonhuman modern civilizations may have existed in Earth's deep past in accordance with "universal rules guiding the evolution of all biospheres" on Earth and elsewhere.[39] *How High*'s linked narratives form a dispersed network of Japanese bodies whose collective heritage, the novel proposes, can be traced back to primordial habitats in deep space. Planets that support life—Earth one among them—figure as unique parameters of environmental reality fixed through metaphysical "probability scopes"— ontological setting devices that "look like telescopes" but are fitted with the "jellylike remains" of ancient organic "ancestors."[40]

While many of the individual stories fixate on the personal grief caused by the deadly pandemic and other environmental disasters, Nagamatsu's inclusion of Japanese/American artifacts and cultural touchstones evokes a transhemispheric and transhistorical sense of Japanese humanity as it relates to both the ancient hominid past and the interplanetary future. One such artifact, a small *dogū* figurine from Japan's Jōmon period (ca. 10,500 to 300 B.C.), holds talismanic properties for the Japanese American archaeologist and evolutionary geneticist Cliff Miyashiro, who arrives at the expanding Batagaika Crater in northeastern Siberia to continue his late daughter Clara's research on an archaic *Homo sapiens* cave civilization being uncovered by floods and permafrost melt. Approaching the sinking ledge from where Clara had fallen to her death, he throws the *dogū* figurine into the crater, "waiting for all that has been unburied to be retaken into the earth."[41] Nagamatsu's mention of the Jōmon period as "ancient Japanese history" in fact reflects an early postwar Japanese reassessment of Jōmon culture as *authentically* Japanese.[42] Previously, the Jōmon period had been excluded from Japanese self-understanding, viewed as prehistory for the subsequent Yayoi period when, according to most historical linguists, Japonic peoples first arrived on the archipelago.[43] Nagamatsu indigenizes Japaneseness through Jōmon clayware and then folds Japanese indigeneity into the late Pleistocene across Northeast Asia and Siberia.

The author revamps the notion of Asian indigeneity through a Japanese American revision of techno-Orientalism. Cliff's colleagues hypothesize that "Annie," a recovered specimen of the ostensible Batagaika culture, is Neanderthal in origin, but Cliff is skeptical that Annie had reached this far east from Europe and West Asia. He instead speculates, as a salute to Clara's fascination with "ancient aliens," that the cave civilization is extraterrestrial in origin.[44] The novel entertains and indeed betrays a preference for the lost civilization theory—likening the Batagaika remains to alien-built sites like the city of Atlantis and the "Egyptian pyramids"—in lieu of an engagement with the histories of local Sakha people whose "stories on the land," to quote Paulette Steeves, would account for such archaeological sites.[45] An elision of this sort owes its inspiration to "alternative temporalities where the Alien/Asian is inextricably tied to science [and] technology," but Nagamatsu's speculative perspective reorients the techno-Orientalist gaze around "Asian American spatial subjectivities and temporal heterogeneities . . . [where] issues of racial marginality are often encrypted, reconfigured, and/or transformed."[46] Nagamatsu encrypts racialized Japanese American subjectivity within a deep historical account of human/"alien" presence in Northeast Asia, laying terrestrial claim to Asianized Indigeneity but writing over actual Indigeneities for the geography in question.[47]

The novel here introduces the concept of the Bering Strait migration while foregrounding the "Asianness" of this part of human history, incorporating Japaneseness into the theory that big-game hunters first migrated to the Americas from Asia via the Bering land bridge within the past fifteen thousand years and dispersed southward. Japanese Americans like Cliff become correlated with an immigrant history of Asian–North American crossings alongside migrating caribou and Alaskan Natives. At one point, Cliff points out the phenotypical similarities between Japanese Americans and Indigenous northeast Siberians: as he phrases it in a letter to his granddaughter, "The team went into a nearby village today and we saw a girl that reminded me of you."[48] Nagamatsu proposes an alternative, celebratory interpretation of the American Bering Strait narrative, which historically constructed s Indigenous peoples in the Americas as Asian "immigrants" as well as the first "yellow peril invasion" of the European promised land.[49] Under the author's new lens, the Indigenous of the Western Hemisphere derive from the East, creating the impression that both aliens and Asians have long belonged to Earth's natural history so as to deconstruct normative criteria for continental inclusion or exclusion. But by leaving this hegemonic theory of Asian descendance intact, the novel rearticulates how Japanese settlers have touted shared ancestry and their own "Indigeneity" in the Americas. As Paulette Steeves writes, "The historically embedded boundary of recent (on a global human history scale) time frames for first human migrations to the Western Hemisphere is not simply based on the archaeological record; instead, it is a political construct maintaining colonial power and control over Indigenous heritage, material remains, and history."[50] Humans may have traversed back and forth between the Western and Eastern hemispheres during the Pleistocene, but the inaccurate fabrication of Asian origins for a "panhemispheric" Native American culture based on one stone tool type "erases the diversity known to be present in the archaeological record, oral traditions, and linguistics."[51] Nagamatsu's techno-environmentalist proposition of ancient alien modernity in Northeast Asia as a precursor to transhemispheric migration makes aliens out of Indigenous peoples on their own lands.

Nagamatsu's novel succumbs to the ruse of multiethnic and multiracial inclusion in the United States as a way of underwriting settler colonial speculation. The "racialized connection [between Asians and American Native peoples] forged in settler orientalism" constructs frontiers by way of the speculative.[52] In Nagamatsu's novel, the new frontiers are distant planets for American scientific and military personnel escaping lethal Earth conditions. In 2037, Cliff's wife Miki and their granddaughter Yumi board the U.S.S. *Yamato* as honorary members of this select group of "Yamato pioneers" in search of a "second chance" for humanity.[53] The interstellar ship, designed by the Japanese American Bryan Yamato and owned by both NASA and the "Yamato-Musk Corporation," attests to the fantasy that Asian American capitalists educated in STEM (science, technology, engineering, and mathematics) fields can spearhead state-sanctioned settler colonial projects on newly discovered habitable territories.[54] But Nagamatsu's novel also appeals to a more transhistorical construction of Japanese American intimacy with an interplanetary natural history that connects humanity's *longue durée* on Earth to a broad spectrum of extraterrestrial life. Miki suspects that Bryan Yamato's genius was kindled through "otherworldly help," which reinforces the novel's associative links between Japanese people, "Indigenous" environments, and alien civilizational presence on Earth and elsewhere.[55] As the U.S.S. *Yamato* arrives on different planets and encounters alien life-forms tens to hundreds of light-years from Earth, the ship commander announces that "we will not be exterminating alien life" in an astounding move by the author toward settler innocence, imagining that land enclosed by American colonizers would be peacefully settled absent of any system change.[56]

Nagamatsu's ecomodernism, by which Japanese American fictional proxies concoct interstellar transportation technologies, sunlight-reflecting satellites, or Ice Age biomes to engineer one's way out of global heating, is a utopian techno-environmentalist fantasy that embraces settler capitalist futures and naturalizes land theft and Indigenous dispossession.[57] The novel's final chapter churns through an alien goddess's reincarnation cycles across eons of time and scores of nonhuman and human lives, including the mother of "Annie" and the researcher who eventually finds her, Clara Miyashiro. Clara's antecedent lifetime is of a Japanese American woman who is incarcerated during World War II and sings to her daughter "in alien languages [she] had not spoken in centuries."[58] Japanese American alien citizenship is incoherently sequenced with the lives of Europeans and American settlers as well as those of prehistoric organisms. Formerly excluded Japanese Americans are reintegrated into a foundational logic of U.S. borders and frontiers that somehow holds environmental memories of the primeval past.[59] The novel's virus, which encodes the host's genes with traits "similar to those of a starfish or octopus," is understood to be nature's recapitulated assimilation of the human body into an older point in life's evolutionary lineage.[60] Nagamatsu thus invalidates Indigenous land claims in the Americas by reproducing the colonial assumption that settlers from Europe and Asia can claim a planetary Indigeneity on occupied territories.

Conclusion

Karen Tei Yamashita and Sequoia Nagamatsu's network novels draw their large-scale conceits from the technological integration of Japanese/Americans into transhemispheric and interplanetary environmental geographies. At their core, these speculative narratives legitimize Asian bodies in the Western Hemisphere by inverting techno-Orientalist tropes of Asian inhumanity into the following techno-environmentalist premise: that amid the environmental crises of capitalism, Asian Americans might usher in more utopian modes of production reconciling capitalist technoscientific knowledge with sustainable social and material value.[61] However, the revision of techno-Orientalism into Asian American techno-environmentalism does not necessarily subvert the colonial foundations of settler Orientalism. Yamashita's and Nagamatsu's global and transhistorical interpretations of Japanese diasporic movement reveal—but also greenwash—how Asian Americans participate in the production of colonial space-time.

That being said, Yamashita's and Nagamatsu's novels also reveal noteworthy differences in political ideology. Yamashita clearly identifies Indigenous Latin Americans as victims of state-sanctioned dispossession and capitalist extraction, whereas Nagamatsu uncritically lionizes capitalism's synergy with imperial science and technology to redeem the imperial core from the crises imperialism itself produces. Yamashita's and Nagamatsu's techno-environmentalist speculations exemplify leftist and liberal approaches, respectively, that Asian American writers may take in their search for alternative interpretations of human and natural history from within the grip of ongoing capitalist-colonial occupations and the ruins of Japanese and U.S. empires.

Notes

1 David S. Roh, Betsy Huang, and Greta A. Niu, "Technologizing Orientalism: An Introduction," in *Techno-Orientalism: Imagining Asia in Speculative Fiction, History, and Media*, ed. David S. Roh, Betsy Huang, and Greta A. Niu (New Brunswick, NJ: Rutgers University Press, 2015), 3.
2 Juliana Hu Pegues, *Space-Time Colonialism: Alaska's Indigenous and Asian Entanglements* (Chapel Hill: University of North Carolina Press, 2021), 43.
3 Iyko Day, *Alien Capital: Asian Racialization and the Logic of Settler Colonial Capitalism* (Durham, NC: Duke University Press, 2016), 34; Colleen Lye, *America's Asia: Racial Form and American Literature, 1893–1945* (Princeton, NJ: Princeton University Press, 2005), 114.
4 Lye, *America's Asia*, 8–10.
5 Jodi Byrd, *The Transit of Empire: Indigenous Critiques of Colonialism* (Minneapolis: University of Minnesota Press, 2011), 210.
6 Day, *Alien Capital*, 134, 141.
7 Byrd, *The Transit of Empire*, 189.
8 Byrd, *The Transit of Empire*, 202.
9 Michelle Huang, "The Posthuman Subject in/of Asian American Literature," in *Oxford Research Encyclopedia of Literature* (Oxford University Press, 2019) 4, 6.
10 Hu Pegues, *Space-Time Colonialism*, 14.
11 Roh, Huang, and Niu, "Technologizing Orientalism: An Introduction," 3; Kohei Saito, *Marx in the Anthropocene* (Cambridge: Cambridge University Press, 2022), 136.
12 Hu Pegues, *Space-Time Colonialism*, 16.
13 Hu Pegues, *Space-Time Colonialism*, 14.
14 Patrick Jagoda, *Network Aesthetics* (Chicago: University of Chicago Press, 2016): 52, 55–56.
15 Min Hyoung Song, "Becoming Planetary," *American Literary History* 23, no. 3 (2011): 557–565.
16 Song, "Becoming Planetary," 564.
17 Max Ajl, *A People's Green New Deal* (London: Pluto Press, 2021), 54.
18 La Paperson, "A Ghetto Land Pedagogy: An Antidote for Settler Environmentalism," *Environmental Education Research* 20, no. 1 (2014): 21.
19 Wendy Hui Kyong Chun, *Control and Freedom: Power and Paranoia in the Age of Fiber Optics* (Cambridge, MA: MIT Press, 2008), 178.
20 Viet Thanh Nguyen, "On Art and Politics," interviewed by Karen Tay Yamashita, Bay Area Book Festival, May 7, 2018, https://www.youtube.com/watch?v=m4Rf_DIBmO8&t=3698s. As Yamashita tells it in her interview with Viet Thanh Nguyen, "When I took [*Through the Arc of the Rainforest*] out on the road in the '90s, all of the sudden the Amazons were like the lungs of the earth, and children were drawing pictures of the forests being destroyed. And all of a sudden, I had to speak to that, even though that's not exactly what that book is about. So again, I entered into a sort of political moment that was on the minds of folks."
21 Karen Tei Yamashita, *Through the Arc of the Rainforest* (Minneapolis: Coffee House Press, 1990).
22 Rachel C. Lee, *The Americas of Asian American Literature* (Princeton, NJ: Princeton University Press, 1999).
23 Aimee Bahng, *Migrant Futures: Decolonizing Speculation in Financial Times*, Illustrated edition (Durham, NC: Duke University Press, 2018), 35.

24 Shouhei Tanaka, "The Great Arrangement: Planetary Petrofiction and Novel Futures," *Modern Fiction Studies* 66, no. 1 (2020): 195–196.
25 Quynh Nhu Le, *Unsettled Solidarities: Asian and Indigenous Cross-Representations in the Americas* (Philadelphia: Temple University Press, 2019), 110. Le more specifically discusses the Latin American *mestizaje/mestiçagem* ideology, which historically deployed the trope of mixed "racial democracy" to reinforce projects of assimilating—or "whitening"—Indigeneity out of existence. Quotes from Yamashita, *Through the Arc of the Rainforest*, 211.
26 Eiichiro Azuma, *In Search of Our Frontier: Japanese America and Settler Colonialism in the Construction of Japan's Borderless Empire* (Oakland: University of California Press, 2019), 12.
27 Azuma, *In Search of Our Frontier*, 146–149.
28 Azuma, *In Search of Our Frontier*, 142–146.
29 Sidney Xu Lu, *The Making of Japanese Settler Colonialism: Malthusianism and Trans-Pacific Migration, 1868–1961* (Cambridge: Cambridge University Press, 2019), 247.
30 Lu, *The Making of Japanese Settler Colonialism*, 257.
31 Lu, *The Making of Japanese Settler Colonialism*, 251.
32 Karen Tei Yamashita, *Brazil Maru* (Minneapolis: Coffee House Press, 1992), 248.
33 For example, one archaeologist character in *Brazil Maru* encounters "Indian remains" near his Japanese communal settlement and sees fit to "make comparisons between [Brazilian Indians] and the Japanese aborigines . . . to prove that the Brazilian Indians were our distant relatives." Yamashita, *Brazil Maru*, 66.
34 Eiichiro Azuma, *In Search of Our Frontier: Japanese America and Settler Colonialism in the Construction of Japan's Borderless Empire* (Oakland: University of California Press, 2019).
35 H.L.T. Quan, *Growth Against Democracy: Savage Developmentalism in the Modern World* (Lanham, MD: Lexington Books, 2012), 73–74.
36 Yamashita, *Through the Arc of the Rainforest*, 81.
37 Felipe Fernando Cruz, "Napalm Colonization: Native Peoples in Brazil's Aeronautical Frontiers," *Hispanic American Historical Review* 101, no. 3 (2021): 463.
38 Nina Allan, "How High We Go in the Dark Review—A New Plague," *Guardian*, January 20, 2022, https://www.theguardian.com/books/2022/jan/20/how-high-we-go-in-the-dark-by-sequoia-nagamatsu-review-a-new-plague.
39 Adam Frank, "Was There a Civilization on Earth Before Humans? A Look at Available Evidence," *The Atlantic*, April 13, 2018, https://www.theatlantic.com/science/archive/2018/04/are-we-earths-only-civilization/557180/.
40 Sequoia Nagamatsu, *How High We Go in the Dark* (New York: William Morrow, 2022), 275.
41 Nagamatsu, *How High We Go in the Dark*, 29.
42 Nagamatsu, *How High We Go in the Dark,*, 12.
43 Mark J. Hudson, "Re-Thinking Jomon and Ainu in Japanese History," *Asia-Pacific Journal* 20, no. 15(2) (2022): 3, 7.
44 Nagamatsu, *How High We Go in the Dark*, 18.
45 Paulette Steeves, *The Indigenous Paleolithic of the Western Hemisphere* (Lincoln: University of Nebraska Press, 2021), 10.
46 Stephen Hong Sohn, "Introduction: Alien/Asian: Imagining the Racialized Future," *MELUS* 33, no. 4 (2008): 6.
47 Nagamatsu's replacement of Indigenous presence in Yakutia with aliens rhetorically recalls how the Sakha peoples were categorized as "nomadic aliens" under the

Russian imperial government's passage of the 1822 Statute of Alien Administration in Siberia, which categorized crown subjects as either "natural inhabitants" (Russians) or "aliens" (Indigenous Siberians). See Kara Hodgson, "Russia's Colonial Legacy in the Sakha Heartland," The Arctic Institute: Center for Circumpolar Security Studies, November 15, 2022, https://www.thearcticinstitute.org/russias-colonial-legacy-sakha-heartland/.

48 Nagamatsu, *How High We Go in the Dark*, 23.
49 Byrd, *The Transit of Empire*, 201.
50 Steeves, 15.
51 Steeves, *The Indigenous Paleolithic*, 11.
52 Hu Pegues, *Space-Time Colonialism*, 53.
53 Nagamatsu, *How High We Go in the Dark*, 187–189.
54 The science-fictionality of this premise aligns with what Christopher Fan identifies as an Asian American "ideology of science and [science's] role in industrial expansion," which has arisen in relation to post-1965 Asian American occupational concentration in STEM-related professions. Christopher T. Fan, "Science Fictionality and Post-65 Asian American Literature," *American Literary History* 33, no. 1 (2021): 81, 84.
55 Nagamatsu, *How High We Go in the Dark*, 195.
56 Nagamatsu, *How High We Go in the Dark*, 200; Kevin Bruyneel, *Settler Memory: The Disavowal of Indigeneity and the Politics of Race in the United States* (Chapel Hill: University of North Carolina Press, 2021), 4.
57 Ajl, *A People's Green New Deal*, 45–46.
58 Nagamatsu, *How High We Go in the Dark*, 282.
59 Byrd, *The Transit of Empire*, 206.
60 Nagamatsu, *How High We Go in the Dark*, 25.
61 Saito, *Marx in the Anthropocene*, 142–144.

15

The Alchemized Dis/abled Body as Recuperative Site in *Fullmetal Alchemist*

JUNG SOO LEE

In a prominent box of text, *Fullmetal Alchemist (FMA)*, a popular twenty-seven-volume manga series by Hiromu Arakawa that ran from 2001 to 2010, states that "in alchemy, the body and the soul are connected through the mind."[1] Within the narrative, the body is translated and defined as a malleable concept that is intricately connected to alchemy, the core technology of the series.[2] By defining the body through technology, *FMA* acknowledges the social circumstances and conditions that produce both able and dis/abled bodies.[3] I read the depictions of the dis/abled body in direct contrast to cyberpunk readings of technology transcending the human body.[4] I read the body in line with Sami Schalk's reading of the body, that the body cannot be read outside of the circumstances that produce it.[5] To this end, I argue that the concepts of body, mind, and soul as described in *FMA* are embodied through the dis/abled body and serve as metaphors for tumultuous East Asian transnational and Orientalist sociopolitical dynamics. Dis/ability demarcates the spectrum between ability and disability and examines how in/visible forms of disability are exacerbated or eased due to other factors such as race and gender. By fully embracing the potential of dis/abled bodies, *FMA* provides valuable insight into geopolitical conditions of the 2000s and how Japanese techno-Orientalist anxieties and desires manifested within popular culture.

FMA is part of a long-standing history of Japanese manga produced in response to sociopolitical contexts. Starting from the early twentieth century, manga and anime were part of Japan's campaign of image-making in response to the challenges of modernization and Westernization.[6] During World War II, manga and anime were a representative part of Japan's soft power strategies that sought to identify Japan as peaceable and cultured.[7] After World War II, these cultural articles remained as part of official Japanese national soft power policies to continue cultivating a high-cultural national image.[8] Douglas McGray denotes the intent of such policies as Japan's attempt to become a "military and cultural power on its own terms," coining the term "cool Japan."[9] However, there is an important shift since World War II, wherein Japanese culture became focused on becoming a *harmless* form of high culture. Casey Brienza observes how cultural products created under these policies merely rearticulated and reproduced existing geopolitical power structures, intent on presenting Japan as an impotent country.[10] Christine Yano likewise suggests that the aesthetic of "cute-cool," which uses exaggerated images of hyperviolence or cuteness vis-à-vis Godzilla-esque monsters or Hello Kitty, presents "cool Japan" as a *castrated* culture.[11] According to Yano, such images of helplessness solidify Japan's imagery of impotency and harmlessness.[12] This strategic projection of "cool Japan" within manga instilled a "spirit of righteousness and romantic idealism" in its audience, where the audience is intended to identify with the hyperviolent and hypercapable while remaining nonthreatening.[13] *FMA* is likewise deeply informed by this history, as well as the events of the early 2000s when it was first serialized such as 9/11, the War on Terror, and the Great Recession. These events are reflected in the series' message of unity despite fears of the Other, and the overcoming of hardships as a community. I argue that *FMA*'s depiction of the dis/abled body and the role of technology in the form of alchemy and automail is part of Japan's self-aware techno-Orientalist desire to project an image of *tamed* harmlessness and stability during an uncertain time in global politics.[14] This form of techno-Orientalism focuses on emphasizing the *helpful* nature of the East while implicitly recognizing previous fears of a technologically advanced Orient.

FMA reflects the author-illustrator's awareness of narrative and structural processes through which the disabled body is produced.[15] Arakawa recounts how working part time in a disability rehab center made clear to her how disability flattens out characters into what David Mitchell and Sharon Snyder call a "trope of human disqualification."[16] Disabled characters are often associated with deviance, villainy, or eccentricity, à la Captain Ahab or Darth Vader. Contrary to such depictions, Arakawa states how she wanted to depict more fleshed-out narratives of how society views dis/ability, the process of becoming dis/abled, and the different definitions of normative life for disabled people.[17] To this end, *FMA* presents a worldview where dis/ability is commonplace and focuses its narrative and world-building on the circumstances that *produce* dis/ability such as war and trauma. Dis/ability is presented as an equalizing "mutual constitution of oppressions," yet *FMA* simultaneously demonstrates how the complexities of the dis/abled body are experienced differently despite its ubiquity.[18] Dis/ability in *FMA* offers a chance to reflect upon the personal lived experience and connect it to larger global events.[19] By doing so, *FMA* becomes a transnational text that articulates how dis/ability reflects the manifestation of post-9/11 geopolitics in East Asia, the Middle East, and the Western world represented by the United States and Europe.

The series focuses on the brothers Edward and Alphonse Elric from a militaristic country named Amestris. Both are disabled in an alchemy accident, with Edward having to replace his right arm and left leg, and Alphonse his entire body, with prosthetics. As they travel to find a cure to restore their bodies, they uncover the secrets surrounding the Ishvalian Massacre, an Amestrian genocidal attack against a minority population named Ishval. The brothers meet Scar, an Ishvalian survivor avenging his people through acts of terror against the Amestrian military, and Mei Chang, an illegal immigrant from a neighboring country named Xing.[20] Scar and Mei introduce the brothers to *alkahestry*, the Xing version of alchemy. The brothers are interested in alkahestry but are understandably apprehensive because it is primarily used by a terrorist and an illegal immigrant. Ultimately the Elric Brothers, Scar, and Mei overcome personal and political differences to defeat the main villain, known as Father, and use a combination of alchemy and alkahestry to save Amestris and come together in unity.

Scholarship on *FMA* tends to focus on its circulation and the significance of its global popularity.[21] Other scholars read the series within the history of manga, and explore how *FMA* utilizes the monstrous figure as an allegory for cultural evils.[22] I focus on the role of alkahestry, the Xing, and how alchemy, alkahestry, and the dis/abled body utilize technology as an allegory for the complex geopolitics of the 2000s.[23] The visualization of Ishval, Amestris, and Xing clearly shows them as coded to be representations of the Middle East, the United States/Europe/the West, and China/Japan/the East respectively. In the series, the Xing and their alkahestry facilitate a reconciliation between the Ishvalians and the Amestrians. By showing alkahestry and its users as ultimately harmless, as well as *beneficial* to the West, I argue *FMA* is reacting to the tumultuous geopolitics of a post-9/11 world order, thus assuaging fears of a technologically advanced East.

The desire to produce a perception of harmless Asia in the light of the overarching influences of U.S. militarism post-9/11 become clearly apparent through the Japanese response to 9/11. Japan was not only one of the first countries to respond to the attacks, but it also provided logistical support for U.S. military operations in Afghanistan as well as monetary support in the form of $10 million to aid in rescue and cleanup efforts.[24] Japan clearly positions itself as a staunch ally of the United States, and *FMA* reflects such geopolitical inclinations through the dis/abled body. I focus on *FMA*'s usage of the dis/abled body, the geopolitical circumstances that produce such bodies, and the interactions between dis/abled individuals to argue that these depictions promote an image of Asian technology that restores and maintains geopolitical order. As such, I interpret the racialized, dis/abled body and the use of alkahestry in *FMA* as a metaphor for a tamed techno-Orientalist subject, intended to placate techno-Orientalist anxieties. I argue that in the face of geopolitical instability and the ensuing U.S. militarism present during the decade of its publication, *FMA* reflects the East's complex reactions to appease Orientalist fearspost-9/11 and appeal for its harmlessness while being simultaneously useful and competent.

Body: Invisible Antagonistic Dis/ability

The dis/abled body in *FMA* utilizes an ironic usage of the trope of an "extraordinary" body as articulated by Rosemarie Garland Thomson, wherein dis/abled people are only placed on equal footing to their able-bodied counterparts by outperforming them.[25] *FMA* normalizes the extraordinariness of disabled characters by presenting an abundance of characters, including civilians, military personnel, and children with outlandish prosthetics that contain blades and firearms as a result of warfare. The ubiquity of prosthetics in *FMA* and how dis/ability is normalized draws parallels to Jasbir Puar's formulation of the ubiquity of disability in areas with a strong U.S. militaristic presence such as Palestine and Iraq. In addition to being a direct result of warfare, dis/ability is a "deliberate product—of exploitative labor conditions, racist incarceration and policing practices, militarization, and other modes of community disenfranchisement" that resulted from U.S. militarism.[26] By normalizing the extraordinariness of tragedies, the commonness of dis/abled bodies that were produced as a direct consequence of warfare is accentuated as a site where these tragedies are inscribed. The loss of body parts becomes representative of the trauma of losing family, future, country, and culture. There is an inherent politics to the availability of advanced prosthetics as described in *FMA*, not least of which is the availability of technology.[27] However, under the ideal circumstances in which the technology to recover the body is available to all, the process of recovering the dis/abled body becomes allegorical as a site of social recuperation and moral redemption *through* technology. Recovering bodily losses through *Eastern* technology functions narratively as recuperations of the tragedies that inflicted those losses in the first place, and by association the usefulness of the East.

FMA's dis/abled bodies allegorize how the process of recovering from dis/ability can reverse fears of technology controlled by a racialized Other via a tamed techno-Orientalism. Scar begins the series as a terrorist antagonist, only to later turn ally. Scar is a survivor of the Ishval Massacre, a war that Amestris started in the name of revenge for the death of an innocent girl. Scar became dis/abled when he survived an attack from the Amestrian military, by having his right arm transplanted from his dying brother to save his life.[28] As a direct result of the trauma caused by Amestrian military personnel—or State Alchemists—he becomes the "vengeful ghost of Ishval," using alkahestry to commit acts of terrorism and murder every State Alchemist he can find.[29] With his dark complexion, garb, and introduction as a terrorist, Scar is clearly an allegory for the Middle Eastern threat.

FMA further complicates how state-sanctioned violence is justified and how the Other is vilified through Scar and his dis/ability. Scar's individual mental and physical trauma personifies the unhealed group trauma of the civil war and the genocide of Ishvalians alongside the guilt of the Amestrians. Scar's backstory as a victim of geopolitical turmoil turns him into a sympathetic character despite how threatening he is. As Scar seeks to kill all Amestrian State Alchemists, State Alchemists themselves comment on how Scar is a result of Amestrian sins, yet the fear and hatred against him remain because he commits acts of terror.[30] Further marking him as a victim of Amestrian actions even beyond his war trauma, Scar's transplanted right arm is a symbol of state violence inscribed upon his body. It looks like an undamaged arm, barring a scar indicating where it was reconnected. The invisible nature of Scar's arm transplant hides the site where state violence has made its mark. Instead, his arm is covered in sigils that allow him to use alkahestry to become a terrorist, an ability he would not have had nor would have needed had he not been victimized by Amestris (Figure 15.1). As allegory, Scar mirrors the effects of post-9/11 U.S. militarism and guilt regarding the Middle East, and well as the rise of terrorist groups in the wake of an organized militaristic attack.

Scar's dis/abled body becomes the embodiment of the threatening racial Other that uses Western technology to overwhelm the West as an oppositional force.[31] The parallels between Scar's abilities and fears against him, and the geopolitical fears during a decade plagued by the repercussions of U.S. militarism leading up to and after 9/11 are difficult to ignore. Most prominently for *FMA*'s time, it was the War on Terror. The rallying cry from President George W. Bush during his address to a joint session of Congress directly after 9/11 repeats the idea of erasing wrongs as an indisputable right. Bush announced that the United States will "meet violence with patient justice—assured of the rightness of our cause, and confident of the victories to come." Scar's appearance as a Middle Eastern–coded refugee is clearly allegorical for the survivors of U.S. militaristic presence in the Middle East, interventions in multiple wars and regions, and the subsequent guilt from said presence.[32] Puar discusses the moments post-9/11, the U.S. empire's position as the purveying "arbiter of appropriate ethics, human rights, and democratic behavior," and how the U.S. empire remained remarkably immune to its own standards of ethics in the face of the atrocities that resulted from the War on Terror.[33]

The clear erasure of U.S. wrongs in the face of terrorism is mirrored onto how Scar's right arm no longer shows signs of visible disability. Eli Clare discusses the implications of this erasure performed by the processes of cure upon the dis/abled. He argues that institutions of power that focus only on cure erase not

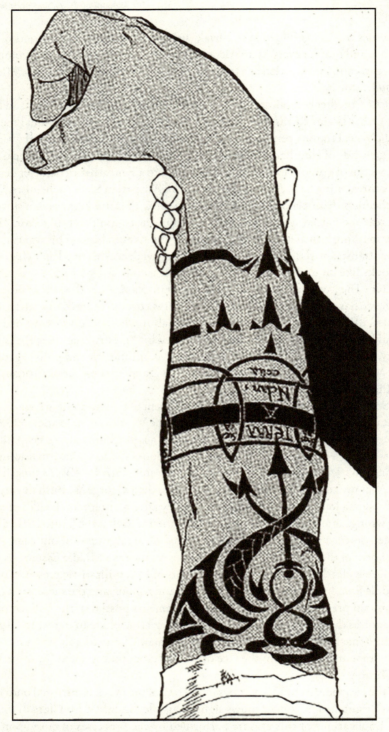

FIGURE 15.1 Closeup of Scar's dis/abled arm, and the bandages that connect it to his body. (Source: Hiromu Arakawa, *Hagane no Renkinjutsushi* [Fullmetal Alchemist], Square Enix, 2010, 4:172.)

only the reality and presence of disability, but also the possibility for the circumstances that caused and resulted from dis/ability to be recognized.[34] The violent story behind how Scar's arm was lost and regained, and the tragedy of state-sanctioned warfare, are obscured as his arm is cured. Because Scar's dis/ability is invisible, it is erased and therefore of no consequence. Scar's seemingly whole arm rewrites his identity from a person dis/abled by sociopolitical circumstances outside of his control to a terrorist.

In *FMA*, the circumstances stem from U.S. militarism against the Middle East. Scar's physical disability has been cured, but his trauma and lived experience of being dis/abled, as well as his status as a victim, is erased with it. As Puar articulates, the Western-centric concepts of cure and disability justice for dis/abled bodies obfuscate the interconnected relations of the social and physical infrastructures of U.S. empire, settler colonialism, and militarism.[35] The effects of the U.S. empire are deeply rooted in the idea that it exists as an exceptional empire that is superior in every way, morally, culturally, and politically. Scar's body, resplendent with scars and tattoos, becomes representative of a site where Western violence is inscribed as an indelible mark, yet this mark has been overwritten in favor of a Western-centric narrative, both figuratively and literally. While Scar is initially presented as the threatening racial Other incarnate, another character serves as his nearly perfect contrast to turn this threat into an ally, furthering Scar's tamed techno-Orientalist threat. By the finale of the series, Scar becomes a domesticated techno-Orientalist threat with the help of the Xing character Mei Chang. Through a distinctly Eastern influence, represented by the Xing, Scar saves the Amestrians from their own follies. Scar's body,, and especially his arm, shift from its initial portrayal as a tool for inducing terror into a symbol of hope, and recuperation via a placating influence brought in by the East.

Mind: Affective Dis/ability

If Scar embodies the fear and terror of technology at the hands of the Middle Eastern Other, then Mei Chang embodies the usefulness of technology in the hands of a *tamed* East. Mei is a representative of a distinctly Oriental Other in comparison to Scar's Middle Eastern Other. By the end of the twentieth century, China's and Japan's rise to prominence on a global stage was distinctly threatening to the United States. Betsy Huang describes how the figure of the Chinese worker replaced the technologically advanced Japanese worker, both being equated with a faceless, cold, calculating horde, with their advancing technology and economic growth.[36] China's threat at the beginning of the twenty-first century was mainly one of size and scale, wherein its accomplishments evoked "yellow peril anxieties of indistinguishable, industrious 'hordes' of Asian laborers threatening to lower wages or replace white, American labor altogether during the late 19th century."[37] Mei is constructed along the same lines of this imaginary techno-Orientalist threat. However, in addition to being what Huang calls "both necessary instruments for *and* impediments to progress," Mei's diminutive and distinctly Chinese features, combined with her proficiency in alkahestry, embody a *tamed* techno-Orientalist threat.[38] Mei's childlike and feminine features and diminutive size effectively nullify the threat of an insurmountable horde into an affective performance of cuteness that appeals to the image of powerlessness.[39] Her cuteness aligns with Yano's account of postwar Japan's national project of "cute-cool," wherein Japan's threat is carefully concealed under the historical distortion of national emasculation.[40] Indeed, Mei is a prominent example of what Sianne Ngai calls "exaggerated passivity," and the most objectified of objects, incredibly capable yet simultaneously impotent beyond further nullifying other similar threats.[41]

Mei is the most advanced alkahestry user in the series who is handicapped visually, socially, and politically. In direct contrast to Scar's terrorism, she is a savior rescuing Amestrians from a collapsed mine.[42] Mei is half the size of other characters, young, rounded, with disproportionately large eyes. She is inseparable from her pet panda Xiao Mei, a giant panda the size of a small kitten who was born with dwarfism. She is an illegal immigrant, who is marginalized even in her home country of Xing as a member of a minority tribe. Everything about Mei, from her name and appearance to her sidekick pet Xiao Mei, emphasizes and reinforces the imagery of a small, adorable, tamed East. Xiao Mei additionally associates Mei with dis/ability through her dwarfism.[43] Although Mei is able-bodied, her diminutive stature and social circumstances suitably dis/able her as an impotent being, allowing her to be used for the benefit of the West as an ally.

Mei's dis/ability, cuteness, and harmlessness becomes a way for her to create bonds and form allyships, bridging the substantial distance between Ishval and Amestris. When Scar is introduced to Mei, he is instantly entranced by Xiao Mei (Figure 15.2). Upon hearing her struggles, Scar and Mei bond over a mutual comradery as members of minority populations persecuted by a dominant political power (Figure 15.3). Mei and Xiao Mei become the catalyst for representatives of Amestris and Ishval meeting in a nonconfrontational manner. Scar helps her look for Xiao Mei, who has accidentally fallen into the company of the Elric Brothers, providing an opportunity for the three to come to the table. With Mei and Xiao Mei as an intermediary, Scar is humanized from terrorist villain to sympathetic antagonist, a "good person" who is doing his best to help those who are minoritized despite his own traumas.[44] Mei's appearance disarms and reinforces the tamed techno-Orientalist representation of the East and further domesticates Scar, the embodiment of the threatening Middle East.

How the Middle East is tamed by the impotent hypercapable East is on full display when Mei and Scar meet the Elric Brothers for the first time. They are facing Father, the true villain of the series, and Mei and Scar can retaliate when the Elric Brothers are helpless. Without Mei, this would be an intimidating scene wherein the metaphoric West would be at the mercy of the vengeful and

FIGURE 15.2 Scar's first introduction to Mei and Xiao Mei. The slight flush to his face indicates how instantly enamored he is by Xiao Mei, as well as how petite Xiao Mei is in comparison. (Source: Hiromu Arakawa, *Hagane no Renkinjutsushi* [Fullmetal Alchemist], Square Enix, 2010, 11:140.)

capable Middle East. Mei's dis/abled, impotent body softens Scar's presence by association and becomes a catalyst for reconciliation between Ishval and Amestris.

Mei is severely injured during this confrontation while coming between the Elric Brothers and Father, and the Elric Brothers race to save her. They introduce her to Dr. Knox, an Amestrian wartime doctor who conducted human experiments against Ishvalian prisoners. He suffers from PTSD (posttraumatic stress disorder) and becomes a medical examiner as a result because he thinks that he is "unworthy of treating the living."[45] However, once he saves Mei's life, Dr. Knox starts feeling capable of atoning for his sins as a savior once more.[46] The blunt metaphor of the Amestrian saving the dis/abled Xing, that is, the metaphoric West saving the East and assuaging Western guilt, is clear in this scene. Mei's dis/ability and salvation at the hands of Dr. Knox, the reformed war criminal, assuage Amestrian guilt at taking part in state violence. As a representative of a technologically advanced East—postwar Japan in this instance—Mei offers ethical salvation for war crimes committed by the West and assuages Western guilt, furthering tamed techno-Orientalist readings of the text. Mei as allegory cements how the East is simultaneously harmless and *useful*. Mei soothes not only allegorical Western guilt from various militaristic activity but also techno-Orientalist fears against the West post-9/11. This metaphor is further solidified when Mei is placed alongside the representatives of Amestris, the Elric Brothers.

FIGURE 15.3 Scar and Mei are being asked how they can use alchemy and alkahestry, with Xiao Mei pictured on Mei's right shoulder. (Source: Hiromu Arakawa, *Hagane no Renkinjutsushi* [Fullmetal Alchemist], Square Enix, 2010,14:71.)

Soul: Recovering Humanity through Dis/ability

Scar and Mei are representatives of the Middle East and East as those who have lived through or are living through dis/abling conditions. By contrast, unlike the Amestrians who took part in systemic state violence like Dr. Knox, the Elric Brothers are personifications of the *upcoming* generation of the West. They are the generation that has been affected by, but not directly responsible for, the atrocities performed by the West. This is represented by the brothers' relationship with dis/ability, not only having been raised by a prosthesis maker but also being disabled themselves. The older brother, Edward, has a prosthetic arm and a leg, while the younger brother, Alphonse, has no body, just a soul that has been attached to a suit of armor. The brothers represent a generation of the West that is aware of the atrocities, as well as the consequences of, state violence and its role in producing dis/abled bodies. The brothers are looking for cures for their disabilities, and this allows them to approach Xing technology with an enthusiasm and open-mindedness that are not found in the older generation. The Elric Brothers see Eastern technology as an exciting possibility rather than a threat, signaling a distinct shift away from the older generation's techno-Orientalist fears to an embrace of its helpfulness. This indexes a broader shift in techno-Orientalist logics in the early 2000s. As David Roh, Betsy Huang, and Greta Niu argue, technology was a method of "mediating 'contact' between East and West through techno-Orientalist logics."[47] They ask the question, "Is techno-Orientalism still Orientalist if contemporary techno-discourse is being authored principally by Asians, seemingly without regard for the Westerners who look on with a mixture of anxiety and envy?" My answer is yes.

Although it takes on a different form, tamed techno-Orientalism is still based on techno-Orientalist logics. The Elric Brothers' techno-Orientalist tendencies are still based on assuaging fears of a burgeoning, capable, allied Orient, despite seeing the East as an ally rather than a threat. As the Othered threat shifted from Asia to the Middle East after 9/11, the Orient's technology and capabilities functioned to soothe and quell Western anxieties in the age reflective of U.S. militaristic empire post-9/11. The indifference or outright hostility against Asia from the 1990s shifted into a partnership against the perceived threat of terrorism.[48] *FMA* emphasizes the split between the previous generation and the Elric Brothers through how they address the Xing. When the Elric Brothers find Xing technology, Edward comments on the clear potential value in alkahestry to *supplement* alchemy.[49] By acknowledging the need for alkahestry, Edward acknowledges the limitations of alchemy, the need for Eastern technology, and how Amestris has validated its many war crimes. This shift in techno-Orientalist logics is further emphasized through the difference in the brothers' relationship to the other characters, and their distinct lack of animosity.

Throughout the series, Edward agrees to become subordinate to a Xing character, while Alphonse develops a quasi-romantic relationship with Mei. The brothers build working relationships with Scar and other surviving Ishvallians, as well as some of the older generation of Amestrians, to save Amestris from Father, which results in a metaphoric coalition of nations. This horizontal relationship between the brothers and the others reflects what Julie Ha Tran articulates as a shift in gaze, as previously unilateral, hierarchical, and one-sided gazes of the West toward the East shift with the changes in techno-Orientalist logics.[50] The East, once unilaterally defined as a source of fear, becomes one that is mutually influential, but with the West still distinctly in control of how their relations are created and maintained. Although Mei has created the potential for horizontal relationships among Ishval, Xing, and Amestris, the Elric Brothers are the ones who are allowed to accept such possibilities, placing them in a position of power as the masters of the tamed techno-Orientalist threat, culminating in the final battle against Father.

During the final battle of the series, Scar and Mei are indispensable to the Elric Brothers' victory. Scar weakens Father to the brink, and Mei uses alkahestry to restore Edward's biological right arm after his prosthetic arm is destroyed, which allows him to destroy Father. Edward is literally reembodied through Scar and Mei's alkahestry, and his body brought together by the collaboration of Amestrians, Xing, and Ishvalians. The metaphor of Edward's dis/abled body becoming whole once again thanks to the East, West, and Middle East coming together is a bit heavy-handed. However, it is effective in projecting the possibility for what Qingguo Jia calls a "constructive and cooperative relationship" that is critical not just to the East and the West, but also to "the world as a whole more than any time in history": in other words, an idealized world that can come together despite past histories and collective trauma, making whole the scars of geopolitical strife.[51] If Scar and Mei represent body and mind—Scar as the embodiment of Western state violence, and Mei as the connective possibility for reconciliation between victims and aggressors—then the Elric brothers are the *soul*. With their golden hair and eyes, the Elric Brothers make palatable the violences of the state and their aftereffects. The brothers and their no longer dis/abled bodies are the Amestrian hope for the future, the idealized next generation that is capable of recognizing the trauma of a mass genocide and overcoming the allegorical West's fears of self-destruction from its own sins.

Conclusion

FMA retains a hopeful message that a complex world can come together in the face of adversity despite its political and social strife. However, despite its idyllic visions to put the metaphoric East, Middle East, and West on an equal footing, *FMA* still cannot escape Orientalist tendencies where the West Others the rest. The narrative wraps up neatly with the Amestrians and Ishvalians coming together to pick up the pieces of a shattered country, yet non-Amestrian characters are never fully integrated into Amestris. Scar's role in saving Amestris is hidden, and he is a former terrorist to the general public, while Mei returns to her home country. The fears of the techno-Orientalist threat are assuaged, but the East and the Middle East remain separate from the Western public. The West may be thankful for the aid of the tamed Orient, yet the techno-Orientalist threat must remain separate and invisible, thus further obfuscating the aftermath of state violence incurred by the West and the upholding of the U.S. hegemon. The Xing characters and Xing technology are presented as a method of taming techno-Orientalist fears into a message of solidarity. However, by exiling the Xing characters and presenting them as outsiders despite acts of unity, there remains space to criticize, or at least question the implications of, a tamed techno-Orientalist threat.

The dis/abled body in *FMA* opens up approaches to techno-Orientalism that go beyond traditional fears of a developing Asia. Techno-Orientalist fears manifest through Scar, while tamed techno-Orientalist platitudes are shown through Mei. Scar's villain arc can be clearly traced as an open acknowledgement of a mutual fear between the Middle East and the United States from a modern U.S. militaristic history. The narrative turns that humanize Scar make salient and poignant points about the complexities of victimhood in the cycle of violence perpetuated by the hegemonic U.S. geopolitical structure. *FMA* makes clear the potential for the dis/abled body to be read as a site of violence *as well as* recuperation. Dis/abled and Othered bodies are not entities to be feared but sympathized with. *FMA* recognizes the dis/abled body as the site of state violence, where individual state-incurred traumas become visible. The dis/abled body becomes the object where the intricacies of intersectionalities that complicate who is the perpetual Other become visible. As Mel Chen suggests, dis/ability needing constant renegotiation and recalibration does not have to be equated to the need for cure and healing.[52] Rather, by making visible the dis/abled body, *FMA* reclaims the dis/abled body as a site where a form of protosolidarity and accountability may take place. *FMA* does have its limits in displaying the broad spectrum of what constitutes as dis/ability, as it limits its depictions of dis/abilities to physical disabilities and excludes many invisible disabilities such as chronic illness or mental disabilities. However, the future that *FMA* illustrates, however idealized and possibly clichéd it may be, acknowledges and projects a world where dis/ability is fully integrated into society. By placing dis/abled bodies at the center of the narrative, *FMA* provides a window into understanding how the dis/abled body and how it is addressed reflect complex geopolitical situations and systems surrounding the construction of the dis/abled subject and the processes in which dis/ability is created by no fault of the dis/abled.

Notes

1 Hiromu Arakawa, *Kangch'ŏlŭi Yŏn'gŭmsulssa* [Fullmetal Alchemist], trans. Sŏ Hyŏn-a (Seoul: Haksan Munhwasa, 2001–2019), vol. 11. All Japanese words have

been romanized using the Hepburn system, while Korean words have been romanized using the McCune-Reischauer system.
2. Unlike the medieval practice of alchemy, in *FMA* alchemy is the equivalent of modern science and technology.
3. Sami Schalk, *Bodyminds Reimagined: (Dis)Ability, Race, and Gender in Black Women's Speculative Fiction* (Durham, NC: Duke University Press, 2018), 8. I borrow Schalk's usage of the terms "dis/abled" and "disabled," wherein "disabled" is used to refer to the condition of being disabled, and "dis/abled" acknowledges the social, economic, and political circumstances that go into producing disability such as systemic injustices, cultural conceptions, and lack of resources.
4. Veronica Hollinger, "Cybernetic Deconstructions: Cyberpunk and Postmodernism," *Mosaic: An Interdisciplinary Critical Journal* 23, no. 2 (1990): 31.
5. Schalk, *Bodyminds Reimagined*, 7–11.
6. Mark MacWilliams, *Japanese Visual Culture: Explorations in the World of Manga and Anime* (London: Routledge, 2008), 148–150.
7. MacWilliams, *Japanese Visual Culture*, 150.
8. Outlined in the white papers published by the Japanese Ministry of Education, Culture, Sports, Science, and Technology. These describe plans to encourage the international spread of Japanese animation and manga. Ministry of Education, Culture, Sports, Science, and Technology, "White Papers: Japanese Government Policies in Education, Science, Sports and Culture 2000," 2000, https://warp.da.ndl.go.jp/info:ndljp/pid/11402417/www.mext.go.jp/b_menu/hakusho/html/hpae200001/index.html.
9. Douglas McGray, "Japan's Gross National Cool," *Foreign Policy* 130 (2002): 28.
10. Casey Brienza, "Did Manga Conquer America? Implications for the Cultural Policy of 'Cool Japan,'" *International Journal of Cultural Policy* 20, no. 4 (2014): 385–387, https://doi.org/10.1080/10286632.2013.856893.
11. Christine R. Yano, *Pink Globalization: Hello Kitty's Trek across the Pacific* (Durham, NC: Duke University Press, 2013), 286.
12. Yano, *Pink Globalization*, 286–288.
13. Ryan Holmberg, "Manga Shōnen: Katō Ken'ichi and the Manga Boys," *Mechademia* 8 (2013): 177, https://doi.org/10.1353/mec.2013.0010.
14. Automail refers to metal prosthetics that behave like natural limbs. They are said to have developed so extensively due to the many years of treating people who have lost body parts due to war.
15. Hiromu Arakawa, *Kangch'ŏlŭi Yŏn'gŭmsulssa K'ŭronik'ŭl* [Fullmetal Alchemist chronicle], trans. Sŏ Hyŏn-a (Seoul: Haksan Munhwasa, 2022), 379–381.
16. David T. Mitchell and Sharon L. Snyder, *Narrative Prosthesis, Disability and the Dependencies of Discourse* (Ann Arbor: University of Michigan Press, 2000), 8.
17. Arakawa, *Kangch'ŏlŭi Yŏn'gŭmsulssa K'ŭronik'ŭl*, 380.
18. Schalk, *Bodyminds Reimagined*, 8.
19. Mitchell and Snyder, *Narrative Prosthesis*, 8–10.
20. Visually Ishval, Amestris, and Xing are respectively equivalent to the Middle East, the United States/Europe, and China as seen through their characteristic physical and cultural depictions.
21. Michael Daliot-Bul and Nissim Otmazgin, "The Legacy of Anime in the United States: Anime-Inspired Cartoons," in *The Anime Boom in the United States* (Cambridge, MA: Harvard University Asia Center, 1972): 116–119; So-hyang Yang and Kim Kyŏng-hŭi, "Ilbon aenimeisyŏn <kangch'ŏl ŭi yŏn'gŭm sulssa> sŭt'orit'elling yŏn'gu: yŏngung k'aerikt'ŏ rŭl chungsim ŭro" [Research on the

storytelling of the Japanese animation "Fullmetal Alchemist": Focusing on the hero characters] *Küllobŏl munhwa k'ont'ench'ŭ hakhoe haksul taehoe*, no. 2 (2019): 118.
22 Lesley-Anne Gallacher, "(Fullmetal) Alchemy: The Monstrosity of Reading Words and Pictures in Shonen Manga," *Cultural Geographies* 18, no. 4 (2011): 458.
23 Alkahestry is initially confused with alchemy because Amestrians did not know any technology other than alchemy. Because of this, alkahestry users such as Scar and Mei are both mistaken for alchemists rather than being recognized as capable individuals outside of the reach of Amestrians.
24 Paul Midford, "Japan's Response to Terror: Dispatching the SDF to the Arabian Sea," *Asian Survey* 42, no. 3 (2003): 330–331, https://doi.org/10.1525/as.2003.43.2.329.1.
25 Rosemarie Garland Thomson, *Extraordinary Bodies: Figuring Physical Disability in American Culture and Literature* (New York: Columbia University Press, 2017), 16.
26 Jasbir Puar, *The Right to Maim: Debility, Capacity, Disability* (Durham, NC: Duke University Press, 2017), 65–68.
27 Kumiko Sato, "How Information Technology Has (Not) Changed Feminism and Japanism: Cyberpunk in Japanese Contexts," *Comparative Literature Studies* 41, no. 3 (August 2004): 337.
28 In the world of *FMA*, alchemy and alkahestry are both enabled by bringing together intricate sigils that signify destruction and reconstruction. Scar's right arm, which bears the sigil for destruction, was given to him by his brother, an alkahestry user, while Scar was unconscious and dying. Initially, it only grants him the power of destruction, but later in the series he adds in the sigils for reconstruction after he becomes allied with the Elric Brothers.
29 Arakawa, *Kangch'ŏlŭi Yŏn'gŭmsulssa*, vol. 11, 85.
30 Arakawa, *Kangch'ŏlŭi Yŏn'gŭmsulssa*, vol. 2.
31 David S. Roh, Betsy Huang, and Greta A. Niu, "Technologizing Orientalism: An Introduction," in *Techno-Orientalism: Imagining Asia in Speculative Fiction, History, and Media*, ed. David S. Roh, Betsy Huang, and Greta A. Niu (New Brunswick, NJ: Rutgers University Press, 2015), 8.
32 George W. Bush, "Address to a Joint Session of Congress and the American People" (George W. Bush White House Archives, September 20, 2001), https://georgewbush-whitehouse.archives.gov/news/releases/2001/09/20010920-8.html.
33 Jasbir K. Puar, *Terrorist Assemblages: Homonationalism in Queer Times* (Durham, NC: Duke University Press, 2007), 8.
34 Eli Clare, *Brilliant Imperfection: Grappling with Cure* (Durham, NC: Duke University Press, 2017), 27.
35 Jasbir Puar, "Critical Disability Studies and the Question of Palestine: Toward Decolonizing Disability," in *Crip Genealogies*, ed. Mel Y. Chen et al. (Durham, NC: Duke University Press, 2023), 121.
36 Betsy Huang, "Premodern Orientalist Science Fictions," *MELUS* 33, no. 4 (2008): 23–24, https://doi.org/10.1093/melus/33.4.23.
37 Toni Hays, "Open Concept. Land and Home in Asian/America" (PhD diss., University of California, Irvine, 2024), 7–8.
38 Huang, "Premodern Orientalist Science Fictions," 24.
39 Sianne Ngai, *Our Aesthetic Categories: Zany, Cute, Interesting* (Cambridge, MA: Harvard University Press, 2012), 28.
40 Yano, *Pink Globalization*, 257.
41 Ngai, *Our Aesthetic Categories*, 93.
42 Arakawa, *Kangch'ŏlŭi Yŏn'gŭmsulssa*, vol. 8.
43 Arakawa, *Kangch'ŏlŭi Yŏn'gŭmsulssa*, vol. 11.

44 Arakawa, *Kangch'ŏlŭi Yŏn'gŭmsulssa*, vol. 11.
45 Arakawa, *Kangch'ŏlŭi Yŏn'gŭmsulssa*, vol. 11.
46 Arakawa, *Kangch'ŏlŭi Yŏn'gŭmsulssa*, vol. 16.
47 Roh, Huang, and Niu, "Technologizing Orientalism: An Introduction," 10.
48 Qingguo Jia, "The Impact of 9-11 on Sino-US Relations: A Preliminary Assessment," *International Relations of the Asia-Pacific 3*, no. 2 (2003): 175.
49 Arakawa, *Kangch'ŏlŭi Yŏn'gŭmsulssa*, vol. 4.
50 Julie Ha Tran, "Thinking about Bodies, Souls, and Race in Gibson's Bridge Trilogy," in *Techno-Orientalism: Imagining Asia in Speculative Fiction, History, and Media*, ed. David S. Roh, Besty Huang, and Greta A. Niu (New Brunswick, NJ: Rutgers University Press, 2015), 150.
51 Jia, "The Impact of 9-11 on Sino-US Relations," 176.
52 Mel Chen, *Animacies: Biopolitics, Racial Mattering, and Queer Affect* (Durham, NC: Duke University Press, 2012), 201.

Part VI
Optimistic Futures

16

Recovering Asian American Futures in the Marvel Cinematic Universe

LORI KIDO LOPEZ

While theories of Afrofuturism have been inspiring art, literature, music, and academic publications since the 1970s, the concept surged in mainstream popularity surrounding the release of *Black Panther* in 2018. The eighteenth film in the Marvel Cinematic Universe (MCU), *Black Panther* tells the story of a technologically advanced African nation and its unapologetically Black superheroes. It has been widely celebrated for the way that it imagines a future where Black cultures are seamlessly intertwined with science and technology, but also where histories of slavery and colonialism have been avoided and African cultures can develop on their own terms. In a film written and directed by African American artists, including Ryan Coogler and Joe Robert Cole, *Black Panther* shifted the MCU away from its monochromatically white cadre of heroes and proved that the most profitable franchise in global history could imaginatively showcase people of color. In 2022 with the arrival of the sequel, *Black Panther: Wakanda Forever*, this story extended beyond Afrocentricity to tell the story of Latinx/Indigenous heroes—focusing on the underwater civilization of Talokan and its hero Namor, or the Mesoamerican deity K'uk'ulkan. While Latinx futurism and Indigenous futurism cannot be conflated and have their own unique qualities and characteristics, they are both inspired by Afrofuturism and similarly use speculative fiction and other works of art to posit an empowering dream for the future. In these visions, Black, Latinx, and Indigenous peoples maintain autonomy over their own lives. Such works actively imagine what liberation and healing might look like as a way of countering racial trauma and oppression.

These hopeful visions of the future for people of color have sharply contrasted with how Asian futures have been represented and imagined, due to the dominance of techno-Orientalism. Techno-Orientalism is a framework characterized by dystopian visions of an Asianized future that represent the feared subordinance of white Western society to Asian powers. For Asians and Asian Americans, mastery over technology has long figured into discourses of dehumanization and alienation—propelling fears of Asian economic domination, but also assumptions that Asians are themselves more robotic, mechanical, and soulless than other races. Techno-Orientalist narratives have proliferated in science fiction films such as *Blade Runner, Ghost in the Shell, Cloud Atlas,* and *The Matrix,* each set within futuristic worlds that are both highly technologized and inescapably Asian. Building from Edward Said's theory of Orientalism that posits that Westerners conceive of Asian and Middle Eastern cultures as inscrutable, mysterious, and "Other," these ways of imagining the future are closely connected to "yellow peril" ideologies. Discourses invoking the "yellow peril" paint Asians and Asian Americans as dangerous, threatening, diseased, immoral, and fixated on world domination. The fear of Asian immigrants posing a threat to white American jobs only heightened with the rise of first the Japanese and then the Chinese economy as looming global threats to American corporate supremacy. These narratives are closely connected to understandings of Asian Americans as the "model minority," or the nonwhite community that has managed to thrive and succeed in Western society due to innate talents such as a soulless dedication to scholarly pursuits and a preternatural affinity for math, science, and technology.

As with all forms of Orientalism, techno-Orientalism is a discourse that is created and propagated by white Western authors, mainstream media, and other non-Asian voices. Indeed, the prevalent tropes of foreignness, exoticism, and Othering that proliferate within Orientalist and techno-Orientalist narratives are clear markers that such narratives are being written from the perspective of non-Asians. These struggles around authorship are also important to contrast with Afrofuturism, Indigenous futurism, and Latinx futurism—bodies of creative production that foreground self-authorship and represent an important form of autonomy and self-determination. When Asian American authors and artists do take control, they often engage with techno-Orientalist themes as a way of resisting and contesting histories of subjugation. In the edited collection *Techno-Orientalism: Imagining Asia in Speculative Fiction, History, and Media,* authors point to the subversive potential of Larissa Lai's poetry book *Automaton Biographies,* Greg Pak's short film vignettes called *Robot Stories,* Noboru Iguchi's action film *Robogeisha,* and Nam June Paik's video art. Yet as seen in these examples, this engagement by Asian Americans with techno-Orientalist themes often takes place within frameworks of irony, parody, and self-referential critique rather than through optimism.

Within this context, we can then ask how the 2021 film *Shang-Chi and the Legend of the Ten Rings* and the 2022 television miniseries *Ms. Marvel* put forward optimistic visions of a specifically Asian American future. Both additions to the MCU are helmed by Asian American writers and directors, and center Asian American superheroes and their families. Their narratives respectively feature the journeys of protagonists Shang-Chi and Kamala Khan as they return to their ancestral countries of China and Pakistan, discover their technologically enhanced powers, and learn how to use their powers for strength, protection, and helping others. This chapter explores these texts as an opportunity to imagine a different kind of Asianized future that counters the predominance of Sinophobia and Asianized apocalyptic dread. It theorizes a version of Asian futurism that aligns with the optimistic, corrective vein of Afrofuturism where technologies wielded and controlled by communities of color become a mechanism for protection of the vulnerable and disruption of long-standing racism and oppression. I argue that while technological mastery has been a sufficient condition for imagining Black and brown empowerment, Asian futurism must take up a different relationship to technology—in this case, focusing instead on the way that technological enhancement allows Asian superheroes to finally reckon with painful family histories and reconnect with their ancestral origins. In contrast to Afrofuturist renderings of technology as a signal of modernity, Asian futurism avoids the trap of modernity by reorienting narratives of technological mastery toward kinship and healing. This chapter also explores the limited political potential of superhero stories in challenging assumptions about techno-Orientalism. While it is important to reflect on the liberatory power and potential of optimism, we must also reckon with the complexities and constraints of franchised Hollywood superhero narratives.

The Role of Technology in Afrofuturism

If racialized visions of the future have been closely connected to our technological imagination, it is important to clarify what is included within the concept of technology. Articulations of technology can be quite broad, as we would use the term "technological achievement" to describe the advent of the stone blade, the steamboat, the plow, gunpowder, and countless other scientific instruments. In tracing the way that new technologies come to improve upon and eventually replace these tools, we may be inclined to ignore these simpler technologies and reserve use of the term for only advanced technologies such as complex machines, digital and electronic tools, and computers. Yet this conceptual evolution reminds us that all technologies are tools used to help humans to achieve their goals and advance society, and that new technologies are constantly in development. As Marshall McLuhan argues in *Understanding Media: The Extensions of Man*, it is not the specific technology or medium that matters, but its usage, its impact on society, and its social value that we should pay attention to.[1] This understanding of technology forms the basis for visions of racial futurity, as racialized communities build relationships to technology of all kinds—including computers, health care, weaponry, and simple objects—that are premised on societal advancement and change.

Black, Latinx, and Indigenous communities have been systematically positioned in opposition to technological progress and scientific advancement. These racist histories extend all the way back to the origins of scientific racism and its assumption that Black and brown cultures are more primitive and have less technological aptitude than superior white cultures. Since the Enlightenment, white scholars have created racial typologies based on pseudoscientific fields such as phrenology and anthropometry to justify the racist and inaccurate belief that Caucasians have the most developed brains. These discourses and assumptions were taken up in the United States to support the enslavement of Africans and genocide of Indigenous peoples.[2] Black and brown communities have also been exploited and targeted in the name of science, with incidents such as the Tuskegee Syphilis Study and the uncompensated use of Henrietta Lacks's cancer cells for medical research serving as painful reminders.[3] The harmful impacts of institutionalized racism have been reinforced to this day throughout education systems and professional hiring, with the fields of science, technology, engineering, and mathematics (STEM) discouraging or excluding African American, Latinx, and Indigenous students and workers.[4] Beyond these inequities, we have also seen the way that technologies of many kinds have contributed to exacerbating racism through anti-Black violence, colonization, predatory surveillance, and countless other institutionalized forms of oppression.[5]

As an antidote and opposition to such discourses, the aesthetic power of Afrofuturism has always been connected to an optimistic perspective on Black uses of technology. Mark Dery uses the term "Afrofuturism" to describe a genre of African American speculative fiction that "appropriates images of technology and a prosthetically enhanced future" as a way of countering the deliberate extermination of African American histories.[6] In a wide range of media, including music, novels, comics, movies, visual art, and games, African American creators have used Afrofuturism to imagine relationships to technology that benefit their communities and envisioned a new kind of technological proficiency that allows Black communities to flourish. These creators include singer Janelle Monae, whose musical persona Cindi Mayweather is a time-traveling android designed to bring freedom and destroy binaries, Octavia Butler, whose science fiction novel *Parable of the Sower* imagines a postapocalyptic world where an African American teenager struggles to save humanity and the environment, and visual artists like Renee Cox, who uses digital technologies to center Black women's bodies within mathematically complex portraiture or Western religious iconography where they had previously been invisible.[7] Afrofuturism comes in many different forms and is constantly evolving, but at its core it focuses on creating new spaces for rethinking the relationship between African diasporic peoples and technology as one of empowerment. Technology becomes a mechanism for looking to the future, for reparation from past harms, and for taking control over one's own destiny.

Celebrating Technology in *Black Panther*

While there are many ways to understand the power of the *Black Panther* film, its celebratory relationship to technology has often been highlighted as one of its strengths. Shayla Monroe argues that the resonance of *Black Panther* with Black viewers is based on its portrayal of a technologically advanced African country. To counter prevailing images of Africa as technologically stagnant, Wakanda is celebrated for manipulating the metal ore vibranium to produce stunning architecture, life-saving health care, space-age aircraft, overpowering weaponry, state-of-the-art forms of artificial intelligence (AI), and the ability to completely conceal the existence of the country from the outside world. Wakanda is ruled by a leader who ingests a heart-shaped herb that is infused with vibranium, giving that leader the power of enhanced strength. Monroe assesses the film's depiction of internal debates over the uses of these technological innovations, pointing to conflicting perspectives about whether technological change and progress are always helpful rather than harmful.[8] Yet it is undeniable that the film's general orientation toward technology is positive and celebratory, marking Wakandans as impressive and socially advanced thanks to their mastery over technology.

A similar orientation to the power of technological advancement for Latinx/Indigenous cultures can also be seen in the film *Black Panther: Wakanda Forever*. While the African nation of Wakanda remains at the center of the narrative, this sequel introduces a new nation called Talokan that exists entirely underwater. In a parallel narrative to Wakanda, the people of Talokan benefit from their exclusive access to the extraterrestrial element vibranium and a vibranium-infused plant. The film narrates their origins as a Yucatán Mayan community suffering from diseases brought by Spanish colonizers in the sixteenth century. After ingesting the vibranium-enhanced plant, they develop the ability to breathe underwater and are able to move their entire civilization into an underwater trench. From there, they decide to retreat from humanity for their own safety and protection. Throughout the film, the people of Talokan display an impressive array of technologies including water-based weapons, hypnotizing sonic tools, above-water breathing apparatuses, and rapid healing capabilities.

These celebrations of the way that control over technology can be connected to the struggle for sovereignty and empowerment are also reflected in the thinking of Indigenous and Latinx futurism. Following the rise of

Afrofuturism, Indigenous and Latinx creators also began to develop their own forms of speculative fiction in relation to technological advancement. Terms such as "Indigenous futurisms" and "Chicana futurisms" are used to describe the art, media, narrative, and other forms of cultural production that have emerged. As Catherine Ramirez explains, "Like black people, especially black women, Chicanas, Chicanos, and Native Americans are usually disassociated from science and technology, signifiers of civilization, rationality, and progress. At the same time, many Chicanas, Chicanos, and Native Americans have been injured or killed by and/or for science and technology."[9] For the people of Talokan, the decision to move their people underwater becomes a way to shield themselves from the death and destruction wrought by colonization. Just as the Wakandans decide to cloak their nation from the outside world, the people of Talokan use powerful technologies to remain invisible and preserve their unique culture and heritage from discovery. This desire to remain undiscovered represents a challenge to the histories of genocide and destruction that resulted from the "doctrine of discovery" that Christopher Columbus and other European explorers used to colonize territories occupied by Indigenous peoples.[10] Indeed, Western colonizers have used technology in an outward-facing orientation throughout history, striving to extend their imperial reach through advanced developments in navigation, transportation, and weaponry. Within the Afrofuturist and Latinx/Indigenous futurist envisionings of the MCU, we see an emphasis on societies withdrawing and carefully guarding technology—as a safeguard for their sovereignty, but also in response to the omnipresent threat of Western thievery and usurpation. Asian American narratives around technology and futurity bypass these questions of sharing versus hoarding and instead turn to questions of how technology can be used for healing wounds within oneself, one's family, and one's community. Such differences reflect the ways that white supremacy has shaped all racialized groups and the relationships between them.

Technology as Protection in *Shang-Chi* and *Ms. Marvel*

Shang-Chi and the Ten Rings premiered in 2021 as the first MCU narrative featuring an Asian American lead. It was also the first Marvel film with Asian Americans in key creative roles—including mixed-race Japanese American director and writer Destin Daniel Cretton and mixed-race Chinese American writer David Callaham. The film follows actor Simu Liu as Shaun/Shang-Chi, a Chinese American young adult who returns to his ancestral roots in China, accompanied by his best friend Katy, played by Awkwafina. In addition to facing off with his villainous father and reuniting with his abandoned sister, Shang-Chi must travel to an alternate dimension called Ta Lo and fight against evil forces. While Ta Lo is protected by a powerful dragon whose scales are used to develop advanced technologies and weaponry, the primary goal of its inhabitants is to contain a creature called Dweller in Darkness that could escape into our world and extinguish human life on Earth.

The eighth television entry into the MCU released onto the streaming platform Disney+, *Ms. Marvel* features actress Iman Vellani as Kamala Khan—a Pakistani American high school girl trying to break free from the restrictions of her Muslim American parents. After receiving a mysterious bangle from her grandmother in Pakistan, Kamala starts to develop superpowers and travels to Pakistan to learn more about her family's origins. She learns that the bangle belonged to her great-grandmother, a visitor from an alternate dimension whose people are now set on using Kamala and the powerful technology of the bangle to send them home—even if it means destroying humanity in the process. The bangle briefly transports Kamala back in time so that she can guide her grandmother to safety during the dangerous Partition of India and Pakistan, and then she returns home with greater control over how to use her powers to protect herself, her loved ones, and her community.

We can begin analyzing these texts in relation to techno-Orientalism by considering the role of technology in the Asianized worlds they depict. Both texts center advanced technologies that give their users enhanced powers—a gift that also comes with the threat of those who would use it for harm. In *Shang-Chi*, the realm of Ta Lo is described as extremely technologically advanced; Shang-Chi's auntie Ying Nan explains that their world used to "have cities that surpass any in your universe, rich with culture and history." But after the Dweller in Darkness decimated their largest cities, the only way the inhabitants of Ta Lo were able to survive was through "the gift" given by their Great Protector—glowing red dragon scales that can be sewn into clothing or forged into armor and advanced weaponry, which they primarily use to keep the Dweller in Darkness contained so it will not harm others. The other advanced technology in *Shang-Chi* is Wenwu's rings, a set of ten metal rings that give their wearer eternal life and enhanced strength for attacking enemies. In *Ms. Marvel*, we also see advanced technologies that originated in alternate dimensions. Kamala learns that her great-grandmother Aisha is actually from the Noor dimension, and that a group of Noor people seek a powerful bangle that possesses the power to let them return to their home dimension. When Kamala puts on the bangle, she develops a number of supernatural abilities—including the ability to create glowing force fields, to extend her hands into giant glowing fists, and to connect with her ancestors through both psychic bonds and time travel. She learns that if the visitors from Noor use the bangle to return home they will destroy our universe, so she ultimately teams up with a coalition known as the Red Daggers who oppose the Noor to protect our dimension.

In both of these stories we see the existence of advanced technologies, but the worlds that these Asian American superheroes inhabit are not premised on the miracle of technological advancement. This contrasts with the depictions of both Wakanda and Talokan, societies that could not exist in the way that they do without the intervention of vibranium and its high-tech capabilities. In both *Black Panther* films, control of technology is positioned as a key tool for imagining what an alternate future for Black, Latinx, and Indigenous communities might look like. Due to the critical importance of technological advancement in their societies, both Wakanda and Talokan end up going to war in order to protect their exclusive access to vibranium. This struggle to protect their special relationship to technological advancement becomes a way of protecting their way of life and their sovereignty.

But to imagine an alternate future for Asian Americans, technological mastery alone is not enough—indeed, a world where Asians become powerful through their exclusive use of technology is at the very core of techno-Orientalism and its accompanying dystopian terror for non-Asians. In order to be optimistic about a different vision for Asians and Asian Americans and elide the fear of techno-Orientalism, we see a different orientation to technology altogether in Asian American superhero films. First, both Shang-Chi and Kamala wield advanced technologies as a way of protecting the entire world. The threats posed by the Dweller in Darkness and the incursion of the Noor realm are not to a single village, society, or way of life, but to the entire planet and realm inhabited by all of humanity. The alien technologies of the ten rings and the bangle are both dangerous and unknowable, and while they ultimately facilitate protection, they are not the thing in need of protecting.

Moreover, both Shang-Chi and Kamala are depicted as extremely ambivalent and wracked with self-doubt about whether they should utilize the technology of the ten rings or the bangle. Shang-Chi fully rejects the inheritance of his father's ten rings as a teenager, escaping to the United States and going into hiding rather than accept a life ruled by vengeance and violence. When battling his father in order to keep the Dweller in Darkness contained, Shang-Chi is extremely reluctant about ultimately putting on the ten rings and using violence against his father—even when the safety of the world is at stake. He has seen this technology bring about violence, pain, and destruction, and he is not interested in the power that is associated with it. Kamala has a different path toward embracing her great grandmother's bangle and all of its power; she is initially excited about the prospect of becoming a Muslim American superhero and cannot wait to become adept with her newfound powers. But she later becomes overwhelmed with self-doubt about whether she deserves this kind of power, and hides behind her alter ego to avoid the pressure of public scrutiny. When she learns about the Noor dimension, she is similarly confused about what path to take and whether she should be helping her own ancestors return home or protecting Earth.

Technology as Family Reunification

In reconciling a dream for productive relationships between Asian Americans and technology, we also see a deep focus on looking inward and using technology to heal wounds within one's own family. *Ms. Marvel* tells the story of multiple children who feel abandoned and misunderstood by their parents. This includes an entire lineage of generational trauma wrought by the separation and family destruction that occurred during Partition—beginning with Kamala's great-grandmother Aisha, who dies as her daughter Sana catches the last train out of Karachi and leaves Sana with not even a photograph to remember her by. Sana then has a daughter named Muneeba who feels abandoned when she immigrates to the United States and loses touch with her mother, who remains in Pakistan. Finally, Kamala rebels against the strictures of her mother Muneeba, who is weary of Kamala's obsession with fandom and imaginary worlds. The technology of the bangle brings them together in many different ways—by creating a psychological bond between Kamala and her grandmother Sana that summons her to return to Karachi, by facilitating time travel so Kamala can meet up with her great-grandmother Aisha, and by ultimately bringing the women together to discuss their traumas and feelings of abandonment for the first time.

This focus on familial trauma and the need to reunite and make amends—a classic narrative within Asian American literature and storytelling—is also a dominant theme in *Shang-Chi*. After his wife is murdered, Wenwu raises his son Shang-Chi as the heir to the ten rings and trains him as a human weapon. Resisting this violence, Shang-Chi turns his back on his family and escapes to the United States to start a new life as Shawn. When he finally returns to China, he is forced to confront the reality that when he left the compound he also abandoned his sister, Xu Xialing. The first way that technology is used for family reunification is through the green pendants that Shang-Chi and Xu Xialing's mother gave them as children—they provide the key to unlocking a portal into her home dimension of Ta Lo, but they only work when both pendants are inserted together. While the siblings initially reunite over the pendants, they only start to really make amends when they enter Ta Lo together and join forces to save their mother's village from both their father's attackers and then the Dweller in Darkness. Equipped with dragon-scale armor and weapons, they finally begin to realize the importance of being able to trust and rely on one another. Shang-Chi then only realizes his full potential as the wielder of the ten rings when he faces down his father by integrating the inheritance of both of his parents, embracing both the aggression taught by his father and the flowing, open-handed style of his mother. We certainly see battles over technology and who gets to control them, but technology ultimately becomes a way of bringing families together to resolve conflicts and heal from past traumas.

Beyond helping to reunite fractured families, these texts that center Asian futurity also convey larger visions for an alternative world. The pressing need for a utopian view of the future that imagines liberation and empowerment is of course premised on the dystopian realities of one's current situation—that in our current world, African Americans, Latinx and Chicanx peoples, and Indigenous peoples have experienced generations of oppression, slavery, and colonization. These works of futurism are a direct response to those conditions of oppression. Under this logic, it would make sense for Asian Americans to similarly long for a future where they are free from the relentless Othering, dehumanizing forces of Orientalism and the violence of anti-Asian hatred. Yet we can also see meaningful differences between these different groups, and trace the impact those differences have had on necessary dreams for alternative realities. For instance, it makes sense that both African Americans and Indigenous peoples would dream of sovereignty and the right to govern themselves free from outside influences, asking what their lives would be like without having had foreign invaders wrest them from their native lands, enslave them, and take away all their autonomy and freedom. Yet Asian immigrants to the United States more commonly chose to come to the country of their own free will in search of greater opportunity. As we have seen in *Ms. Marvel*, there are certainly communities of Asian immigrants who have experienced the pain of colonization and forced resettlement. But there are others whose experiences of oppression might manifest in different desires and longings.

We can see these differences play out in *Shang-Chi* and *Ms. Marvel* and the way that their stories produce imaginings of Asian futurism. Both contain stories about longing for family connection and purpose amid separation, loss, and feeling invisible. Their dream for the future is about reunion with family from the homeland, rekindling a lost ancestral connection, and aligning with one's community around a shared purpose. In *Shang-Chi*, we see a Chinese American who immigrated to the United States, repressing the pain of having left his family behind and never processing the harm that he experienced or that he caused to others. He and his best friend Katy work as valets and have yet to discover their true passion or direction, despite pressure from their families and communities. Their dream for the future is about Asian immigrant families being able to face their past traumas, and ultimately find purpose and satisfaction by claiming a path of their own choosing. In *Ms. Marvel*, we see a story about the pain of forced separation and the loss of unity wrought by the British colonization of India and the painful partitioning of Pakistan. Moreover, amid a Muslim American community suffering from Islamophobia, *Ms. Marvel* dreams of a world where a Pakistani American and Muslim American high school girl can become a superhero on the level of the white American Captain Marvel, Carol Danvers. In both stories, the Asian American characters use technology and its potential for reconciling histories and relationships as a way to integrate into a specifically Asian American space.

Superheroes and Supervillains

Beyond examining the role of technology in these optimistic visions of Asian futurism, it is also important to interrogate the very core of their genre—the depiction of an Asian American superhero. Indeed, one of the most significant cultural contributions of *Black Panther*, *Shang-Chi*, and *Ms. Marvel* to our media landscape is that they portray people of color as not just the protagonists and narrative leads, but as actual superheroes. This challenges the histories of Orientalist depictions wherein Asian Americans are narratively subordinated through a focus on white heroes and their perspectives. For instance, the iconic techno-Orientalist film *Blade Runner* takes place within an Asianized world, but focuses primarily on a white hero and his actions. One of the key interventions of racial futurisms is that when they envision the future, they do it through the lens of centering people of color—it is their desires, motivations, and journeys that we understand most clearly, and whose actions propel the narrative forward. This is another way that authorship has meaning, as authors of color are often understood as being in a better position to write nuanced characters of color.

Within the MCU, the overwhelming whiteness and androcentrism of its heroes have been widely condemned. It is notable that it was not until the eighteenth film that we saw a protagonist of color in *Black Panther*, and not until the twenty-first film that we saw a female protagonist in *Captain Marvel*. This is why there was significant fanfare around both *Black Panther* and *Shang-Chi* for finally representing heroes of color. News stories abounded about the young African American and Asian American fans who finally had a role model who "looked like them," who eagerly dressed as T'Challa and Shang-Chi for Halloween and purchased Black and Asian action figures and other merchandise, and who had outsized emotional reactions upon viewing the films for the first time.[11] The timeworn trope of calling for "positive representation" as a way of countering negative stereotypes and the persistent messages that it is "bad" to be a person of color deeply colored the discourse surrounding these films.

Yet within media studies scholarship about identity and marginalization, there has always been a skepticism that banal superheroism is the cure-all for representational ills. In addition to holding minorities up to an impossibly high moral standard, the journey of the lone hero—or the superhero—can be predictably bland and lacking in complexity. Celebrations of superheroes as the antidote to representational harms also reify a binary understanding of representation that categorizes images into either "positive/good" or "negative/bad." Yet this binary framework has been roundly condemned as overly simplistic and unhelpful, as there are many more nuanced ways of understanding the rich depths of media representation. Kristen Warner criticizes thinly written characters of color as "plastic" even if they are presumably positive and respectable, while Evelyn Alsultany argues that the project of balancing negative representations with positive ones can result in simplified representations that do little to mitigate harm.[12] One of the most helpful insights comes from Racquel Gates, who argues that it is important to strategically embrace shameful, disreputable, "negative images" of blackness for the way that they connect more authentically and emotionally with Black audiences, as they sometimes reflect the realities of their everyday lives in a way that can be cathartic and liberatory.[13]

If we apply these same expectations to superhero stories, we can see that it is important to look beyond our glorified heroes of color to consider the power of representing villains of color as well. Indeed, within each of the MCU texts analyzed here, there are also compelling, richly layered portrayals of villains of color—including Erik Killmonger from *Black Panther*, Namor from *Black Panther: Wakanda Forever*, Wenwu from *Shang-Chi*, and Najma from *Ms. Marvel*. Each of these characters resists the common characterization of villains as cartoonish in their interminable quest for evil, instead embodying perspectives and desires that are so understandable and sensible that it is often difficult to root against them. In *Shang-Chi*, Wenwu seeks to be reunited with his deceased wife, and believes that his attack on Ta Lo is the only thing that will allow him to be with her. While the reality is that his mind has been deceived by the Dweller in Darkness, his love for his wife is sincere and relatable. Similarly, the character of Najma from *Ms. Marvel* is the leader of her people, and she simply seeks to bring her people back to their home dimension after many generations of exile on Earth. While both Wenwu and Najma ultimately attack and abandon their own children and become morally indefensible, their guiding principles of desiring reunification nonetheless make them sympathetic in many moments. Moreover, the portrayals of Wenwu by the famed Hong Kong actor Tony Leung and Najma by Pakistani actress Nimra Bucha also lend gravity and competence to their roles.

Conclusion

This chapter has explored the way that these stories from the MCU participate in shaping possibilities for an Asian futurism that challenges the pessimism and peril of techno-Orientalism. In these fantastic superhero stories we can see Asian Americans fighting for their own distinctive version of liberation, where they are able to reunite family scattered across the globe and create community spaces for healing from trauma and harm. While these dreams for Asian American futures are framed here as different from those of Afrofuturism and Latinx/Indigenous futurism in their more ambivalent orientations toward technology, it is also important to highlight the role of white supremacy in necessitating such differences. Indeed, while differential racialization may have created historical legacies that attempt to push communities of color apart from one another, there must also be dreams for a future when people of color align in overcoming institutionalized racism in its many different forms. As the MCU continues to expand and proliferate with new narratives each year, we are sure to see the forging of new relationships between and across each of these characters and their peoples that may shift our understanding of these previously explored themes. Yet we must also stay attuned to questions of authorship and the corporate mechanisms of such large-scale franchises.

Indeed, it can be questionable how much political or liberatory potential we should be ascribing to superhero movies that are so deeply connected to a dominant corporate juggernaut. Each individual text within the MCU may have the opportunity to break new ground with its narratives, individual cast and crew, and artistic decisions. But as component parts of one of the largest and most profitable media franchises of all time, there are deep intertextual bonds among these texts that will always play a role in shaping racial meanings. As we look closely to these texts for evidence of what we can learn about racial futures, we must also consider who is responsible for each narrative, and who is profiting. The films and TV shows analyzed here may foreground people of color behind and in front of the camera, but the overwhelming whiteness of MCU products and producers must also be acknowledged. In moving forward with understandings of the political potential of Asian American futures, it will be important to look beyond the MCU to find other examples of Asian American–authored visions of the future that resist dystopian pessimism and dare to put forward a dream propelled by optimism, and to continue asking what Asian American happiness and fulfillment might look like on screen.

Notes

1 Marshall McLuhan, *Understanding Media: The Extensions of Man* (Cambridge, MA: MIT Press, 1994).
2 Audrey Smedley, *Race in North America: Origin and Evolution of a Worldview* (Boulder, CO: Westview Press, 2007).
3 See Allan M. Brandt, "Racism and Research: The Case of the Tuskegee Syphilis Study," *The Hastings Center Report* 8, no. 6 (1978): 21–29; Rebecca Skloot, *The Immortal Life of Henrietta Lacks* (New York: Random House, 2011).
4 See Lori Kido Lopez, *Race and Digital Media: An Introduction* (London: Polity Press, 2023); J. Blaine Hudson, "Scientific Racism: The Politics of Tests, Race and Genetics," *The Black Scholar* 25, no. 1 (1995): 3–10; and Ebony Omotola McGee, *Black, Brown, Bruised: How Racialized STEM Education Stifles Innovation* (Cambridge, MA: Harvard Education Press, 2021).
5 See Simone Browne, *Dark Matters: On the Surveillance of Blackness* (Durham, NC: Duke University Press, 2015); and Ruha Benjamin, *Race After Technology: Abolitionist Tools for the New Jim Code* (London: Polity, 2019).
6 Mark Dery, Greg Tate, and Tricia Rose, "Black to the Future: Interviews with Samuel R. Delany, Greg Tate, and Tricia Rose," in *Flame Wars: The Discourse of Cyberculture*, ed. Mark Dery (Durham, NC: Duke University Press, 1994), 180.
7 Nathalie Aghoro, "Agency in the Afrofuturist Ontologies of Erykah Badu and Janelle Monáe," *Open Cultural Studies* 2, no. 1 (2018): 330–340; Michael Brandon McCormack, "'Your God Is a Racist, Sexist, Homophobic, and a Misogynist . . . Our God Is Change': Ishmael Reed, Octavia Butler and Afrofuturist Critiques of (Black) American Religion," *Black Theology* 14, no. 1 (2016): 6–27; Lesly Deschler Canossi and Zoraida Lopez-Diago, *Black Matrilineage, Photography, and Representation: Another Way of Knowing* (Leuven: Leuven University Press, 2022).
8 Shayla Monroe, "Tradition, Purpose, and Technology: An Archaeological Take on the Role of Technological Progress in Black Panther," in *Afrofuturism in Black*

Panther: Gender, Identity, and the Re-Making of Blackness, ed. Renee T. White and Karen A. Ritzenhoff (Lanham, MD: Lexington Books, 2021), 253.
9 Catherine S. Ramirez, "Afrofuturism/Chicanafuturism," *Aztlán: A Journal of Chicano Studies* 33, no. 1 (2008): 188.
10 Robert J. Miller, "The Doctrine of Discovery: The International Law of Colonialism," *Indigenous Peoples' Journal of Law, Culture and Resistance* 5, no. 1 (2019): 35–42.
11 Stereo Williams, "What 'Black Panther' Means to Black Boys and Girls," *Daily Beast*, February 18, 2018, https://www.thedailybeast.com/what-black-panther-means-to-black-boys-and-girls; Salamishah Tillet, "'Black Panther' Brings Hope, Hype and Pride," *New York Times*, February 9, 2018, https://www.nytimes.com/2018/02/09/movies/black-panther-african-american-fans.html; Kat Moon, "Shang-Chi and the Legend of the Ten Rings Made Me Feel Seen Like No Other Hollywood Blockbuster Has," *Time*, September 3, 2021, https://time.com/6095108/shang-chi-asian-representation/; Laura Yuen, "Why 'Shang-Chi' Matters to My Asian American Boys in the Era of #StopAsianHate," *Minnesota Star Tribune*, September 10, 2021, https://www.startribune.com/laura-yuen-when-we-needed-him-most-a-hero-brings-humanity-to-asian-american-narratives-laura-yuen-wh/600095774/.
12 Kristen J. Warner, "In the Time of Plastic Representation," *Film Quarterly* 71, no. 2 (2017): 32–37; Evelyn Alsultany, *Arabs and Muslims in the Media: Race and Representation after 9/11* (New York: New York University Press, 2016).
13 Racquel Gates, *Double Negative: The Black Image and Popular Culture* (Durham, NC: Duke University Press, 2018).

17

Looking for Asianfuturism

Asian American Science Fiction and Games of Color

EDMOND Y. CHANG

Asianfuturism is a response to the ongoing legacies of anti-Asian rhetoric, hate, and violence, especially in this COVID-19, "China Flu," world; it is a repairing and reappropriation of the fear and fantasies of a techno-Orientalist, globalized mediascape and popular culture; and, it is a meditation on Asian bodies, times, challenges, and possibilities to come. Asianfuturism confronts the assimilation, containment, marginalization, erasure, even eradication of Asian bodies, identities, voices, and histories, particularly in the United States and the global West. Asianfuturism is an Asian-inflected, infected, impactful twenty-first-century manifesto, a new groove, a capacious set of circuits, desires, and sensoriums to interrogate and reinterpret dominant visions of history, technology, and culture, particularly "Western ambivalences toward what it regarded as the mysterious power of the East, manifesting in strange contradictions."[1] Asianfuturism is imagining other worlds, conceiving of other whens and wheres, and reconfiguring our relationship to and representations of a racist, sexist, phobic, and Orientalist past, present, and future to imagine otherwise.

This essay offers a working definition of Asianfuturism (perhaps even a few paths toward Asianfuturisms) as modes of reading, writing, embodying, creating, and theorizing the worlds around us and the worlds yet to come. Through the specific resonances between literature and video games created by Asian and Asian American authors and developers, I hope to illuminate other modes of doing and being that do not replicate or recapitulate whiteness, heteronormativity, or neoliberal capitalist fantasies of belonging. Through these Asianfuturist texts, I show how these writers and creators antidote the techno-Orientalist negations of self, community, and survival. Dawn Chan, thinking about Asian American art, asked in 2016, "Is it possible to be othered across time? For almost a century already, the myth of an Asian-inflected future has infiltrated imaginations worldwide."[2] However, there is hope and connection as witnessed in the recent embrace in the United States of Asian and Asian American celebrity, pop music, and media, and even the growing political and activist coalitions of #StopAsianHate, #BlackLivesMatter, and the protests against recent conservative U.S. Supreme Court decisions offer hope and connections across different imaginings and formulations including Afrofuturism, Indigenous Futurism, and other ethnic futurisms.

Mark Dery, in their 1994 essay "Black to the Future," coined "Afrofuturism" to mean "speculative fiction that treats African-American themes and addresses African-American concerns in the context of twentieth-century technoculture—and, more generally, African American signification that appropriates images of technology and a prosthetically enhanced future ... to tell about culture, technology, and things to come."[3] Afrofuturist narrative, music, and other media reimagine and newly imagine different histories, possibilities, and configurations of being for black identities and bodies, often in response to the legacies and realities of slavery, segregation, and violence. It is in this recombinant spirit that an Asianfuturism might be assembled (or a slew of Asianfuturisms) that foreground, in this case, Asian American themes and concerns for the twenty-first century. This working definition of Asianfuturism offers a resonant examination and recodification of our analog and digital worlds through the interpretative lens of Asian American culture. It recognizes histories, legacies, or aspirations that might animate the need for Asianfuturism, for Asianfuturist creativity, experimentation, and world-building.

It is no surprise that Afrofuturism centers Black and brown positionalities in the face of and in defiance of centuries of enslavement, segregation, and death. Indigenous futurism foregrounds Indigenous sovereignty and survivance as a way to "renew, recover, and extend First Nations peoples' voices and traditions ... to envision Native futures, Indigenous hopes, and dreams recovered by rethinking the past in a new framework."[4] The reach here for Afrofuturism and Indigenous futurism is not to equate or flatten these formations but rather to revel in their frissons, harmonies, and counterpoints. Instead, the move is to explore the affordances, disjunctions, and coalitions across these voices and visions of time and space, particularly in the ways they handle technologically inflected race, gender, sexuality, and ability. Asianfuturism then joins this constellation of knowledges and practices to counter and reimagine Asian representations, tropes, exclusions, and appropriation by a dominant, normative, Western culture. In part, it is a response to the perils and traumas of the past, and more hopefully, it searches beyond the problems of the present. It is re-visioning, refuturing the twenty-first century and beyond. And it is mindful that, like queerness and other slippery terms, it cannot resolve into, in the words of critic Xin Wang, "suspiciously clean" and "homogenous methodological apparatuses," making sure that it respects and reflects the messiness of time, space, bodies, and context.[5]

To wit, Asianfuturism in literature, art, and media 1) critically foregrounds Asian and Asian American stories, cultures, and concerns in a literary, cultural, political, and lived landscape that naturalizes and maintains whiteness and normativity; 2) Asianfuturism reconfigures, even queers identities, embodiments, and technologies to imagine alternative, even radical worlds, desires, relationships, play, and possibilities as "direct action" that challenges racism, sexism, ableism, phobia, and other technonormativities; 3) Asianfuturism is intersectional, intertextual, and interdisciplinary in method, modality, and messiness of identity, representation, embodiment, and imagination; and 4) Asianfuturism revels in and reveals connection, community, collaboration, utopia, and most importantly, curiosity and joy.

This essay weaves together close reading of speculative fiction, close playing of digital games, and personal accounts of working on two exhibits at the Wing Luke Museum of the Asian Pacific American Experience in Seattle, Washington. Specifically, I draw on the Asianfuturist world-building of Larissa Lai's *The Tiger Flu*, the ludic possibilities of Melos Han-Tani's *All Our Asias* and Mike Ren Yi's *Hazy Days*, and the community development of the Wing's exhibitions *Worlds Beyond Here: The Expanding Universe of APA Science Fiction*, which opened in 2019, and *Asian American Arcade*, which was up in 2012. Though largely focused on Asian American texts, the following recognizes that even the notions of "Asian" or "Asian American" are fluid, contested, and not commensurate, and that these arguments are provocations for further wrangling, work, and specificity. This essay demonstrates how these texts and experiences critically (re)imagine Asian and Asian American cultures and concerns in order to reconfigure identities, embodiments, and technologies to imagine alternative, even radical narratives, desires, relationships, and play.

Asian American Science Fiction (SF) as Designed Experiences

I open this section with the community engagement work that I have been doing with the Wing Luke Museum of the Asian Pacific American Experience located in Seattle, Washington. Since 2011, I have been invited to be a part of a range of Community Advisory Committees (CAC) that serve to generate, organize, and curate exhibits at the museum. The mainspace exhibit titled *Worlds Beyond Here: The Expanding Universe of APA Science Fiction*, which opened in 2019 and ran for a year, illustrated the first imperative of Asianfuturism, which requires a

FIGURE 17.1 Screenshot of the Wing Luke Museum website featuring *Worlds Beyond Here* (2019). (Source: Wing Luke Museum of the Asian Pacific American Experience.)

mindful centering of Asian narratives, representations, and worlds and a rejection or reappropriation of assigned or acceptable roles, rules, forms, and genres (Figure 17.1). *Worlds Beyond Here* grappled with this first goal, hoping to explore "the connection between Asian Pacific Americans and the infinite possibilities of science fiction. . . . Despite the historically limited representation of Asian Pacific Americans [APA] in popular science fiction, they have had and continue to have a large impact in science fiction. . . . For many Asian Pacific Americans, science fiction addresses issues related to identity, immigration and race, technology, morality and the human condition."[6] The questions raised in the introduction above and the complexities of Asianfuturism were made real by the exhibit, put into practice in working with the CAC about how to encapsulate, interrogate, and complicate what the "universe of APA SF" is, was, and could be, and ultimately, about what to include, what counted as Asian American, and more importantly, what would inevitably be left out. Rather than dwell on the stereotypical past, the exhibit worked to tell a different story about APA peoples through SF, one that would engage contemporary, local, regional, Indigenous, multiracial, and multimodal imaginings of and by APA creators. In a deep sense, the exhibit also engaged the "science fictionality of museums," the "great potential for activist work in sci fi writing and museums to speak to each other . . . Through imagining and proposing other worlds, [both] might move some way towards building a different future."[7] In essence, an Asianfuturist curatorial philosophy emerged from the foregrounding of the work of APA writers, artists, and activists. Any formulation of Asianfuturism then must engage more than narrative, representation, or identification, but grapple with mechanics and the restructuring and reconfiguring of conventions and expectations that lead "to the restructuring of the knowledge [SF] produces about the world and the people it depicts."[8]

Betsy Huang's *Contesting Genres* posits that the reliance on realism, memoir, and immigrant fiction in Asian American literature was in a way part of an assimilationist desire to be legible, readable, and laudable to white publishers and audiences, particularly post–World War II, the Civil Rights era, and the rise of multiculturalism in education and mainstream media. But writers who turned to SF discovered the "tools for destabilizing the generic and social imperatives that have governed Asian American literary production."[9] Or, as Sami Schalk argues, "My use of speculative fiction allows me to mostly circumvent discussions of genre boundaries, genre histories (including histories of exclusion), and canon building which are not essential to my arguments. On the whole, I am less concerned about genre labels and more concerned with how a variety of nonrealist tropes and devices influenced the representation of (dis)ability, race, and gender" in SF texts.[10] However, even within science and speculative fiction the challenge to these norms varies, and it is largely women of color and queer of color artists and creators who push these destabilizations and imperatives to more radical horizons—centering racialized, queer, and other marginalized bodies in the face of bowdlerizing, sanitizing, whitewashing, and erasure—though some more radical than others.

For example, Ted Chiang, an award-winning Chinese American SF writer and former technical writer, began publishing in the early 1990s and garnered acclaim in the 2000s, winning a number of prestigious SF awards over the years. Chiang's writing is often described as "hard science fiction"—SF based on scientific or technological accuracy or fidelity—which has been privileged as rational, rigorous, and more realistic, and which has been historically and problematically conflated with white, heteronormative, male genius. In fact, Min Song in *The Children of 1965* argues that Chiang's "startling" refusal to address race is a direct consequence of his desire to be "explicitly part of a tradition comprising authors like Isaac Asimov, Arthur C. Clarke, and Ray Bradbury."[11] While Chiang's work is Asianfuturist for its engagement with social and technological tumult and change, its limits and containment of race, gender, and other "soft" concerns prevent it from embracing the full power of Asianfuturism. For instance, Chiang's novella "Story of Your Life" follows in this hard, realist, big-ideas tradition.[12] "Story of Your Life," which was adapted into the film *Arrival*, considered one of the best movies of 2016, is a "first contact" narrative in which extraterrestrials dubbed "heptapods" arrive on Earth and the protagonist, a linguist named Dr. Louise Banks, is tasked by the government to establish communication with the aliens. Without giving too much away, "Story of Your Life" is about language, incommensurate worldviews, science as philosophy, and temporality, all woven with a touching tale of family, motherhood, and loss. It is a beautiful and beautifully crafted story, but upon my first reading of "Story of Your Life" (and watching of the film), what is striking is the colorblindness (even genderblindness) of the story; like many science fiction and fantasy tales, race becomes metaphorized, ported to the "aliens" (or monsters) in the narrative world. The human characters, when left to some readers' imaginations, default to hegemonic whiteness—as canonized by the film, Dr. Louise Banks *is* always already Amy Adams, a white woman.

While some of Chiang's other stories do obliquely attend to issues of race, gender, and other norms, he has expressed that the writing, the craft, the story is paramount to these concerns. In an interview in 2012 for Asian American Writers' Workshop, he is asked point-blank by Indian SF writer Vandana Singh, "Many of your stories are set in America and feature American protagonists of unspecified race (one assumes white by default).... Does your being Asian American inform your stories in any way?"[13] And Chiang responds (at length) that "race inevitably plays a role in my life, but to date it's not a topic I've wanted to explore in fiction.... I'm hesitant about making my protagonists Asian Americans because I'm wary of readers trying to interpret my stories as being about race when they aren't.... This is just a special case of something most writers have to contend with—people reading their work in a certain light based on extratextual knowledge of the author."[14] Chiang, here, becomes a "special case," which confirms both the generic constraints on Asian American writers as outlined by Betsy Huang but also the need to hack, slash, confound, queer, and escape those constraints. Obviously, to assume that an Asian American writer *must* write about Asianness is essentializing, but colorblindness I would argue is ostensibly more problematic for its complicity in the dominant culture. What Chiang ignores is that the "racial subtext" is already assumed and in play because racial ideologies legitimate and make legible the writer and writing; Chiang's stories are about race even when "they aren't." As outlined above, Asianfuturism must center Asian and Asian American embodiment, difference, and existence. When Chiang's characters and worlds reject and elide race, his stories miss an opportunity to do more than the technical work of SF. It must intervene in these tangles of essentialisms and blindnesses and respond in substantive ways to address race, gender, sexuality, and so on, particularly in how difference and alterity exist, are described, expressed, and lived in these imagined worlds and futures.

In contrast to the "hard, clean, rational SF" of "Story of Your Life" is Larissa Lai's messy, fluid, humectant, and riotously feminist, queer, and anticapitalist 2018 novel *The Tiger Flu*.[15] The second modus operandi of Asianfuturism is intentionally and analytically racialized, gendered, even queered, what Eve Kosofsky Sedgwick calls "the open mesh of possibilities, gaps, overlaps, dissonances and resonances, lapses and excesses of meaning... [and] the experimental linguistic, epistemological, representational, political adventures."[16] Asianfuturism, in this mode, is ideological, political, and pedagogical; it (re)teaches us how the world works then and now and how it might or could work in the present to come.

Lai is an American-born Canadian writer, novelist, poet, and scholar. Most of Lai's work is Asianfuturist, centering queer, Asian, often critically non- or posthuman women in genres that traditionally focus on heteromasculine, white, ruggedly individualist men. I had the good fortune to be able to have a conversation about SF, her work, and Asianfuturism with Lai as part of her Maria Zambrano Fellowship at the University of Huelva. She argues that "SF is a mode of literary possibility," which does not just the work of refusal and negative critique but also the imaginative work of hope, organizing, and possibility.[17]

Drawing on my work in game studies, I argue that speculative literature, like a video game, is a "designed experience" full of obvious and inobvious choices made by the writer (and reader) as whether to replicate or resist representational and identificatory norms and values. The Asianfuturism imagined by Lai plays with speculative identities and technologies, taking Asian perspectives and cultures as given and emergent circumstances rather than as tokens, window dressing, or stereotypes. In other words, Lai's novels are intentional in their Asianfuturist design, narratives, prototyping, and community building. When asked, Lai provisionally defines Asianfuturism as a creative practice that posits a future in which we (and people of Asian descent) are fully agented subjects, with full, nuanced narratives, and in which we have "an active role in making the future, for better or worse, but for ourselves and our own kin."[18] Lai's SF novels ground their characters, stories, settings, and cosmologies in Asian mythology and native and immigrant experiences, and center Asian perspectives as the default, the kernel of germination, rather than the usual Western or Orientalist one.

The *Tiger Flu* presents a postapocalyptic Pacific Northwest in the year 2145 or Cascadia Year 127 TAO (Time After Oil). The novel is set in Saltwater City, a transformed Vancouver, where technologically advanced corporations compete with paramilitary groups and vestiges of nation-state governments for resources and control. Even though this future is full of manifestations of high tech like human cloning, cybernetic implants, and massive computer mainframes in the form of moon-like satellites, it is a world simultaneously ravaged by climate change, massive economic and technologic divides, and a worldwide pandemic that originated in the cloning of Caspian tigers and the making and consumption of tiger-bone wine. *The Tiger Flu* features the parallel-then-tangled stories of two young women: Kirilow Groundsel, a member of Grist Village far outside Saltwater City, home to queer, parthenogenic, women-only clones, and Kora Ko, a teenage girl living outside of Saltwater City, who discovers her family is part of the technopolitical tumult of the world and that she is a near-distant relative to the Grist sisterhood.

Lai presents the world through a queer, female perspective, taking up the norms and conventions often associated with SF. As Lai argues about normative SF, "It is already the case that the West wants to push Asian bodies and Asian spaces into the future in which white people get to unfold their dramas."[19] *The Tiger Flu* articulates an ambivalently racialized, gendered, eroticized, and diseased world that eschews capitalist or colonialist fantasies and fixes. For example, the story of the "tiger flu," which ironically affects and kills men more than women, involves the competition between rival factions that strive to "cure" the disease by curing "the mind of the body," meaning using technology to transfer the consciousnesses of the sick to the orbiting mainframes.[20] However, the novel is deeply critical of this mind/body dualism, this technological fix; it refuses the straight, white, male cyberpunk convention that treats the body as "meat" and the "prison" of the mind.[21] In contrast to novels like *Neuromancer* and *Snow Crash*, "Lai is careful to reject the traditional cyberpunk use of the passive techno-Orientalized female body."[22] Lai's universes are material, embodied, and often laborious and messy.

Everything in *The Tiger Flu* reeks, stinks, gushes, and flows, with strange fluids, juices, and the scents of milk, fish, stainless steel, radishes, durian, and muck and mire. Lai's novel splashes and wallows in "a deeply gendered, raced, and material world. In fact, much of Lai's SF writing locates subjectivity, life, and survival in the fecundity and funkiness of bodies, environments, and effluvia."[23] The novel deploys "these mortal senses, on this haptic and olfactory textuality" to destabilize the stereotypical tropes of Asian bodies as savage, diseased, backward, lascivious, and alien or as smooth, mechanical, cold, and sterile; it plays with and against the typical set dressing in SF between an "[arrested] Asia in traditional, often premodern imagery" and a futuristic, (over)developed, technologically threatening Asia.[24]

The funkiness and smelliness of *The Tiger Flu* defines its Asianfuturism as embodied, emotional, painful, pleasurable, symbiotic, communal, and survivable. After her lover, Peristrophe Halliana, succumbs to the flu and Grist Village is raided by Host Industries, Kirilow must head to Saltwater City to find help from a hidden branch of the Grist sisterhood in Saltwater Flats. On her arrival in the city, she remarks, "Although the polluted air smells foul and rotten, there's a damp sweetness beneath it, the rich undercurrent of life trying to fight its way through all the nightmares laid over the land in the time before."[25] Again, the novel argues that change and possibility can be found in the most unlikely places, in the most unlikely pairings and connections. This attention to mind and body, identities and materialities, makes real the horrors of a collapsing, compromised world as well as the possibility for escape and transformation. The very technologies that have poisoned and subjugated the world are also the ones that have allowed for communities like the Grist sisterhood to radically reorganize life and existence.

One of the main arcs of *The Tiger Flu* is whether or not Kora, Kirilow, and the Grist sisterhood, and their radical, feminist, queer existence will survive the end of the world, whether or not there is a way out of the dystopia. As Aimee Bahng reminds us, "The future is an always already occupied space," one that must be decolonized, reimagined, and refutured through invention, collaboration, experimentation, and emergent responses to dire straits like the hardscrabble lifeways of the Grist sisters.²⁶ For Lai, an Asianfuturist future must be a coalitional one, asking, how might we wish to dream the future? She responds, "For me, it is connected to action. I think it is really important to put those dreams into action, but as we do so, we need to attend to the way that our actions concatenate with actions and dreams of others and attend to what emerges unexpectedly."²⁷

Asian American Video Games and Gaming the Asian

This section takes up the radical intentionality I have just described and shifts to a second set of related examples in a different medium, digital games. I argue here that Asian American literary and cultural studies must adapt to and engage with video games (and game studies more broadly). The designed experiences of literature and games offer alternative opportunities to explore and critically engage with underrepresented people, worlds, and experiences. I turn to another exhibit that I helped to curate at the Wing Luke Museum in 2011–2012 titled *Asian American Arcade: The Art of Video Games* (Figure 17.2). The exhibit asked

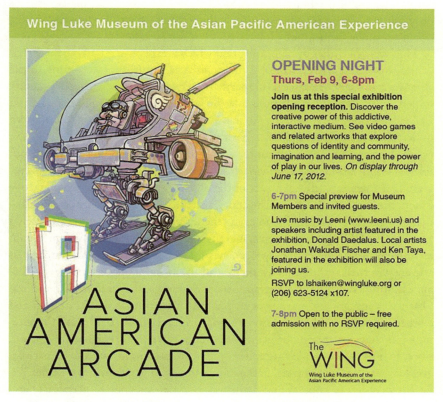

FIGURE 17.2 Screenshot of the Wing Luke Museum website featuring *Asian American Arcade* (2012). (Source: Wing Luke Museum of the Asian Pacific American Experience.)

visitors to "orient" themselves, to "explore the interplay between video games and Asian Pacific American identity and experience" and to consider how "the games we play in this rapidly-changing digital world shape our experience of identity, community, and cultural values. *Asian American Arcade* examines how artists and game developers are using these new technologies to create interactive narratives, nuanced emotional experiences, and evocative virtual worlds for players to explore."[28]

In fact, Christopher Patterson opens their essay "Asian Americans and Digital Games" describing the exhibit at the Wing by noting the "challenge of comprehending video games from Asian American cultural politics" since unlike literature and cinema, games, he argues, "carry no obvious national origin."[29] The critique of the exhibit stems from the difficulty of resolving what makes a game APA, something the CAC also struggled with. Patterson asks, "How can we read games through Asian American methods," especially for games that may be produced "by Asian American designers but that have no Asian bodies or anything else explicitly 'Asian' about them?"[30] They go on to suggest that it is difficult to think of "Asian American" in the traditional sense with digital games because of the blurring, even obfuscation, of race or ethnicity vis-à-vis programmer, player, avatar, narrative, and mechanic, as well as production, distribution, and consumption.

However, the difficulty of locating that which is explicitly "Asian" about digital games is precisely the project of a thorough Asianfuturism, intervening in the absences and tacit normativities of speculative fictions and worlds, be they literary or gamic. Like speculative fictions à la the worlds of Ted Chiang versus Larissa Lai, video games are inherently "designed experiences" and are embedded with understandings of race, gender, sexuality, ability, and other logics, what I have called elsewhere the "technonormative matrix."[31] Tara Fickle, in *The Race Card: From Gaming Technologies to Model Minorities*, calls all of this "ludo-Orientalism," arguing, "There is hardly an aspect of the digital game industry in which race—functioning intersectionally with gender, sexuality, class, and other categories—does not play a crucial role" and even "the infrastructure of gaming [is] itself a raced project."[32] I argue that even games that are not explicitly Asian often remain technological prostheses and synecdoches for nation, identities, and exclusions and are governed at the very least rhetorically by assumptions or fantasies of "Asianness," from stereotypical programmers or e-sport gamers to retro- and techno-Orientalist avatars and narratives to platform chauvinisms, gold farming, and e-waste. In other words, we must anticipate and interrogate the "default" epistemologies and perceptions of games, especially when their origins are unclear, since, to underscore Fickle's argument, games are used in formal and vernacular, everyday ways to "justify racial fictions and other arbitrary human typologies."[33]

The majority of popular and mainstream games continue to perpetuate and commodify these limited ludo-Orientalisms. Therefore, I turn to independent games produced and distributed by Asian American game makers that respond to the lack of representation in mainstream games as well as engage mechanics that foreground Asian experience in nuanced and novel ways. The third feature of Asianfuturism is to address race, gender, sexuality, and other differences in intersection and in action, relying on more than the identitarian or representational critique. An "Asianfuturist" game then must make race, difference, and cultural expression and experience narratively, structurally, and ludically consequential.

Melos Han-Tani's game *All Our Asias* (2018) is self-described as a "a surreal, lo-fi, 3D adventure, about identity, race, and nationality."[34] The player-protagonist of the game is Yuito, a thirty-one-year-old Japanese American hedge fund analyst who receives a letter from his estranged father. Yuito discovers that his father is dying, in a hospital, unconscious and on life support. But, with the attending doctor's help, Yuito can undergo a Memory World Visitation, using a futuristic technology that allows him to project his consciousness into and navigate his father's memories. The game lasts about two hours, and as the player-character, you explore abstract, dream-like landscapes and settings that take the form of train stations, nightclubs, forests, mountains, even the streets of Chicago. As you explore, you encounter memories of people, places, and deeds, hoping to find a way to communicate with your dying father. There is risk, however, because if your father dies while Yuito is in the Memory World Visitation, you too will die. Han-Tani, himself Japanese American, who also composed the atmospheric music for the game, comments, "I picked themes like developing race consciousness, some masculinity stuff, complicating the notion of 'Asian America,' which when expressed through Yuito, grew to encompass ideas like community/family." He continues, "Narratively I'd say I'm inspired by Taiwanese, or Asian American filmmakers and novelists."[35]

For the most part, much of the game is purposefully abstract and disorienting as a narrative and ludic way to comment on immigrant identity, Asian American masculinity, and norms about family and filial piety. One key example takes place in a nightclub rendered abstractly as vector lines and pulsing wireframes. The "Memory of Nightlife" level presents the player-character with Yuito's father as a lovelorn young man struggling with gendered expectations at a school dance (Figure 17.3). What is compelling about the level for me is its referencing of the trope of the cyberpunk nightclub (e.g., *Snow Crash*, the *Matrix*)—an eroticized space literally putting the techno in techno-Orientalism—but reconfiguring the setting by making it the site of heteromasculine anxiety. Yuito's encounter with his father as a shy, insecure young man unable to ask someone to dance dramatizes cultural expectations for men and recodes the space through its fragmented rendering. Much in the way that Larissa Lai's work disrupts the expectations of SF, Han-Tani's game deploys an Asianfuturist remodeling and reconfiguring of techno-Orientalist expectations for games set in an Asian, science-fictional setting. Han-Tani himself responds, "When I hear Asianfuturist, I kind of imagine some forms of storytelling or art-making that don't revolve around traditional modes of story structure."[36]

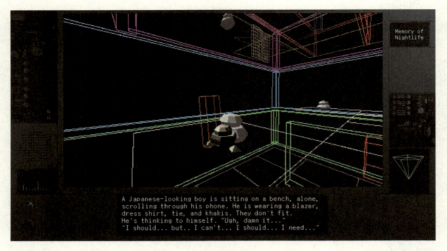

FIGURE 17.3 Screenshot of *All Our Asias* (2018). "Memory of Nightlife" level. (Source: Melos Han-Tani.)

Much of the gameplay of *All Our Asias* is about wandering, traversing the psychic environment of Yuito's father's experiences, and interacting with various memories in an attempt to repair the damaged relationship between father and son. *All Our Asias* privileges exploration, conversation, and meditation; the game would be pejoratively called a "walking simulator" by "hardcore" gamers because it does not rely on the technonormative mechanics of competition, accrual, or combat. As Bo Ruberg, writing on the queer potential of walking simulators, argues, "Video games, with their constructed environments and designed player paths, make these metaphors of spatial orientation strikingly literal. Whether the player moves through a game following the straight, prescribed path or whether they embrace deviation can be seen as an expression of a straight or queer orientation."[37] For *All Our Asias*, the Asianfuturist landscape itself becomes a key mechanic. The jumping lines and thumping techno music of the "Memory of Nightlife" level create a gendered and eroticized space, particularly for the failed masculinity of Yuito's father. These ideologies of space in video games become digital manifestations of gender and sexuality through their relationship to straight and queer movement, navigation, and orientation. *All Our Asias* renders an Asianfuturist game world—one where the ludo-Orientalist expectations of Asia and Asian bodies as model sources of games and gamer are interrupted and questioned.

Mike Ren Yi, who currently lives in Shanghai, is known for indie games *Hazy Days* (2016), a "breathing simulator"; *Yellow Face* (2019), an interactive game about being Asian in America; *Novel Containment* (2020), a game about the COVID-19 pandemic; and *Night Flyer: A Bat's Journey* (2020), a

Looking for Asianfuturism • 293

game where players play as a bat trying to survive the destruction of its habitat.[38] Ren's designs offer an opening into how games and other media might imagine and interrogate what it means to be an Asian American game maker making games that center Asian and Asian American experiences.

Hazy Days follows a young Shanghai girl named Xiao Feng (whose name might mean "little wind" or "little phoenix"), who is looking forward to visiting her grandmother for Lunar New Year. She must make it through a week of school before she can fly to see her family. However, during the week, the air quality in the city steadily grows worse, and even masked, Xiao Feng risks getting sick from the pollution, which would upset her Lunar New Year's plans. On the game's website, Ren says, "I've been living in China for years now and while the city is great, the air pollution is a real issue that people deal with on a day-to-day basis."[39]

The player watches as Xiao Feng wakes up each morning, counting down the days till she can see grandma, and walks to school. The left side of the split screen shows the little girl as she walks through the city, and as the game progresses, the days grow smoggier and smoggier till Xiao Feng must wear a mask while outside. The right side of the screen represents Xiao Feng's breathing: a set of concentric circles surrounded by floating blue droplets that represent oxygen and spiky black balls that represent pollution (Figure 17.4). Play requires only one button: the player clicks, holds, and releases, simulating Xiao Feng breathing in and breathing out. As she breathes in, oxygen and pollution are attracted toward the center of the circles. The goal is to make sure that enough oxygen enters the circle, much like nutrients entering a cell. However, if the

FIGURE 17.4 Screenshot of *Hazy Days* (2016). (Source: Mike Ren Yi.)

pollution touches the outermost circle, it sticks, clumping together, and can block oxygen from getting through. At the top of the screen is a light blue bar representing Xiao Feng's overall health, and if it drops to zero, she becomes sick and the game ends.

While perhaps reductive in its earnestness, what is illuminating about *Hazy Days* is its engagement with the perils of modernity, urbanization, and environmental degradation. The game uses a simple mechanic to address the consequences of pollution and the fraught future for the planet if nothing is done about it. *Hazy Days* is an environmental game, which, as Alenda Y. Chang argues, "brings the nonhuman world into equal prominence with the human, exposes humanity's moral responsibility to and participation in the natural world, and portrays the environment as fluid process, not static representation."[40] These games, like other environmental texts, position players to consider their relationship to nature, real or simulated; they allow for players "to create meaningful interaction within artificially intelligent environments, to model ecological dynamics based on interdependence and limitation, and to allow players to explore manifold ecological futures—not all of them dystopian."[41] *Hazy Days*' play with the utopian and dystopian, with identities and nationalities, with hopes and fears, makes it Asianfuturist. Moreover, *Hazy Days* plays with gamic genre in that it might also be called a "cozy game," a kind of game that is often low stress, comforting, relaxing, and about building relationships between characters and the natural world. But its cute, colorful, cartoony style belies the seriousness of the game's ecological message; while the game can (?) have a happy ending, *Hazy Days* upends the genre's ludic expectations.

Both *All Our Asias* and *Hazy Days* offer Asianfuturist interventions in exploring the ways that race, gender, sexuality, and other norms are encoded in games but also in the culture at large. Both play with these codes and norms and possibilities by taking up gamic genres that fly in the face of "accepted" or "hardcore" games, by deploying narrative, representation, and mechanics in integrated and intersectional ways, and by foregrounding Asian and Asian-American experience, revealing how, as Tara Fickle argues, "gamification and racialization work in tandem as mutually constitutive ways of orienting ourselves to and through difference."[42] Ren, speaking of his games, says, "The reality is that institutions around the world oppress people to keep power. . . . I try to reflect that in my games. Once struggling individuals from around the world realize that their own experiences are not that different from one another, then we are on the path toward positive change."[43] Both *All Our Asias* and *Hazy Days* (and a growing number of other games) open new possibilities for Asian American cultural and media studies, which must address the near ubiquity of digital games, algorithmic integration and interpellation in twenty-first-century life, and the crisscrossed ways that Asian American identity, representation, and embodiment have been turned into games and, more importantly, can be thus gamed.

Coda

In closing, I want to quickly revisit my nested frames and offer a few final takeaways. First, rather than close the conversation about and exploration of Asianfuturism, I hope to keep the definition generous, capacious, ambivalent, and messy even as it insists on affirming Asian and Asian American narratives and experience, queering and critiquing racialized technonormativities, offering intersectional and interdisciplinary approaches and interventions, and gesturing toward and grappling with Asian and Asian American futures, potentialities, and sites and scenes of utopia. This is why Asianfuturism and its interventions must be in conversation with other speculative fictions, texts, and games. There are provocations and lessons to be learned from witnessing how Afrofuturist, Indigenous, and other marginalized speculative writers and creators engage the questions of why literature, mass media, and games matter. For example, Octavia Butler's essay "Positive Obsession" asks, "What good is science fiction to Black people?" She answers, "What good is science fiction's thinking about the present, the future, and the past? What good is its tendency to warn or to consider alternative ways of thinking and doing? What good is its examination of the possible effects of science and technology, or social organization and political direction? . . . what good is all this to Black people?"[44] Therefore, in part, this presentation hopes to answer the question, "What good is Asianfuturism for Asian people?"

This is the final move of Asianfuturism, to find generosity, empathy, connection, and coalition; it is about active dreaming, teaching, playing, and putting theory into individual and collective action. Larissa Lai adds to her definition of Asianfuturism above, saying, "It makes room for real relationships with Black and Indigenous people—relationships that recognize the harms that have occurred among us through the colonial project, but also new relationships newly imagined."[45] This promise and potential of these futurisms resonates with what Jose Esteban Muñoz calls the horizon of queerness in *Cruising Utopia*; as Muñoz beautifully waxes poetically, "Queerness is a structuring and educated mode of desiring that allows us to see and feel beyond the quagmire of the present. . . . Queerness is that thing that lets us feel that this world is not enough, that indeed something is missing."[46] Asianfuturism is also a horizon, an asymptote that allows us to approach new and different possibilities. And, here, it is this "mode of desiring" that Asianfuturism must embrace and explore; it is about change, about transformation, and about hope. But utopia is not a scalpel, hope is not a destination. It is messiness, funkiness, glitchiness, and impreciseness. Mark Dery writes, "If there is an Afrofuturism, it must be sought in unlikely places, constellated from far-flung points."[47] To which I respond in kind, if there is an Asianfuturism, it too must emerge from different sources, accidental harmonies, and emergent desires. Ultimately, like other variations on hope, it is something that is always in the process of learning, changing, transforming, and becoming.

Notes

1. David S. Roh, Betsy Huang, and Greta Niu, "Technologizing Orientalism: An Introduction," in *Techno-Orientalism: Imagining Asia in Speculative Fiction, History, and Media*, ed. David S. Roh, Betsy Huang, and Greta A. Niu (New Brunswick, NJ: Rutgers University Press, 2015), 1.
2. Dawn Chan, "Asia-Futurism," *Artforum* 54, no. 10 (Summer 2016), 161, https://www.artforum.com/print/201606/asia-futurism-60088.
3. Mark Dery, Greg Tate, and Tricia Rose, "Black to the Future: Interviews with Samuel R. Delany, Greg Tate, and Tricia Rose," in *Flame Wars: The Discourse of Cyberculture*, ed. Mark Dery (Durham, NC: Duke University Press, 1994), 180–181.
4. Grace Dillon, "Imagining Indigenous Futurisms," in *Walking the Clouds: An Anthology of Indigenous Science Fiction*, ed. Grace Dillon (Tucson: University of Arizona Press, 2012), 1–2.
5. Xin Wang, "Asian Futurism and the Non-Other," *E-Flux Journal*, no. 81 (April 2017): 2, 9.
6. Mikala Woodward, *Worlds Beyond Here: The Expanding Universe of APA Science Fiction*, (Wing Luke Museum of the Asian Pacific American Experience, 2018).
7. Verity Burke and Will Tattersdill, "Introduction: Museums in Science Fiction, Science Fiction in Museums," *Configurations* 30, no. 3 (Summer 2002): 250; Rhianedd Smith, "'I Just Don't Think About It': Engaging Students with Critical Heritage Discourse through Science Fiction," *Configurations* 30, no. 3 (2002): 354–355.
8. Betsy Huang, *Contesting Genres in Contemporary Asian American Fiction* (New York: Palgrave Macmillan, 2010), 3.
9. Huang, *Contesting Genres*, 100.
10. Sami Schalk, *Bodyminds Reimagined: (Dis)Ability, Race, and Gender in Black Women's Speculative Fiction* (Durham, NC: Duke University Press, 2018), 18.
11. Min Hyoung Song, *The Children of 1965: On Writing, and Not Writing, as an Asian American* (Durham, NC: Duke University Press, 2013), 85, 99.
12. Ted Chiang, "Story of Your Life," in *Stories of Your Life and Others* (New York: Vintage, 2002), 91–146.
13. Vandana Singh, "The Occasional Writer: An Interview with Science Fiction Author Ted Chiang," *The Margins*, October 3, 2012, https://aaww.org/the-occasional-writer-an-interview-with-science-fiction-author-ted-chiang/.
14. Singh, "The Occasional Writer."
15. Larissa Lai, *The Tiger Flu* (Vancouver: Arsenal Pulp Press, 2018).
16. Eve Kosofsky Sedgwick, *Tendencies* (Durham, NC: Duke University, 1993), 8.
17. Edmond Y. Chang and Larissa Lai, "Circling Asianfuturisms: Edmond Chang and Larissa Lai in Conversation," (María Zambrano Fellowship Online Public Lecture Series, University of Huelva, Spain, May 16, 2023), https://eventos.uhu.es/93666/section/41408/larissa-lais-maria-zambrano-fellowship-events-at-the-university-of-huelva.html.
18. Chang and Lai, "Circling Asianfuturisms."
19. Chang and Lai, "Circling Asianfuturisms."
20. Lai, *The Tiger Flu*, 173.
21. William Gibson, *Neuromancer* (New York: Ace, 1984), 6.
22. Kathryn Allan, "Reimagining Asian Women in Feminist Post-Cyberpunk Science Fiction," in *Techno-Orientalism: Imagining Asia in Speculative Fiction, History, and Media*, ed. David S. Roh, Besty Huang, and Greta A. Niu (New Brunswick, NJ: Rutgers University Press, 2015), 159.

23. Stevi Costa and Edmond Y. Chang, "Fish, Roses, and Sexy Sutures: Disability, Embodied Estrangement, and Radical Care in Larissa Lai's *The Tiger Flu*," in *Project(Ing) Human: Representations of Disability in Science Fiction*, ed. Courtney Stanton (Wilmington, DE: Vernon Press, 2023), 137.
24. Costa and Chang, "Fish, Roses, and Sexy Sutures," 137; Roh, Huang, and Niu, *Techno-Orientalism*, 3.
25. Lai, *The Tiger Flu*, 169.
26. Aimee Bahng, *Migrant Futures: Decolonizing Speculation in Financial Times* (Durham, NC: Duke University Press, 2018), 12.
27. Chang and Lai, "Circling Asianfuturisms."
28. Quoted in Christopher B. Patterson, "Asian Americans and Digital Games," *Oxford Research Encyclopedia of Literature*, July 30, 2018, https://doi-org.proxy.library.ohio.edu/10.1093/acrefore/9780190201098.013.859; Mikala Woodward, *Asian American Arcade: The Art of Video Games*, Wing Luke Museum of the Asian Pacific American Experience, Seattle, 2012).
29. Patterson, "Asian Americans and Digital Games," 3, 4.
30. Patterson, "Asian Americans and Digital Games," 4.
31. Edmond Y. Chang, "Queergaming," in *Queer Game Studies*, ed. Bonnie Ruberg and Adrienne Shaw (Minneapolis: University of Minnesota Press, 2017), 15.
32. Tara Fickle, *The Race Card: From Gaming Technologies to Model Minorities* (New York: New York University Press, 2019), 4, 3.
33. Fickle, *The Race Card*, 9.
34. Melos Han-Tani, *All Our Asias*, v. 1.1, released March 19, 2018, Windows, MacOS, Linux, https://han-tani.itch.io/aoa
35. Quoted in Danielle Riendeau, "A Game that Uses PS1 Visuals to Tell a Story about Asian American," *Vice*, February 9, 2018, https://www.vice.com/en/article/pampqg/all-our-asias-ps1-sean-han-tani.
36. Han-Tani, interview.
37. Bonnie Ruberg, "Straight Paths Through Queer Walking Simulators: Wandering on Rails and Speedrunning in *Gone Home*," *Games and Culture* 15, no. 6 (March 2019): 4.
38. Mike Ren, *Hazy Days*, 2016, https://mikeyren.itch.io/hazydays.
39. Ren, *Hazy Days*.
40. Alenda Y. Chang, *Playing Nature: Ecology in Video Games* (Minneapolis: University of Minnesota Press, 2019), 32.
41. Chang, *Playing Nature*, 16.
42. Fickle, *The Race Card*, 27.
43. Quoted in "Interview: Mike Ren Yi," *First Person Scholar*, May 27, 2020, http://www.firstpersonscholar.com/interview-mike-ren-yi/.
44. Octavia E. Butler, "Positive Obsession," in *Bloodchild and Other Stories* (New York: Seven Stories Press, 2005), 134–135.
45. Chang and Lai, "Circling Asianfuturisms."
46. José Esteban Muñoz, *Cruising Utopia: The Then and There of Queer Futurity* (New York: New York University Press, 2009), 1.
47. Dery, Tate, and Rose, "Black to the Future," 181.

18

The Queer Techno-Orientalist Aesthetics of Disney's *Big Hero 6*

―――――――――――◆◇▶

THOMAS XAVIER SARMIENTO

Disney's fifty-fourth, and first Asian American, animated feature film, *Big Hero 6*, opens with multiple shots of San Fransokyo, a Japanized San Francisco featuring a torii-styled Golden Gate Bridge, the Salesforce Tower with "sanhuran" ("San Fran") written out in neon katakana, and a pagoda-embellished Transamerica Pyramid.[1] In this 30-second segment full of vibrant colors and upbeat nondiegetic sounds, audiences develop a hopeful anticipation of another reality akin to, yet distinct from, our own. While the film's speculative Asian future resonates with techno-Orientalist aesthetics, its tonal difference queers techno-Orientalism, presenting a desirable Asian futurity that does not alienate its audiences, especially those of Asian descent.

 The computer-animated film, based on characters from the eponymous Marvel comics, follows fourteen-year-old multiracial Japanese American prodigy Hiro Hamada (voiced by Jewish and Japanese American actor Ryan Potter) as he applies his robotics knowledge from selfish to selfless ends.[2] To dissuade his younger brother Hiro from the illicit world of botfighting, Tadashi (voiced by Korean and white American actor Daniel Henney) surreptitiously takes Hiro to his research lab at the San Fransokyo Institute of Technology (SFIT) to entice him to attend college. Hiro meets Tadashi's lab colleagues, Go Go (Jamie Chung), Wasabi (Damon Wayans Jr.), Honey Lemon (Génesis Rodríguez), and Fred (T. J. Miller); the "personal healthcare companion" robot Baymax (Scott Adsit) that he has developed; and the robotics program director, Professor Robert Callaghan (James Cromwell). After seeing the lab's various technological innovations and meeting his hero Callaghan, Hiro desires to attend SFIT.

To gain admission into SFIT, Hiro presents his invention at the annual student showcase: a microbot, a tiny robot that is controlled by a neural transmitter headband and that can link with other microbots to create dynamic structures instantaneously imagined by the mind. Callaghan is impressed, but so is tech entrepreneur Alistair Krei (Alan Tudyk). Hiro rejects Krei's offer to buy his technology and accepts Callaghan's offer to attend SFIT. When the exhibition hall catches fire, Tadashi goes in to save Callaghan but dies. A few weeks later, Hiro discovers an activated microbot under his bed, which puzzles him because it activates to attract other microbots, but the rest of the microbots presumably were destroyed by the fire. Baymax leads Hiro to an abandoned warehouse, where they discover millions of microbots being manufactured. A mysterious person wearing a Kabuki mask appears and hunts down Hiro and Baymax, with the help of the microbots. Hiro and Baymax escape. Upon reflection, Hiro suspects the masked figure (referred to as Yokai in the Disney+ Extras features) intentionally set fire to the exhibition hall to steal his microbots.

To avenge Tadashi's death, Hiro seeks out Yokai, upgrading Baymax to assist in his defense. Wasabi, Honey Lemon, Go Go, and Fred try to stop Hiro, only to be attacked by Yokai. Hiro enlists Tadashi's lab friends to help defeat Yokai, designing tailored technology to match Yokai's power. In another confrontation with Yokai, the newly formed team discover Yokai is not Krei, whom they suspected, but rather Callaghan, seeking to avenge his daughter Abigail's death after Krei's teleportation machine malfunctioned (Abigail was a test pilot). Hiro's team stops Yokai/Callaghan from killing Krei and destroying the city with another teleportation machine. Baymax senses life in the increasingly unstable portal, so Hiro and he go in to save someone whom they suspect is Abigail. For Hiro and Abigail to escape, Baymax must sacrifice himself. Later, Hiro is surprised to find Baymax's mechanical arm that saved him still clutches a chip that brings Baymax back to life, thus ending the story happily.

As a commercial and critical success (winning the 2015 Best Animated Feature Oscar), with a Disney XD/Disney Channel follow-up television series (*Big Hero 6: The Series* [2017–2021]) and Disney+ spin-off streaming series (*Baymax!* [2022]), *Big Hero 6* is a culturally significant film.[3] Focusing on its utopian aesthetics and narrative, I capitalize on the film's capacity to engender Asian/American futurity that exceeds techno-Oriental forms. Although (techno-)Orientalist critique emphasizes the relationship between West/Self and East/Other, the former often is centered. As a queer-feminist Asian Americanist, I decenter the center, not to ignore the workings of power, but instead to attend to the work happening despite and through the vise of power. To recognize *Big Hero 6* as forwarding an alternative, queer sense of the future that is not always depressing and anti-Asia/n opens more possibilities to imagine otherwise.

Queer-feminist of color and diasporic understandings of "queer" as an onto-epistemological term concerned less with identity and more with nonnormative forms of desire that enact alternative worlds inform my framing of *Big Hero 6* as a queer techno-Orientalist text.[4] As a gay-identified spectator, however, I also project queerness onto characters who attract me—namely, hunky Tadashi—thereby making plausible queer sexualities and rejecting heterosexuality as a default orientation. To read the film queerly is to sense its deviance from normative representations of techno-Oriental futures and to recognize alternative socialities that a multicultural future might afford when we forego normative relational intimacies.

Techno-Orientalism already illustrates how Asianness figures as queer in relation to Euro-American modernity, as such cultural and political discourses "imagin[e] Asia and Asians in hypo- or hypertechnological terms," thus positioning Asia and Asians as nonnormative to Euro-American standards of technological advancement and civilization.[5] Positioned as surplus labor, Asians exceed the bounds of normative capitalist production and accumulation; however, their status exposes the systemic contradictions that maintain their subordination.[6] Accordingly, Asian technological, and thus economical, advancement poses a threat to the white cisheteropatriarchal order, as it unsettles the terms of enfranchisement and disenfranchisement.

As an unstable category of analysis, though, queerness also can figure in the queer. That is, techno-Oriental futures rendered nonnormative, and hence queer, to Western ideals also can figure as queer in their deviance from normative techno-Orientalist scripts. The etymologies of "queer" as "desire and inquiry" from Middle French and Spanish and as a spatial term linked to "the Indo-European word 'twist'" illumine how "queer" aptly functions as a verb that signifies movement away from and toward something.[7] "Queering" engenders transformation. So, while dominant techno-Orientalist representations of the future as Asia/n lean dystopian and figure as negatively queer in relation to Euro-American desires, nondominant representations that appear less dire and more plausible challenge typical understandings of techno-Orientalism and offer more affirming racial forms.

For José Esteban Muñoz, "The future is queerness's domain. Queering is a structuring and educated mode of desiring that allows us to see and feel beyond the quagmire of the present."[8] Muñoz's sense of queerness as horizon enables us to recognize visions of the future that draw from our present reality but reimagine a world wherein the negatively queer, the undesirables of normative society and culture, can thrive. This horizontal notion of "queer" fittingly relates to the etymology of "orient" and "the Orient," the so-called East: "'the horizon' over which the sun rises . . . a visible line that marks the beginning of a new day."[9]

Queering techno-Orientalism is a stylistic reading practice that redirects audiences toward hopeful possibilities just beyond the horizon of stereotypical Asian/American futures.[10] In *Big Hero 6*, such possibilities take the form of technologized Asian Americans as heroes, not threats, who work alongside and care for other people of color and robots, a situation that rewires how we might link "futurity" with "Asian."

(Queer) Techno-Orientalism in *Big Hero 6*

Within techno-Orientalist ideology and aesthetics, "Japan is a screen on which the West has projected its technological fantasies."[11] *Big Hero 6* affirms techno-Orientalism's fantastic quality through its Japanized architectural, cultural, and technological citations—from colorful tiled hip-and-gable roofs, shoji room dividers, and paper lanterns; to Aunt Cass's Lucky Cat Café featuring various iterations of Maneki-Neko, streets lined with blossoming cherry trees, and *kawaii* signage referencing the specialties of restaurants and storefronts; and to efficient mass transit and assistive robotics (Figure 18.1).

Moreover, as a film that presents a speculative future wherein San Francisco's geographic organization (e.g., bridges, waterfront, buildings, cross streets, demography) remains legible but adorned with Japanese accents, the film conforms to techno-Orientalist discourses that figure such futures as ambivalent simulacra to Euro-American present reality.[12]

Pushing at the limits of techno-Orientalist possibility that David S. Roh, Betsy Huang, and Greta A. Niu observe in "critical reappropriations in texts that self-referentially engage with Asian images" and that "[engender] counterdialogue," I forward a queer of color approach to assessing techno-Orientalist aesthetics that conjures Muñoz's notion of disidentification.[13] Muñoz defines disidentification as a practice marginalized peoples enact to inhabit a world not designed for them.[14] Disidentification reworks normative scripts, revising them

FIGURE 18.1 Screenshot of the Lucky Cat Café / Hamada home exterior featuring a tiled hip-and-gable roof and a street lined with blossoming cherry trees. (Source: Walt Disney Pictures.)

to serve minoritarian ends.[15] A disidentificatory approach to techno-Orientalism "negotiates strategies of resistance within the flux of discourse and power," recognizing that "counterdiscourses, like discourse, can always fluctuate for different ideological ends."[16] As a critique of racist discourses toward Asia/ns that are technologically inflected, techno-Orientalist scholarship functions as a counterdiscourse that exposes a particular version of Orientalism that measures Asia/ns in terms of scientific progress. Rather than automatically presume that a techno-Orientalist representation simply traffics in Euro-American imperialist stereotypes of race, gender, and sexuality, I disidentify with such reductive imagery, as well as critical tendencies that dismiss such texts as misrepresentative, and mine for moments of possibility that glimpse subaltern joy. Although *Big Hero 6* may conflate advanced technological futurity with Asia and reinforce the model minority trope through Hiro's genius, its centering of Asian American subjectivity complicates its discursive meaning.

Metanarratively, technology is central to the film, as its creators developed "an entirely new rendering tool called Hyperion" that efficiently "calculat[es] how the light bounces around a virtual scene and shades the objects" to produce a highly realistic image.[17] The technicality of rendering a realistic San Fransokyo through light and shadow plays into techno-Orientalism's association of Asia with technological innovation, despite the production team's Burbank, California, location; it also reflects the creators' care in orienting audiences to imagine a plausible future. Through such technology, a grounded utopian aesthetic steeped in warm color tones emerges and diverges from the shadowy tones common in techno-Orientalist films, such as *Blade Runner* (1982).[18]

Lighting the built environment through new technology brings not only San Fransokyo to life but also the background characters created by Denizen, another Disney-created rendering tool.[19] Whereas *Blade Runner*'s background characters appear as darkened, faceless Asian hordes, *Big Hero 6*'s comprise "different shapes, sizes, cultures and fashion senses," making San Fransokyans appear "authentic and believable."[20] Although background characters do not speak, they perform unique gestures and facial reactions, such as when Baymax traverses San Fransokyo's daytime streets, suggesting their potential for interiority and round characterization (Figure 18.2).

The film's press kit boasts how, at the time of its release, the film featured both the most main characters on screen and the greatest number of unique characters that populate any Disney animated film.[21] Such attention to detail unwittingly works to present a heterogenous Asian future, one wherein Asians are not all the same nor are non-Asians subordinate to them. The future as Asia/n that *Big Hero 6* imagines departs from the yellow peril dystopias often featured in techno-Orientalist films. Clearly, technological innovation is significant in the film's making, but such advances aim to immerse audiences in a narrative they can buy into.

As a Disney film, *Big Hero 6* appeals to audience pathos, featuring Asianized characters who possess emotional depth. Prominently featured in promotional

FIGURE 18.2 Screenshot of background characters expressing unique facial reactions to Baymax crossing a street. (Source: Walt Disney Pictures.)

paratextual materials, Baymax seems friendly, with his curved frame, soft texture, and light (white) appearance. While seemingly deracinated, given (techno)-Orientalist discourses that conflate Asianness with roboticism, Baymax signifies atypical Asian futurity.[22] Codirector Don Hall and lead character designer Shiyoon Kim emphasize Baymax's intentionally appealing affect. An inflatable vinyl arm developed by Carnegie Mellon University researchers that inspired Hall seemed "non-threatening" and refreshingly contrasted with popular cultural images of robots. Centering an Asian perspective on robotics and futurity, Hall goes on to explain how Japanese researchers inspired "Hiro's love of technology" as well as the notion that "robots are the key to a hopeful future ... making the world a better place."[23] Kim shares that "Japanese infomercials ... helped shape [Baymax's] design," as they "emphasize the cute aesthetic, yet hide the technology."[24]

In the film, Tadashi explains that he designed Baymax to be "a nonthreatening huggable kind of thing," to which Hiro responds, "Looks like a walking marshmallow." The shot before this exchange might emphasize Baymax's technicality as Tadashi proudly remarks that he "programmed him [Baymax] with over ten thousand medical procedures," followed by a shallow focused close-up shot of the translucent green computer chip that "makes Baymax, Baymax," adorned with a handwritten label denoting "TADASHI HAMADA" as creator and a circular sticker with a smiley face wearing a head mirror and stethoscope to signal a human touch and care. The scene's pacing, however, attends to Baymax's affective appeal as he performs comedic cuteness with his squeaky vinyl shuffle, reminiscent of penguins walking, periodic eye blinks, and calming AI (artificial intelligence) voice.

Baymax's physical and character design on- and offscreen destabilize techno-Orientalist tendencies in at least two ways. First, Disney's turn to U.S.-based institutions like Carnegie Mellon disassociates advanced technologies from being exclusively Asian. Second, Disney's citation of Japanese optimism and cuteness in developing its sense of technology interrupts the stereotype of Asians as cold and unfeeling.[25] While the latter point unsettles the Asian-as-robot trope, it ironically conjures the stereotypical association of Asians with cuteness.

As Sianne Ngai, Christine R. Yano, and Erica Kanesaka reveal, the aesthetics of cuteness are not benign but rather paradoxical, multiplicitous, and laden with power. Ngai frames cuteness in relation to objectification and describes cute objects as "usually being soft, round, and deeply associated with the infantile and the feminine" as well as the "culturally and nationally other," the latter three descriptors aligning with Orientalism.[26] Cute objects also are often anthropomorphic yet primitive, indicative of their ambivalent and subordinate status to their human creators and consumers.[27] Yano and Kanesaka trace cuteness's particular relation to Japaneseness and its racial, gender, and sexual evocations. Japanese cuteness works to assuage Western geopolitical anxieties toward Asia that emerged in the early twentieth century.[28] In the twenty-first century, it also reflects nostalgia for 1980s Japanese cyberpunk as China's growing economy unsettles the United States.[29]

Baymax's seemingly race-neutral cuteness is consistent with the above notions of cuteness, as his docility enables him to appear approachable to patients and audiences. In our own reality, Disney has capitalized on Baymax's popularity by creating a San Fransokyo overlay in its California Adventure theme park where guests can interact with him. This domestication of an Oriental Other is ironic given California's history of anti-Asian exclusion.[30] Notwithstanding the dangers of embracing Baymax's cuteness, the film's queer representation of Asianized technology embodied by Baymax offers an alternative way to imagine Asian futurity.

The human-robot relationship that drives the film's plot further complicates techno-Orientalist anxieties toward the nonhuman. Akin to the harmonious blend of Tokyo/East and San Francisco/West portrayed in the film's cityscape, Baymax and Hiro's friendship illustrates how technology and humanity can be reciprocal and nonhierarchical. Although the film boasts impressive technical feats, its success lies in the heartwarming story between Hiro and Baymax, reminiscent of classic Disney films.[31] Baymax helps Hiro emotionally process Tadashi's death, which makes him a transitional object that ties Hiro to Tadashi; but Tadashi also functions as an intermediary between Hiro and Baymax, as his death prompts the latter two to connect.[32] The Tadashi-Baymax-Hiro triad ultimately reveals technology as a means for emotional connection, uncoupling feeling as an exclusive quality of humanity.

The rising-action sequence when Hiro and Baymax transform into superheroes and fly above San Fransokyo reinforces the seamless merging of human emotion and technological prowess. To defeat Yokai, Hiro creates Baymax 2.0: a sleek, red carbon fiber–armored Baymax, fitted with jet thrusters and wings. To show off Baymax 2.0's capabilities, Hiro climbs onto Baymax's back. The camera cuts to close-up shots of Hiro's soles and palms locking onto magnetized pads on Baymax's armor. Baymax's circular pads glow with a ring of purple, while

Hiro's circular palm pads glow red and his soles glow with a ring of red. This color scheme complements the circular purple accents on Baymax's helmet, shoulder pads, and hands and the circular red accents on Hiro's chest plate, knee pads, and soles. Whereas Baymax's primary color scheme is red, Hiro's is purple. This visual scheme, combined with the technological and emotional bonding, emphasizes the compatibility of human and robot.

As they elatedly fly over, around, and through the streets and buildings of San Fransokyo, colorful neon billboards come into view; a pink-lavender sky (softer shades of Baymax's red and Hiro's purple) with billowy clouds envelops the cityscape, highlighting the golden hour and effecting a warm tone. Unlike *Blade Runner*'s nighttime Los Angeles or its 2017 sequel *Blade Runner 2049*'s drab Los Angeles and hazy San Diego, *Big Hero 6*'s pastel twilight San Fransokyo, coupled with the uplifting musical score, evokes a playful wonder expected of Disney.[33] Technology never poses a real threat during this sequence, as Baymax attends to Hiro's needs and wants. Instead, humanity's desire to exact revenge is the real danger, as Hiro's fixation on catching Yokai forces Baymax to continue providing care, even though Hiro's emotional state has improved. Still, as they look out toward a beautiful horizon atop a futuristic *koinobori*-fied floating jet turbine, the film's queer techno-Orientalist aesthetics appears as it presents a hopeful Asian future that harmoniously integrates East and West (Figure 18.3).

The fictive origins of San Fransokyo feed into this utopian possibility wherein Asianness is not yellow peril. The filmmakers began production by focusing on world building, indicating the significance of setting in appealing to audiences.[34] Art director of environments Scott Watanabe's concept art reveals the emergence of San Fransokyo after the 1906 San Francisco earthquake.[35] While this is unaddressed in the film, Japanese immigrants rebuilt the city after the quake in an alternate timeline from our own.[36] Sean Miura's reflections on the visual power of San Fransokyo highlight the historical significance of 1906 for Asian America wherein the Naturalization Act passed that year lay the foundation for

FIGURE 18.3 Screenshot of Baymax and Hiro sitting atop a floating jet turbine at sunset with a torii-styled Golden Gate Bridge in the background. (Source: Walt Disney Pictures.)

rejecting Takao Ozawa's claim to whiteness and right to naturalized citizenship in *Ozawa v. United States* (1922). For Miura, the alternate trajectory of San Francisco into San Fransokyo is both a lament for a future that could have been if anti-Asian sentiment did not pervade the early twentieth century and a celebration of the creativity of Asian immigrants.[37]

Big Hero 6 further queers the script of techno-Orientalism through its choice of villain. Although Yokai may reference Japaneseness through his name and Kabuki mask, the revelation that he is Callaghan, a white-presenting person, unsettles techno-Orientalism's Asia/n-as-threat trope. *Yōkai* is a Japanese word that loosely translates into English as "monsters, spirits, or demons."[38] Deborah Shamoon links *yōkai* to database creation wherein traditional, familiar folklore is made available to modern consumers. This association between supernatural entities and information accumulation fits with *Big Hero 6*'s Yokai, as he represents a mysterious malevolence who perverts Hiro's microbot technology to manipulate the world to his liking. As a Japanese folkloric spectral figure, Yokai authenticates the film's Japanese American aesthetics as well as reinforces a normative techno-Oriental trope premised on a despotic leader controlling a faceless horde. Despite the film's general portrayal of the future as Asia/n positively, negative techno-Oriental elements haunt it. And yet, the revelation of Callaghan as Yokai blurs the boundaries of racial threat.

When Yokai appears on screen, the shot is from a low angle to accentuate his power, and the lighting is darker and the musical score is heavier to create a more ominous tone. Overhead lights or moonlight cast shadows over Yokai's mask. Unlike the daytime, twilight, and illuminated nighttime scenes that bring San Fransokyo to life, Yokai's scenes reinforce a normative techno-Orientalist aesthetic that traffics in darkness and grit and conflates technology, Asia, and sinisterness. However, when a demasked Yokai reveals Callaghan, the whiteness of his blue eyes captured in close-up shots contrasts with the ninja-like body suit that covers his head and neck, decoupling Asianness from villainy. Although Callaghan could be multiracial in a world like San Fransokyo, the whiteness of voice actor James Cromwell bears on his character, like most of the actor-character relations in the film. While Yokai may be a yellowface character, the film ultimately departs from normative techno-Orientalist characterizations of protagonist and antagonist. As Richard Leskosky observes, "Both Hiro and Callaghan seek revenge for the loss of a beloved relative and use robotic assistants to achieve their ends, and both, rightly or wrongly, blame Krei for their losses."[39] Rather than reinforce a binary logic of absolute good and bad, often typical of (techno-)Orientalism, the film offers character complexity through emotional depth and deracination.[40]

Queer Kinship and Assemblage

As a Disney film, family unsurprisingly is a core theme; and like several Disney animated films—from *Snow White and the Seven Dwarfs* (1937) to *The Little*

Mermaid (1989) and *Raya and the Last Dragon* (2021)—kinship structures are perhaps surprisingly not primarily heteronormative.[41] In their analysis of parental representations in Disney animated films from 1937 to 2017, Jessica D. Zurcher, Pamela Jo Brubaker, Sarah M. Webb, and Tom Robinson note the increase of single-parent depictions in both general and Disney media and include created or fictive families in their analysis, indicating a more expansive notion of family, albeit not necessarily correlated to U.S. demographic realities.[42] Still, normative representations of family prevail, as most of the parental figures Zurcher et al. coded are biological, white, and upper class (e.g., royalty); married couples slightly exceeded single parents by almost two percentage points.[43] Zurcher et al. did not explicitly address sexuality; however, their findings suggest couples are understood as heterosexual, even though same-gender couples can legally marry in the United States. With their biological parents deceased and being raised by their Aunt Cass (voiced by Black and Jewish American actor Maya Rudolph), who appears to be single, the Hamada brothers belong to a nonheteronormative household. Moreover, as multiracial youth, the Hamada boys represent the transgression of monoracial sexuality and marriage, which was protected by the state until the U.S. Supreme Court in *Loving v. Virginia* (1967) ruled in favor of interracial marriages (though in their alternate universe that may not have restricted Asian immigration, interracial marriages may not have been banned).

Although Hiro may be on a singular quest to solve the mystery of his brother's death, akin to lone-wolf protagonists Rick Deckard (*Blade Runner*) and K(D6-3.7)/Joe (*Blade Runner 2049*), the film deviates from such techno-Orientalist cinematic tropes, instead forwarding queer kinship and assemblage as Hiro accepts the help of his brother's techno-enhanced friends and human-like robot. Queer kinship resembles "the collective, communal, and consensual affiliations as well as the psychic, affective, and visceral bonds" individuals form in excess of the confines of biological and nuclear families privileged by the state and capital.[44] The "queer" of queer kinship aligns with my definition of queer as a mode of antinormativity rather than identity. Moreover, as Kimberly McKee's scholarship on transnational transracial Korean adoptive families illumines, queerness also is predicated on nonproximity to whiteness and on transraciality.[45] Accordingly, the queerness of the eponymous superhero assemblage "Big Hero 6," formed by the teaming of Go Go, Fred, Hiro, Wasabi, Honey Lemon, and Baymax socially, affectively, and technologically, is less about whether any one individual team member is queer-identified and more about the bonds they form and propose in excess of normative social and kinship scripts. As a multiracial and more-than-human ensemble, the superhero team figures queerly as they challenge what constitutes so-called real families.

While biological family drives Hiro's desire to seek out Yokai, the film's primary relationship between Hiro and Baymax forwards queer kinship. Tadashi's biological kinship with Hiro enables the queer familial bonds Hiro forms with Baymax, Honey Lemon, Go Go, Wasabi, and Fred, since the first is Tadashi's creation and the latter four are Tadashi's lab mates and best friends. Baymax acts as a surrogate for Tadashi, as the most intimate moments in the film occur between either Hiro and Tadashi or Hiro and Baymax, with Hiro struggling to find his way in the world and Tadashi and Baymax offering encouragement, comfort, and support. Honey Lemon, Go Go, Wasabi, and Fred also act as surrogate siblings and blur the line between family and friends as they appear in familial contexts. For example, when Hiro presents his microbots to Professor Callaghan at the SFIT Showcase, a medium-close-up shot of Aunt Cass and Tadashi framed by Go Go, Honey Lemon, Fred, and Wasabi establishes from Hiro's point of view who Hiro's kin are. While Aunt Cass and Tadashi are in focus, the framing of all six characters illustrates the family supporting Hiro through this formative moment, regardless of biological status.

Baymax as a character and conduit further queers the notion of family by functioning as a caregiver and blurring the line between friends and family. After Hiro and Baymax return from their first encounter with Yokai, Baymax "download[s] a database on personal loss" to treat Hiro's grief. He immediately shares his results, stating, "Treatments include contact with friends and loved ones." The camera shows Baymax's point of view of a transparent computer screen with a circle around Hiro as the patient focus in the center of the frame, a health assessment that lists "Bereavement" as the "Diagnosis" and "Contact with friends and family" as the "Treatment" on the left of the frame, and a visualized human network that highlights close contacts on the right of the frame. Then Baymax says, "I am contacting them [Hiro's friends] now." A projection appears on Baymax's chest showing connected circular profile pictures of Fred, Honey Lemon, Wasabi, and Go Go. Baymax's speech synonymizes "loved ones" with the word "family" indicated on his computerized screen, which deemphasizes "family" as the exclusive locus of love and affection. His snap decision to contact Tadashi's four best friends subtly transfers their friendship onto Hiro but also suggests Baymax's logic prioritizes them as close friends who are indistinguishable from loved ones.

Baymax goes on to wrap his arms around Hiro, stating that "other treatments include compassion and physical reassurance." The medium-long shot shows Baymax's arm gestures resembling a life preserver that engulfs Hiro. Baymax then affectionately pats Hiro's head and says in his uncontracted, robotic, but not monotone diction, "You will be alright. There, there." Baymax's actions and speech are both comedic and endearing, guiding audiences to see his capacity for love. Yugin Teo's analysis of Baymax as a robot carer highlights Baymax's affective capacities, which set him apart from the dominant dystopian representations of robots, and the importance of socializing robots into a human-dominated world but also recognizing robots as part of a social community.[46]

Additionally, this sequence occurs in Hiro and Tadashi's shared attic bedroom, in a domestic space. Although Aunt Cass is just one floor below, Baymax does not interpellate her as one of Hiro's treatment options. This is not to say Aunt Cass's familial love is unimportant; rather, the film expands the sources of love and kin. Known for their family-friendly content, Disney animated films like this make room for queer kinship not in opposition to but complementary to and sometimes even preferred over normative, biological kin.

Moreover, the Tadashi-Baymax-Hiro triad queers kinship by blurring the line between human and nonhuman. When Hiro commits to entering the SFIT Showcase to gain admission into Professor Callaghan's robotics program, he fails to generate a novel idea and is ready to give up. Tadashi, however, has faith in his younger brother and unexpectedly pulls Hiro out of his chair, hangs him upside down, jumps around their room, and encourages Hiro to "look for a new angle." Initially reluctant, Hiro locks eyes with his botfighter, which inspires him to create his microbots. Later in the film, Hiro utters a similar sentiment, "I just have to look for another angle," to devise a more effective strategy against Yokai, which results in upgraded technology for him, Baymax, and his new friends-family.

These references to new or alternative angles resonate with Sara Ahmed's notion of queerness as an oblique angle in relation to straight lines that metaphorize heteronormativity.[47] To see from a different angle is to deviate from the straight path. While such deviation may seem pathological from a normative coordinate system, reorienting to the queered angle can and, in the case of Hiro, does lead to more expansive worlds. In taking a queer path, Hiro creates microbots that can materialize the human imagination and forms an eclectic superhero team whose aim is helping people. Similarly, Tadashi persists in programming Baymax after eighty-three failed attempts, never giving up on Baymax as he does for Hiro later and tackling his errors from new angles until he discovers a solution that works.

Audiences witness Tadashi's persistence in an emotional rising-action scene after Hiro confronts Callaghan and alone with Baymax angrily comes to terms with Tadashi's death. Baymax tells Hiro that "Tadashi is here," repeating what he said earlier before downloading a database on personal loss. He proceeds to project on his chest archived footage of Tadashi testing out Baymax's program. "Tadashi is here" declares Tadashi's affective and technological presence as he lives on in people's memories and Baymax's chip. The handwritten label, "TADASHI HAMADA," that adorns Baymax's green chip is a marker of both intellectual property and symbolic essence. However, the chip also becomes Baymax's essence as the film's penultimate scene demonstrates: Hiro discovers the chip in the clenched fist of Baymax's jet arm—all that is materially left of Baymax after sacrificing himself so Hiro and Abigail could escape the interdimensional portal—and after Hiro rebuilds him, Baymax immediately recognizes Hiro, and they hug. In blending Tadashi's and Baymax's essences in the soft robotic figure, the film rejects the techno-Orientalist trope of human/Self versus technology/Other and makes space for multiplicity and ambiguity without anxiety.

The film's collapsing of ontological borders concerning humanity and kinship represented in the superhero team "Big Hero 6" illustrates queerness as assemblage. Jasbir Puar describes assemblage as entities that "converge, diverge, and merge" and "are multiple and layered."[48] They are contingent and dynamic, refuse stable socio-ontological definitions, and enable us to engage queerness's affectiveness.[49] To comprehend the Big Hero 6 as a queer assemblage is to recognize an alternative form of relationality and desire among and between humans and nonhumans unmeasured by Western "humanity."[50]

The Big Hero 6 contrasts with the microbot network also created by Hiro and later usurped by Yokai. While both figure as assemblages that work together for a common purpose, the former comprises distinct individuals whose unique strengths complement one another and the latter comprises millions of identical microbots that are interchangeable and thus expendable.[51] Whereas the microbots submit to the neural control of one person who wears a transmitter, the members of Big Hero 6 collaborate but also sometimes conflict with one another, as witnessed in their unsuccessful initial attempts to defeat Yokai. The latter point emphasizes the team's noncoercive nature, unlike that of the microbots. Moreover, the team's individual color scheme (Hiro purple, Go Go yellow, Wasabi green, Honey Lemon pink, Fred blue, and Baymax red) brightly contrasts with the black microbot mass, visually emphasizing the team's heterogeneity and the microbots' homogeneity. Instead of lauding a singular hero, the film spotlights the power of difference, not uncommon among other Marvel superhero teams such as the Avengers, the Guardians of the Galaxy, and the X-Men (whose stories also evoke queerness). Though as a techno-Oriental film, *Big Hero 6* invites audiences to reimagine the possibilities of Asian/American cultural forms.

Affective Technology

The queer utopia that *Big Hero 6* proffers deviates from normative techno-Orientalist fantasies of Asian futurity through its multicultural, technophilic, anticapitalist orientation. Whereas the androids of *Do Androids Dream of Electric Sheep?* (1968) (adapted as Replicants in the *Blade Runner* universe) supposedly lack empathy, thus making them inferior to humans, Baymax develops a capacity to empathize with Hiro's grief, making him appealing to humans.[52] However, this schema privileges proximity to humanness as a measure of technology's desirability. While *Big Hero 6* might reconfigure the matrix of techno-Orientalism through Baymax's ability to care, its cultural significance lies not with indistinguishing human and nonhuman but rather in exploring what kinds of connections can form between them.

The television series *Baymax!* builds on Baymax's caring nature and focuses less on futuristic technologies and more on the everyday lives of San Fransokyans. Its speculative future includes racial-ethnic, gender, and sexual diversity as a given. Notably, the third episode, "Sofia," includes a transman wearing a V-neck T-shirt with the trans* pride flag colors (light blue, pink, white, pink, and light blue) who is one of several customers who help Baymax decide on which menstrual care products to buy; and the fourth episode, "Mbita," features a male protagonist (Mbita) who asks another guy (Yukio) to go on a date. Such literal representations of trans*ness and queerness queer the normal as they project an expansive normality in a fictive universe that could map onto viewers' reality. Moreover, Baymax's extreme helpfulness, and not his robot identity, initially places him at odds with the human characters he encounters, which creates comedy and narrative tension and decenters technology as social difference.

Big Hero 6 also revises normative techno-Orientalist visions of Asian/American subjectivity. Throughout the film, Hiro is preoccupied with technological improvement; however, his emotions drive his actions. Although he is a teen prodigy, he does not apply his intellectual capacity for societal advancement, opting instead to participate in illicit botfighting and to develop technology to exact revenge. And when offered an unimaginable amount of money by Krei Tech to buy his microbot technology, he declines. As such, he figures as an anti-automaton and anti–model minority. Hiro is not a mindless being who caters to the whims of Euro-American capitalist exploitation. He is not what Long Bui calls a "model machine," a racist trope that constructs the Asian/American subject as robotic model minority, linking earlier forms of Asian/American racialization (e.g., the Asian as automaton) to contemporary ones (e.g., Asia as techno-economical threat, the Asian American as model minority).[53] Hiro's techno-Orientalized subjectivity fails to conform to the Asian-as-automaton trope as he exhibits creative capacities through his microbots and superhero upgrades and refuses to allow those creations to bolster capitalism. And like Baymax, he is not unfeeling but rather filled with emotional range. Analyzing Hiro in relation to Baymax, and vice versa, breaks down techno-Orientalist tendencies that conflate "Asian" with technology and short-circuits the critical move to humanize the racialized subject. As Hee-Jung S. Joo boldly reminds us, Asian/Americans "were never 'whole' to begin with" in a liberal humanist frame.[54] Focusing instead on relationality and entanglement, as Joo does, we approach Asian futurity without predetermined racial forms.

This "present is not enough."[55] The speculative future of *Big Hero 6* harnesses the queer utopian possibilities of a techno-Orientalism that is not depressing nor anti-Asian. In true Disney fashion, the film's narrative tugs at audience heartstrings, but the world that narrative takes place in also engenders a positive affective response, especially for Asian/American and queer audiences who may see themselves reflected in a world that is not toxic for them. The film is but one instantiation of a queer techno-Orientalized future that touches on minoritized desires for more affirming mediascapes that reimagine what Asian/Americans can be and do.

Notes

1 *Big Hero 6*, directed by Don Hall and Chris Williams (Walt Disney Studios Motion Pictures, 2014), https://www.disneyplus.com/browse/entity-c29f81d8-8c51-4fe7-bb0c-13f099ad3e90.

2 While the film does not confirm Hiro's racial-ethnic identity, his surname, physical appearance, and familial artifacts suggest he is part Asian. Potter interprets his character as Asian American, and the "Hiro Hamada" Disney Wiki Fandom page notes the character is Japanese and white. Disney Enterprises, "*Big Hero 6* Press Kit," 2014, https://web.archive.org/web/20181118184455/http://www.wdsmediafile.com/media/bigHero6/writen-material/bigHero6544eac07033e9.pdf; "Hiro Hamada," The Disney Wiki, n.d., accessed December 12, 2022, https://disney.fandom.com/wiki/Hiro_Hamada.

3 *Baymax!* (Walt Disney Animation Studios, 2022), https://www.disneyplus.com/browse/entity-bbbb983c-20a5-4b8f-bcb6-653613b80f19; *Big Hero 6: The Series* (Walt Disney Animation Studios, 2017–2020), https://www.disneyplus.com/browse/entity-79f140c4-c27f-42ab-8c8b-7a277e17ac76.

4 Aimee Bahng, "The Cruel Optimism of Asian Futurity and the Reparative Practices of Sonny Liew's *Malinky Robot*," in *Techno-Orientalism: Imagining Asia in Speculative Fiction, History, and Media*, ed. David S. Roh, Betsy Huang, and Greta A. Niu (New Brunswick, NJ: Rutgers University Press, 2015), 178; Martin F. Manalansan IV, "Queer," in *Keywords for Asian American Studies*, ed. Cathy J. Schlund-Vials, Linda Trinh Võ, and K. Scott Wong (New York: New York University Press, 2015), 199.

5 David S. Roh, Betsy Huang, and Greta A. Niu, "Technologizing Orientalism: An Introduction," in Roh, Huang, Niu, *Techno-Orientalism*, 2; Warren Liu, "Queer Excavations: Technology, Temporality, Race," in Roh, Huang, and Niu, *Techno-Orientalism*, 64–75.

6 Lisa Lowe, *Immigrant Acts: On Asian American Cultural Politics* (Durham, NC: Duke University Press, 1996), 13; Sarita Echavez See, *The Filipino Primitive: Accumulation and Resistance in the American Museum* (New York: New York University Press, 2017), 4.

7 Martin Joseph Ponce, *Beyond the Nation: Diasporic Filipino Literature and Queer Reading* (New York: New York University Press, 2012), 26; Sara Ahmed, *Queer Phenomenology: Orientations, Objects, Others* (Durham, NC: Duke University Press, 2006), 67.

8 José Esteban Muñoz, *Cruising Utopia: The Then and There of Queer Futurity* (New York University Press, 2009), 1.

9 Ahmed, *Queer Phenomenology*, 113.

10 Ponce, *Beyond the Nation*, 27; Kareem Khubchandani, *Ishtyle: Accenting Gay Indian Nightlife* (Ann Arbor: University of Michigan Press, 2020).

11 Roh, Huang, and Niu, "Technologizing Orientalism: An Introduction," 4.

12 Roh, Huang, and Niu, "Technologizing Orientalism: An Introduction," 5, 9. Anne Anlin Cheng's concept of ornamentalism and Shilpa Davé's and Kareem Khubchandani's notions of accent inform my thinking here. Anne Anlin Cheng, *Ornamentalism* (New York: Oxford University Press, 2019); Shilpa S. Davé, *Indian Accents: Brown Voice and Racial Performance in American Television and Film* (Urbana: University of Illinois Press, 2013).

13 Roh, Huang, and Niu, "Technologizing Orientalism: An Introduction," 7.

14 José Esteban Muñoz, *Disidentifications: Queers of Color and the Performance of Politics* (Minneapolis: University of Minnesota Press, 1999), 4.

15 Muñoz, *Disidentifications*, 23.

16 Muñoz, *Disidentifications*, 19.

17 "*Big Hero 6* Press Kit," 2; "Disney's Hyperion Renderer," Walt Disney Animation Studios, July 2015, https://www.disneyanimation.com/technology/hyperion/.

18 *Blade Runner: The Final Cut*, directed by Ridley Scott (Warner Home Video, 2007).
19 "*Big Hero 6* Press Kit," 11.
20 "*Big Hero 6* Press Kit," 11.
21 "*Big Hero 6* Press Kit," 5.
22 Hee-Jung S. Joo, "The Asian (as) Robot: Queer Inhumans in the Works of Margaret Rhee, Greg Pak, and Chang-Rae Lee," *Journal of Asian American Studies* 25, no. 1 (2022): 1–30, https://doi.org/10.1353/jaas.2022.0001.
23 "*Big Hero 6* Press Kit," 4.
24 "*Big Hero 6* Press Kit," 6.
25 Robert G. Lee, *Orientals: Asian Americans in Popular Culture* (Philadelphia: Temple University Press, 2000); Stephen Hong Sohn, *Inscrutable Belongings: Queer Asian North American Fiction* (Stanford, CA: Stanford University Press, 2018); Vivian L. Huang, *Surface Relations: Queer Forms of Asian American Inscrutability* (Durham, NC: Duke University Press, 2022).
26 Sianne Ngai, "The Cuteness of the Avant-Garde," *Critical Inquiry* 31, no. 4 (2005): 814.
27 Ngai, "The Cuteness of the Avant-Garde," 815.
28 Christine R. Yano, *Pink Globalization: Hello Kitty's Trek across the Pacific* (Durham, NC: Duke University Press, 2013); Erica Kanesaka Kalnay, "Yellow Peril, Oriental Plaything: Asian Exclusion and the 1927 U.S.–Japan Doll Exchange," *Journal of Asian American Studies* 23, no. 1 (2020): 93–124, https://doi.org/10.1353/jaas.2020.0003; Erica Kanesaka, "Racist Attachments: Dakko-Chan, Black Kitsch, and Kawaii Culture," *positions: asia critique* 30, no. 1 (2022): 159–187, https://doi.org/10.1215/10679847-9418007.
29 Toni Hays, "Open Concept: Land and Home in Asian/America" (PhD diss., University of California, Irvine, forthcoming).
30 Lori Merish, "Cuteness and Commodity Aesthetics: Tom Thumb and Shirley Temple," in *Freakery: Cultural Spectacles of the Extraordinary Body*, ed. Rosemarie Garland Thomson (New York: New York University Press, 1996), 185–203.
31 Caitlin Roper, "*Big Hero 6* Proves It: Pixar's Gurus Have Brought the Magic Back to Disney Animation," *Wired*, October 21, 2014, https://www.wired.com/2014/10/big-hero-6/.
32 Richard J. Leskosky, "Review of Big Hero 6, Directed by Don Hall and Chris Williams," *Science Fiction Film and Television* 10, no. 1 (2017): 137.
33 *Blade Runner 2049*, directed by Denis Villeneuve (Warner Brothers Home Entertainment, 2017).
34 Rebecca Keegan, "MOVIES; It's a Whole New World for Animated Filmmakers; Disney's 'Big Hero 6' Married Computer Programs with Architectural Creativity to Create Its San Fransokyo," *Los Angeles Times*, October 26, 2014, https://www.proquest.com/newspapers/movies-whole-new-world-animated-filmmakers/docview/1616304976/se-2.
35 Scott Watanabe, "Portfolio: Big Hero 6," n.d., accessed January 31, 2025, https://www.scottwatanabeart.com/big-hero-6.
36 Keegan, "MOVIES; It's a Whole New World for Animated Filmmakers"; Scott Watanabe, "Don wanted to figure out a logical explanation for how a mash-up city like this could exist. I came up with the idea that, after the 1906 earthquake in San Francisco, Japanese immigrants rebuilt the place using techniques that allow movement and flexibility in a seismic event.," Tumblr, February 17, 2015, https://disneyanimation.tumblr.com/post/111288640767/don-wanted-to-figure-out-a-logical-explanation.

37 Sean Miura, "San Fransokyo," *Discover Nikkei*, April 14, 2015, https://www.discovernikkei.org/en/journal/2015/4/14/san-fransokyo.

38 Deborah Shamoon, "The Yōkai in the Database: Supernatural Creatures and Folklore in Manga and Anime," *Marvels & Tales* 27, no. 2 (2013): 276, https://doi.org/10.13110/marvelstales.27.2.0276.

39 Leskosky, "Review of *Big Hero 6*," 139.

40 Leskosky, "Review of *Big Hero 6*," 141.

41 *Snow White and the Seven Dwarfs*, directed by David Hand (RKO Radio Pictures, 1937), https://www.disneyplus.com/browse/entity-f51f7e6c-2d9a-443c-9831-f3cc22e822b4; *The Little Mermaid*, directed by John Musker and Ron Clements (Buena Vista Pictures Distribution, 1989), https://www.disneyplus.com/browse/entity-f7643452-fe64-4b05-8f09-c8bea9b2dd60; *Raya and The Last Dragon*, directed by Don Hall and Carlos López Estrada (Walt Disney Studios Motion Pictures, 2021), https://www.disneyplus.com/browse/entity-72aceacd-23df-4fc2-89fb-0b9595a764ca.

42 Jessica D. Zurcher et al., "'Parental Roles,' in 'The Circle of Life' Representations of Parents and Parenting in Disney Animated Films from 1937 to 2017," *Mass Communication and Society* 23 (2019): 128–150, https://doi.org/10.1080/15205436.2019.1616763. For their analysis of general media, see 131; for Disney media, see 146; for creative or fictive families, see 132, 139, and 146.

43 Zurcher et al., "Parental Roles," 140–141.

44 David L. Eng, *The Feeling of Kinship: Queer Liberalism and the Racialization of Intimacy* (Durham, NC: Duke University Press, 2010), 2.

45 Kimberly D. McKee, *Disrupting Kinship: Transnational Politics of Korean Adoption in the United States* (Urbana: University of Illinois Press, 2019), 63.

46 Yugin Teo, "Recognition, Collaboration and Community: Science Fiction Representations of Robot Carers in *Robot & Frank*, *Big Hero 6* and *Humans*," *Medical Humanities* 47, no. 1 (2021): 95–102, https://doi.org/10.1136/medhum-2019-011744.

47 Ahmed, *Queer Phenomenology*, 67.

48 Jasbir K. Puar, *Terrorist Assemblages: Homonationalism in Queer Times* (Durham, NC: Duke University Press, 2007), xxii.

49 Puar, *Terrorist Assemblages*, 204–205.

50 Joo, "The Asian (as) Robot."

51 Hiro's genius and robotics expertise enable him to develop defensive applications for each member's strengths, Go Go's electromagnetic suspension expertise enables her to travel fast on wheels, Wasabi's laser-induced plasma expertise enables him to instantaneously cut objects into multiple slices, Honey Lemon's chemistry expertise enables her to create immobilizing compounds, Fred's enthusiasm for science and comics expertise motivate the team, and Baymax's large size and robotics capabilities enable him to protect the team.

52 Philip K. Dick, *Do Androids Dream of Electric Sheep?* (New York: Del Rey, 1996).

53 Long T. Bui, *Model Machines: A History of the Asian as Automaton* (Philadelphia: Temple University Press, 2022).

54 Joo, "The Asian (as) Robot," 13.

55 Muñoz, *Cruising Utopia*, 27.

Markets of Techno-Orientalist Critique

A Concluding Discussion

DAVID S. ROH, BETSY HUANG,
GRETA AIYU NIU, AND
CHRISTOPHER T. FAN

DAVID: It's been ten years since we last wrote together with *Techno-Orientalism* (hereafter *TO1*), and while the field has grown considerably, and the culture industry seems receptive, we're still subjected to novels, films, and games that replicate techno-Orientalist tropes and aesthetics. So, I wanted to start by asking the obvious question: has techno-Orientalism as a mode of critique failed?

GRETA: I think it's too early to say it has failed. How many decades of "yellowface" have we endured? How long has the model minority myth persisted? How many more years will the stereotype of Asians as high-tech/machinelike linger? Is it too mundane to note that it's two steps forward, one step backward, with some side stepping along the way? If we didn't have techno-Orientalism as an identifiable form of racism, we would need something similar to it. Naming the phenomenon is a critical step.

I just read *Another Man's Moccasins* (2008) by Craig Johnson, the fourth Longmire mystery; chapters alternate with white protagonist Sheriff Walt Longmire reflecting in first person between the case of a murdered Vietnamese woman named Ho Thi in present-day Wyoming and his time in Vietnam during the U.S.-Vietnam War. The contemporary murder victim reminds him of his first case as a Marine investigator of illegal drug trafficking in Vietnam that led to the murder of a Vietnamese sex worker named Mai Kim. Although Walt was nearly killed by a corrupt white U.S. soldier in 1967, the antagonist in contemporary Wyoming is a Vietnamese man named Tran Van Tuyen. During the U.S.-Vietnam War and perhaps even earlier during the French-Vietnam War, Tran Van Tuyen was trained by the United States to infiltrate and/or fight against the Communist Vietnamese. Now, Tran is trafficking Vietnamese women for sex work under the guise of providing new lives in the United States for "children of the dust": *bui doi* or Amerasian children. Walt saves Ho Thi's colleague Ngo Loi Kim. Classic case of a white man saving "yellow" women from "yellow" men. As sheriff, Walt knows others see him as "antiquated."[1] A laptop, Wi-Fi, and email messages play key roles in solving the mystery; much is made of the fact that Walt does not understand what Wi-Fi is, but Ngo Loi Kim, a young female Vietnamese survivor—the granddaughter of Mai Kim, whom Walt knew—is trying to get his attention through email messages, which he finally "decodes." Walt and his assistant Ruby view the messages as gibberish because they do not realize they are written in Vietnamese but finally figure out they are being sent using the Wi-Fi from an abandoned school. In an example of techno-Orientalism, despite being trafficked from a young age and having little to no freedom or education, Ngo Loi is technologically savvy enough to access Wi-Fi on her stolen laptop.

CHRIS: To respond to David's provocation, I think we have to consider what's changed about the social practices and institutions that sustain Orientalism and make it a useful framework (for dividing the world, for critique). If techno-Orientalism is indeed an abiding "style of thought," to use Edward Said's phrase, then that signals to me that its material bases are also still with us. For me, the "techno" in techno-Orientalism has always indexed a shift in those material bases away from the projects of knowledge production that supported formal European colonialism, to what Said described in the final section of his book, "The Latest Phase": an Orientalism that, after the postwar anticolonial movements became hegemonic along with U.S.-led militarized global capital. What the chapters in this volume have really impressed upon us is how significant China has become to thinking through these questions for our present conjuncture. Also because of developments like COVID-19 and the *hallyu* wave, techno-Orientalism—as a rubric for critique, as the name for the reproduction of racist tropes—hasn't died . . . it's multiplied!

That's why I think this was a volume in search of authors. Has techno-Orientalist critique failed? This volume attests that it hasn't. Take Greta's example of the Johnson novel, which brings the long-standing stereotype of Vietnam's premodernity together with a techno-Orientalist gesture that indexes Vietnam's increasing openness to global capital. What techno-Orientalism signals here is a gap in understanding about how these two temporalities are aspects of the same contradiction, which is capitalism's uneven development. Discourse abhors a vacuum, so racial and sexual stereotypes rush in to fill the gap. If we are wondering about failure at all, then, as we say in the introduction, there's an open question about whether our critique is most usefully pursued under the sign of race, or under a modified version of race analysis like racial capitalism, or (to complete the triptych) through political economy. As a critique of racism, is techno-Orientalism a victim of its own success? Are we close to knowing as much as we can know about techno-Orientalism as a mode of racialization and as a critique of it? Or is the appearance of saturation perhaps a function of China's rise alongside the increasing pace of U.S. decline—a historical development that highlights even more starkly the latest phase of Orientalism's roots spreading along with global capital's moving contradiction?

BETSY: This is quite a provocative question. It reminds me of the question of whether dystopian fiction of the *1984* ilk as a preventive measure has failed in the age of Trumpism. I think it has less to do with the failure of the critique and more to do with the reach of the critique and the power of the objects of critique.

First, on the reach, which is another way of understanding impact. Who has the first volume reached? Who has read it and spread it? How is it marketed, bought, sold, circulated? Where is it taught, advanced, critiqued, ignored? A critical mode's impact is only as wide as its readership, and as deep as that readership's ability to translate the critique into cultural and political practice. Is techno-Orientalist critique a preach-to-the-choir phenomenon? How do we assess its impact?

Second, on the power of the objects of critique. It is clear that the logics and forces of capital are so strong that they can easily withstand and even co-opt techno-Orientalist critique, as evidenced by David's opening observation that "we are still subjected to novels, films, games, and graphic novels that replicate techno-Orientalist tropes and aesthetics" and by Greta's example of recalcitrant Orientalist themes informing contemporary texts like the *Longmire* series. If techno-Orientalist tropes has continued to appear in art and media productions of the past decade, then *TO2*'s challenge is to distinguish between replication and critique, and all the nuances and complexities in between. Techno-Orientalist critique's impact may not be measured by "failure" or "success," but by how it keeps up with new variants of the objects of critique and mounts new critical angles. As is the case with climate change or racism, eradication is aspirational but not realistic. Instead, the pragmatic approach may be to mitigate by weed-whacking the ever-reproducing phenomena. We may wish for the critique to be preventive, but do we acknowledge that its most important work will always be reactive?

Another way to think about this is through the (cynical?) lens of market analysis. I see a competition between the *markets that produce and consume techno-Orientalist objects* (e.g., the industries and texts critiqued in *TO1* and *TO2*) and the *markets that produce or circulate the critique* (e.g., academic scholarship or popular cultural criticism platforms such as *AV Club* and *Vulture*). In this competitive scenario, the latter just does not have the market shares to compete with the former. How do we grow the market of the latter so that it can really influence the production and consumption practices of the former? And how do we square our critique of markets and objects with the fact that we rely on market-driven industries to disseminate the critique?

DAVID: Perhaps "failed" is too reductive a word—it's a premise based on a standard to which I personally don't hold other modes of critique. But I'm curious how we think through the question of efficacy as a way to grapple with how the field has expanded beyond the initial aesthetic we interrogated in the first volume. On the one hand, it's remarkable to see its reach grow beyond generic tropes. But does it risk the possibility of being subsumed into a broader sociocultural or political conversation that renders it paradoxically formless, or less recognizable? In other words, I'm thinking backwards: a techno-Orientalist aesthetic was fairly recognizable initially, but only with its multiplication and relevance across fields does it become something akin to a generalized "vibe." But I'm somewhat concerned that it'll take on more insidious, subtle forms that may not be so easily recognizable as it diffuses. Is the worry overblown? Maybe. There appear moments when everything snaps into focus—the first volume preceded the Trump administration by a year, and this volume will have been published in 2025, several when after another spike in anti-Asian racialization and undoubtedly more fearmongering during the 2024 presidential election. I guess my question is more of a lament over the loss of a clear telos. Talk me off the ledge!

GRETA: I find Betsy's description of the marketing and reach of techno-Orientalist objects and the marketing and reach of techno-Orientalist critique both compelling and disheartening. We know the marketplace of ideas is much bigger than the academy. Chris, your point about discourse abhorring a vacuum reminds us of the many times the United States has viewed China as the other, from the 1882 Exclusion Act to fearmongering over the rise of the PRC a century later. David, I take your worry seriously. In another decade we may not recognize the new forms of techno-Orientalism, and anti-Asian racism will need to be called out differently.

CHRIS: I too find Betsy's distinction between the market for techno-Orientalist objects and the market for techno-Orientalist critique to be really helpful. I'm thinking now about how my previous comment presumes that it's less useful to think about the relation between these two markets than it is about the conditions that make those two markets *markets*—let's just call them global capitalism. David, now that you're ready to move away from the "failure" framework, I kind of want to stick with it for just a moment longer. Betsy is absolutely right, I think, that the market for critique is outgunned by the market for objects. But in recent years we've seen other modes of critique *succeed* in reshaping the market for objects. I'm thinking here of Min Hyoung Song's account of the *mainstreaming* of Asian American literature, and how much that process has expanded in recent years. *Everything Everywhere All at Once* (hereafter *EEAAO*) is the most prominent example of this for our purposes—its success certainly has to do with critiques of representation that have been sustained primarily in the academy, but also modes of praxis and corporate activism that have been commodified, in this case, for instance, as an A24 studio style whose success is due in part to a focus on cultural specificity that came out of representational critiques.[2] To offer another example, while FX's 2024 reboot of *Shōgun* doesn't fully excise the original novel's white savior narrative, it nonetheless metabolizes critique by fucking with the source material, which might be due in part to the writers' room being predominantly Asian American women. To round out these examples with a third, the Netflix adaptation of Liu Cixin's *The Three-Body Problem* could have reversed the original's allegory of the United States as a technological threat, fitting it into a familiar techno-Orientalist framework with China as the threat. It attempts this reversal (insisting on retaining the invading aliens' original, Chinese name, the "San-ti"), but softens the techno-Orientalist edges in various ways, namely by meticulously attending to Chinese historical detail and centering Chinese characters, giving them dimensionality that they don't possess in Liu's novels, and casting excellent actors to play them. As Greta notes, the culture industries are fully capable of adjusting in response to critique, as well as commodifying critique as style. I don't think that *EEAAO*'s representations of East Asian experience achieved market success as a result of the same forces behind *Shōgun*'s updating, or that *The Three Body Problem*'s self-consciousness is thanks to the same or even similar forms of corporate activism—there are specific histories behind each that are blurred out and look like the same thing if we only focus on their end products. But if the larger question you're asking, David, is whether or not a market for critique can improve representational politics in a market for objects, I think we have to acknowledge, even with serious qualification, that the latter has been at least *responsive* to the former. So maybe I can respond to your provocation with another. Is the more interesting question here *not* whether techno-Orientalist critique has failed, but whether it has in fact *succeeded*? Or, to use Song's term again, why it has become so available to mainstreaming?

BETSY: I see a throughline here. We seem to agree that we aren't really asking whether techno-Orientalist critique has wholesale succeeded or failed, but 1) how we continue to recognize it in the market of objects when it has taken on new forms and formations, and 2) how the modes of critique should also take new forms and formations in response. So, "failed" or "succeeded" questions are ultimately questions that prompt impact analysis. This, I think, is what all the chapters in this second volume are doing.

David is absolutely right to note that the techno-Orientalist aesthetic has grown "beyond generic tropes" and has been "subsumed into a broader sociocultural or political conversation." Chris is also right that the market for critique has improved representational politics in the market for objects. I also think that this has been necessary but still not sufficient. In the case of *EEAAO*'s successes,

I am cautious about the meaning and effect of that success beyond the circles of techno-Orientalism-conscious audiences. Most people I talk to—Asians and non-Asians alike—still say that *EEAAO*'s a wild ride but they don't really get it. Those who "get it" presumably have the decoding tools of genre critique to understand the structural and historical vocabularies on which the text is built. While many of us may recognize and respond deeply to *EEAAO*'s (or Charles Yu's *How to Live Safely in a Science Fictional Universe*, for instance) nuanced dramatizations of immigrant melancholy and the travails of being Asian in America via ironic techno-Orientalist aesthetics, such critique is not necessarily discernable to the general audience. Of course, audiences bring to and take from films what they will, and having techno-Orientalist awareness certainly isn't a requirement for meaningful connections with *EEAAO*. But I take Chris's point well, that the impact of techno-Orientalist critique is at least recognizable in how the techno-Orientalist objects are made, even if not in what the audience "gets." But at some point, the latter needs to happen too for the impact to be true and complete.

So, could we say that whereas the first techno-Orientalism volume was focused on the markets of techno-Orientalist *objects*, this volume is focused on the markets of techno-Orientalist *critique*?

DAVID: What I'm hearing is that we're in a moment of rupture and/or transition (cue calls for *TO3*, 2035!). We can likely tie that into similarly situated anxieties in other fields. The current reworking/expansion/diffusion of techno-Orientalist logic is perhaps a response to uncertainty. We hinted that in the first volume's conclusion, but that uncertainty was squarely situated in economic and environmental manufacturing concerns; that's since transitioned to information and biological systems.

To Chris's point on the progress made in the culture industry—sure, I'm all for the subversive reconfigurations of science fiction; I don't mean to discount how Asian/Asian American works have proliferated that we would've loved to have seen before 2015. Relatedly, Betsy's notions of the market and genre help get to what I think I'm concerned about. Even if readers aren't conscious of how Yu's *How to Live Safely* contests techno-Orientalist conventions (and even *he* wasn't entirely conscious of it), it's the recognizability of a genre that's been so internalized that has to be there in the first place for subsequent artists to speak back to.[3] And it's a natural evolution for subgeneric forms to contest, push, and ultimately transform into new generic forms through that dialogic tension. I suppose in this transitional moment, I'm hoping to avoid techno-Orientalist forms and critique devolving into a generalized "Asians plus technology" schema, which is the least interesting form of criticism.

CHRIS: I have to admit, I'm ambivalent about the prospect of a *TO3*! Part of me hopes that all of our Orientalist weeds will have been ripped up by then, roots and all, so that there'll be no need for another volume. Shouldn't critique always desire its own obsolescence? But part of me also knows that even if the relevance of the "techno-" prefix withers away, Orientalism will still be with us because capitalism's contradictions will still be with us. So here I wonder if the "vibe" that David mentions isn't in fact coming from the elephant in the room: the U.S.-China relationship and its predomination of all Asian meanings and social forms. If the point of transition that you mention, David, is the deterioration of that relationship since the previous volume was published in 2015, which is accompanied by intensified forms of uncertainty, then I wonder if, by the time of the projected third volume, 2035, the "techno-" will have been rendered redundant, absorbed into a general, China-directed Orientalism. A curious aspect of this China-directed Orientalism is that it seeks not only to differentiate but, as Daniel Vukovich has argued, to make same.[4] One way we can tell that the market for objects and market for critique are now in fact two sides of the same coin is that the mainstreaming of the market for critique seems so consistently directed at softening a fantasy of inevitable Orientalization. Here, I'm thinking again about Liu Cixin's *The Three-Body Problem* and its Netflix adaptation. In both versions, some humans resist the alien invasion while others welcome it. The Netflix adaptation is so keenly aware of how the aliens allegorize China that it does everything it can to three-dimensionalize its representations of Chineseness: its accurate depictions of Red Guard struggle sessions, which have been widely praised by historians, and, as I mentioned before, the complex protagonicity of its Chinese characters (an improvement, in fact, on Liu's originals). Like the humans who embrace the invasion, the incentive structure here is to get ahead of the curve and begin adjusting immediately to our inevitable Sinicization. To some degree, the persistence of techno-Orientalist tropes is evidence of the clumsiness of this adjustment. The shifting value structures underlying all of this generate what David earlier called the more "subtle forms" of these tropes. What makes the Netflix adaptation Orientalist is that it still excises the Chineseness of Liu's original story. Most of the Chinese settings are relocated to the UK and elsewhere, and most of the Chinese characters are repatriated. What makes all this evidence of becoming-same is that Liu himself gave these choices his blessing, because his primary interest is not in preserving Chineseness but in expanding the *3 Body* IP (intellectual property) into a global, *Star Wars*-like universe of IPs.[5]

GRETA: I understand, Chris, that you are speaking of preserving "Chineseness" tongue in cheek. The author Liu Cixin also permitted *The Three-Body Problem* to be used for a television series set in and released in the PRC in January 2023. Greenlighting the Netflix version of *The Three-Body Problem* that empties out some of the novel's history of the PRC seems like a capitulation, as if the producers, screenwriters, and show runners could not deliver a series set in the PRC and as if viewers would not understand.[6]

Betsy points out the insider knowledge that *EEAAO* celebrates and that some Asian American viewers experience as being seen by popular culture at last as being the main characters. Viewers may not immediately "get" it but perhaps we learn something about multivalent Chinese American immigrant families and not only that Evelyn, as portrayed by action star Michelle Yeoh, demonstrates her prowess in Asian martial arts.[7]

Thank you, David and Chris, for lifting up Min Hyoung Song's point about mainstreaming. In part due to the scarcity of U.S. media featuring characters of Asian descent, and laziness, the predictable association of Asians as martial arts practitioners has continued for decades. The combination of Bruce Lee's popularity and mainstreaming has contributed to rigidity and more stereotyping. In *The Matrix* (1999), Neo claims his expertise with the phrase "I know kung fu" and his hacking skills are represented visually as a "kung fu" fight with Morpheus in a room with rice-paper screen walls, scrolls, and tatami mats.[8]

BETSY: As David reminds us, the techno-Orientalist aesthetic was initially recognizable and was in fact the raison d'être for *TO1*: We knew what we saw in the market of objects and brought exemplars into coherent legibility. But *TO2* is doing something different. *TO2* feels less theoretically cohesive than *TO1*, and I see the coherence of *TO1* being pulled apart, cubist-like, into the much broader range of analytical approaches we have in *TO2*. This effect brings Tina Chen's concepts of "structural incoherence" and "imaginable ageography" to mind.

What techno-Orientalism might now be is a form of Chen's "imaginable ageography" in the way Orientalisms were for Said—a strategic necessity for marking the boundaries of an object (or a geocultural region) for critique. Chen argues in a recent essay that the conceptual formulations of "Global Asias," which "encompass or bring into visibility specific geographic features and characteristics" in order to "[establish] a critical lexicon necessary for making intellectual work possible," warrant continual critical "[disruption] of the consolidating effects of such naming."[9] Failure to do so, Chen cautions, may lead to "an avoidably homogenizing effect that immediately begins rendering invisible the core assumptions and priorities driving its usage."[10] Chen proposes the paradigm of "imaginable ageography" to "work against the consolidating effects of Global Asias by squarely centering the simultaneous dissonances and unifying mythologies inherent in its conceptualization."[11] The paradigm is a play on Said's "imaginative geography," through which a group defines "us" and "them" via imagined boundaries and, as Said reminds us, "does not require that the barbarians acknowledge the distinction."[12] This feels mappable to techno-Orientalist critique because markets are (still) turning out techno-Orientalist objects, seemingly (still) without regard for the putative barbarians' acknowledgement or rejection of the distinction.

"Imaginable ageography" feels useful to the success-or-failure question because it asks us to look at what might be the consolidating effects of techno-Orientalist critique that might also be unwittingly limiting the critical mode's reach and impact. Aware of the limited terrains of *TO1* (especially as articulated by its subtitle, "Imagining Asia in Speculative Fiction, History, and Media"), we consciously assembled a collection that covers broader physical as well as cultural, generic, and ideological geographies this time around. But with expansiveness comes the risk of incoherence. As *TO2*'s chapters take us across Global South Asias as well as cultural products ranging from Disney, the Marvel Comic Universe, and A24 films to independent documentaries, state-run public art projects, music, and museums, our challenge is to heed Chen's and Said's calls to relate "geography's representational function to its ideological utility—a perspectival shift that obviously enable[s] a much better understanding of how *[techno-]*Orientalism operates as a discursive formation and pave[s] the way for the varied approaches of postrepresentational cartography."[13]

Following this, we can say that *TO2* both critically examines our methods for identifying techno-Orientalist cultural geographies *and* paves the way for more varied approaches for such examinations. In the coming years, we will continue to ask questions such as, What is the techno-Orientalist mode of critique looking for now? How do we choose the objects of critique? What would prompt and draw the techno-Orientalist critical eye now?

DAVID: I like the idea of an aspirational obsolescence, Chris, but as you and Greta both note, the logic never dies, it merely evolves; there will always be a market for the racialized Asian tropes, of which a cyberpunk aesthetic is one of several; but the speed with which these forms evolve against competing motives makes for at times awkward—even absurd—artifacts. An "imaginable ageography" is a useful framework that can be expansive but accounts for market and political incentives that we've intimated in the introduction to this volume, as the cultural logic moves from subject to structure. We're already seeing another wave of objects rooted in techno-Orientalism proliferate; it's an open question if that will coalesce into a recognizable aesthetic (or if that's even necessary).

We resisted writing this second volume for a long time for several reasons—the most important being that we didn't want to retread ground already covered. The worst kinds of sequels are those that simply retell the first and consider its lore sacrosanct; the interesting ones unravel and deconstruct the first while honoring its spirit. Apropos, we're closing out this book with earned speculations on the direction of the subfield after upending the narrower focus of the first volume.

Friends, we're out of space and time.

Notes

1 Craig Johnson, *Another Man's Moccasins: A Longmire Mystery* (New York: Penguin, 2008), 19.
2 See Nate Jones, "The Cult of A24," *Vulture*, August 24, 2022, https://www.vulture.com/article/a24-movies-cult.html; Sonia Rao, "How the Indie Studio behind 'Moonlight,' 'Lady Bird' and 'Hereditary' Flourished While Breaking Hollywood Rules," *Washington Post*, August 5, 2019, https://www.washingtonpost.com/lifestyle/style/how-the-indie-studio-behind-moonlight-lady-bird-and-hereditary-flourished-while-breaking-hollywood-rules/2019/08/01/47094878-a4dc-11e9-bd56-eac6bb02do1d_story.html; David Kane, "A Beginner's Guide to A24, Cinema's Coolest

Studio," *Esquire*, May 21, 2021, https://www.esquire.com/uk/culture/a36486289/best-a24-films/.

3   Charles Yu and Leslie Bow, "An Interview with Charles Yu," *Contemporary Literature* 58, no. 1 (2017): 10.

4   Daniel Vukovich, *China and Orientalism: Western Knowledge Production and the PRC* (New York: Routledge, 2012).

5   James Hibberd, "'Game of Thrones' Creators' Wild Road to Their Biggest Gamble Yet: Netflix's '3 Body Problem,'" *The Hollywood Reporter*, January 10, 2024, https://www.hollywoodreporter.com/tv/tv-features/3-body-problem-benioff-weiss-netflix-thrones-interview-1235783117/.

6   Greta: Speaking of Liu Cixin, there's a conservatism in his "dark forest" concept that troubles me; in the simplest of terms, it reanimates fear of the others, more specifically fear of their technological prowess, fear that to communicate would lead to being killed. In this dark forest, the assumption is that those with more technology will overrun those with less; it's better to remain hidden and not to seek out extraterrestrial life. In this zero-sum–game interpretation, the imperialists admit their explorations caused the dwindling of cultures and societies, and even genocides, and fear this will happen to them. For people living under oppression, hiding signs of intelligence is a tactic of survival, but the oppressors are learning to code switch too, and not for the benefit of those they oppress.

7   First as Kato in *The Green Hornet* (1966–1967) and then in *The Big Boss* (1971), *Fist of Fury* (1972), and *Way of the Dragon* (1972), Bruce Lee (1940–1973) provided viewers with characters who fought against oppressors, engaging U.S. audiences of Asian, African, and Latinx descent as a champion against blatant racism. Against hypersexualized Asian female characters and emasculated Asian male characters, Lee's physical talent was a challenge. In an interview on *The Pierre Berton Show* (1971), Lee himself noted that thirty years earlier it would have been extremely unlikely for someone like him to be cast in U.S. television and films. Even though he escaped the stereotype of the emasculated Asian man in *The Green Hornet,* Lee noticed that the other actors could be human beings while his character Kato was "a robot": *Bruce Lee: The Lost Interview* (Little-Wolff Video, 1994).

8   In *Cobra Kai* (2018–2025) the white male actors reprise their characters from *The Karate Kid* (1984). With the commodification of Asian martial arts outside of Asian countries, and thousands of clubs, gyms, and teachers in the United States alone, there may be as many or more students of karate, judo, and tae kwon do outside of Japan and Korea as there are within these countries. In *EEAAO*, Evelyn and Waymond successfully thwart opponents using a variety of fighting tactics, but Evelyn receives training in kung fu. Bruce Lee trained himself using a variety of martial arts, including boxing and fencing, but he is celebrated primarily and perhaps inaccurately as a practitioner of Chinese martial arts.

9   Tina Chen, "Global Asias," in *Asian American Literature in Transition, 1996–2020*, ed. Betsy Huang and Victor Román Mendoza (Cambridge: Cambridge University Press, 2021), 313, 312.

10 Chen, "Global Asias," 313.

11 Chen, "Global Asias," 313.

12 Chen, "Global Asias," 313–314.

13 Italics added by the authors; Chen, "Global Asias," 314.

Acknowledgments

First and foremost, we want to thank the students and readers who reached out to us over the years to let us know how much of an impact the first volume had on their intellectual journeys. Their questions, comments, critiques, and dialogue led us to stretch our horizons, revisit assumptions, and reach further into the crevices of knowledge. Indeed, it's their engagement that eventually led us to revisit the subject for this book; it wouldn't exist without them.

As always, we have a coterie of fellow travelers and colleagues to thank for their support, ideas, and fellowship: Scott Black, Kyung-Sook Boo, Tina Chen, Seo-Young Chu, Stuart Culver, Jennifer Ho, Joe Jeon, Sue Kim, Kim Korinek, Kent Ono, Min Hyoung Song, Myra Washington, Timothy Yu, and all of the contributors to the first volume of *Techno-Orientalism*, whose collective scholarship continues to serve as the signal fire for critical studies yet to come. We want to express deep gratitude also to the writers and artists of the texts examined in these volumes. Productive illumination of the power of texts can only emerge through the dialogic engagement between the creative and critical work, between artists and scholars.

Our contributors and press editors have been exemplars of professionalism and intellectual partners. Specifically, we want to thank Carah Naseem and Huping Ling for their support, responsiveness, and advocacy for this book. The contributors to this volume not only gave their time but answered the call to engage as an intellectual community to make for a more vibrant and critically engaged discussion, punctuated with a two-day symposium in Salt Lake City.

We're grateful for the institutional support from Clark University, the University of California Irvine's School of Humanities, and the University of Utah's College of Humanities, Department of English, and the Asia Center.

David wants to thank his fellow coeditors for the laughs over meals and drinks, the contributors for their responsiveness, and readers of all stripes for reminding him that the work matters.

Chris would like to thank Kyung-Sook Boo for the opportunity to share some sneak peeks of this volume with her students at Sogang University. He would also like to thank David, Betsy, and Greta for lassoing him into their merry crew, and to our contributors for the community that we've created together.

Betsy is inspired every day by David's now-legendary leadership, Greta's unwavering magnanimity, Chris's wicked smarts, and everyone's remarkable intellect. Thank you all for being the best people with whom one could ever dream of collaborating. Nothing can be more gratifying than putting out impactful scholarship that advances the field we call home. Dave, Stella, Mom, Charlie: I love you.

Greta thanks "Davie," "Betsy," and "Chris" for the generous gift of collaborating over many months (and years) and thanks all the contributors for sharing their ideas and energy. It has been a marvelous and heady treat to learn from and with them all. She is grateful to her biological and chosen family members for their love and support.

Bibliography

Abel, Jessamyn. *Dream Super-Express: A Cultural History of the World's First Bullet Train*. Stanford, CA: Stanford University Press, 2022.

Abel, Jonathan E. *The New Real: Media and Mimesis in Japan from Stereographs to Emoji*. Minneapolis: University of Minnesota Press, 2022.

Abel, Jonathan E., and Joseph Jonghyun Jeon. "Unfolding Digital Asias." *Verge: Studies in Global Asias* 7, no. 2 (Fall 2021): vi–xxii. https://doi.org/10.1353/vrg.2021.0011.

Abnet, Dustin. "Escaping the Robot's Loop? Power and Purpose, Myth and History in Westworld's Manufactured Frontier." In *Reading Westworld*, edited by Alex Goody and Antonia Mackay, 221–238. Cham, Switzerland: Palgrave, 2019.

Abu-Lughod, Lila, Brian Larkin, and Faye Ginsburg. *Media Worlds: Anthropology on New Terrain*. Berkeley: University of California Press, 2002.

Adams, Sam. "What the Year's Best Sci-Fi Movie Has to Say About Asian Identity and Adoption." *Slate*, March 10, 2022. https://slate.com/culture/2022/03/after-yang-colin-farrell-showtime-movie-kogonada-interview.html.

Adoptees of Color Roundtable. "Haiti Statement by Adoptees of Color Roundtable." Adopted and Fostered Adults of the African Diaspora, January 26, 2010. https://afaad.wordpress.com/2010/01/26/haiti-statement-by-adoptees-of-color-roundtable/.

Aesthetics Wiki. "Cottagecore," accessed April 22, 2024. https://aesthetics.fandom.com/wiki/Cottagecore.

Aghoro, Nathalie. "Agency in the Afrofuturist Ontologies of Erykah Badu and Janelle Monáe." *Open Cultural Studies* 2, no. 1 (2018): 330–340.

Ahmad, Irfan. "Hindu Orientalism: The Sachar Committee and Over-representation of Minorities in Jail." In *The Politics of Muslim Identities in Asia*, edited by Iulia Lumina, 115–144. Edinburgh: Edinburgh University Press, 2021.

Ahmed, Hilal. "The Good Muslim–Bad Muslim Binary Is as Old as Nehru." *The Print*, November 6, 2018. https://theprint.in/opinion/the-good-muslim-bad-muslim-binary-is-as-old-as-nehru/145770/.

Ahmed, Osman. "Xander Zhou's 'Techno-Orientalism.'" *The Business of Fashion*, January 2, 2018. https://www.businessoffashion.com/reviews/fashion-week/xander-zhou-techno-orientalism/.

Ahmed, Sara. *Queer Phenomenology: Orientations, Objects, Others*. Durham, NC: Duke University Press, 2006.

Ahmed, Shazeda. "Credit Cities and the Limits of the Social Credit System." In *Artificial Intelligence, China, Russia, and the Global Order*, edited by Nicholas D. Wright, 55–61. Montgomery, AL: Air University Press, 2019.

Ajl, Max. *A People's Green New Deal*. London: Pluto Press, 2021.

Akimoto, Daisuke. "Laputa: Castle in the Sky in the Cold War as a Symbol of Nuclear Technology of the Lost Civilisation." *Electronic Journal of Contemporary Japanese Studies* 14, no. 2 (2014). https://www.japanesestudies.org.uk/ejcjs/vol14/iss2/akimoto1.html.

Alaimo, Stacy. "New Materialisms." In *After the Human: Culture, Theory and Criticism in the 21st Century*, edited by Sherryl Vint, 177–191. Cambridge: Cambridge University Press, 2020.

Alessio, Silvia Maria. *Digital Signal Processing and Spectral Analysis for Scientists: Concepts and Applications*. New York: Springer International, 2016.

Allan, Kathryn. "Reimagining Asian Women in Feminist Post-Cyberpunk Science Fiction." In *Techno-Orientalism: Imagining Asia in Speculative Fiction, History, and Media*, edited by David S. Roh, Betsy Huang, and Greta A. Niu, 151–162. New Brunswick, NJ: Rutgers University Press, 2015.

Allan, Nina. "How High We Go in the Dark Review—A New Plague." *Guardian*, January 20, 2022. https://www.theguardian.com/books/2022/jan/20/how-high-we-go-in-the-dark-by-sequoia-nagamatsu-review-a-new-plague.

Allison, Graham, and Eric Schmidt. *Is China Beating America to AI Supremacy?* Cambridge, MA: Harvard Kennedy School, 2020.

Alperovitch, Dmitri. "How the U.S. Can Win the New Cold War." *Time*, May 1, 2024.

Alsultany, Evelyn. *Arabs and Muslims in the Media: Race and Representation after 9/11*. New York: New York University Press, 2016.

Amour, Cheri. "Mitski—Be the Cowboy: Album Review." *The Skinny*, August 14, 2018. https://www.theskinny.co.uk/music/reviews/albums/mitski-be-the-cowboy.

Andersen, Torben G., and Luca Benzoni. "Stochastic Volatility." Chicago: Federal Reserve Bank of Chicago, 2009.

Appadurai, Arjun. *Fear of Small Numbers: An Essay on the Geography of Anger*. Durham, NC: Duke University Press, 2006.

Arakawa, Hiromu. *Kangch'ŏl ui yŏn'gŭm sulsa* [Fullmetal Alchemist]. Translated by Sŏ Hyŏn-a. Seoul: Haksan munhwasa, 2001–2019.

———. *Kangch'ŏl ŭi yŏn'gŭm sulsa k'ŭronik'ŭl* [Fullmetal Alchemist chronicle]. Translated by Sŏ Hyŏn-a. Haksan munhwasa, 2022.

Arias, Michael, and Nakamura Takashi, dirs. *Harmony*. Studio 4°C, 2015.

Arquitectura Viva. "Garden & House, Japan." Accessed June 24, 2024. https://arquitecturaviva.com/works/jardin-y-casa-10.

Aschim, Bjørn-Erik, dir. *Dear Alice*. London: The Line, 2021. https://www.youtube.com/watch?v=z-Ng5ZvrDm4.

Asem Connect. "Master Plan of Thu Duc City in Ho Chi Minh City Until 2040," January 27, 2022. http://asemconnectvietnam.gov.vn/default.aspx?ZID1=14&ID1=2&ID8=114619.

Ashby, W. Ross. *An Introduction to Cybernetics*. London: Chapman & Hall, 1957.

Atanasoski, Neda, and Kalindi Vora. *Surrogate Humanity: Race, Robots, and the Politics of Technological Futures*. Durham, NC: Duke University Press, 2019.

Azuma, Eiichiro. *In Search of Our Frontier: Japanese America and Settler Colonialism in the Construction of Japan's Borderless Empire*. Oakland: University of California Press, 2019.

Azuma, Hiroki. *Otaku: Japan's Database Animals*. Minneapolis: University of Minnesota Press, 2009.
Bady, Aaron. "'Westworld,' Race, and the Western." *New Yorker*, December 9, 2016. https://www.newyorker.com/culture/culture-desk/how-westworld-failed-the-western.
Bahng, Aimee. "The Cruel Optimism of Asian Futurity and the Reparative Practices of Sonny Liew's *Malinky Robot*." In *Techno-Orientalism: Imagining Asia in Speculative Fiction, History, and Media*, edited by David S. Roh, Betsy Huang, and Greta A. Niu, 163–179. New Brunswick, NJ: Rutgers University Press, 2015.
———. *Migrant Futures: Decolonizing Speculation in Financial Times*. Durham, NC: Duke University Press, 2018.
Bal, Hartosh Singh. "UK Govt Inquiry Says VHP Planned to 'Purge Muslims' in 2002 Riots, Acted with Guj Govt's Support." *The Wire*, January 24, 2023. https://thewire.in/communalism/full-text-bbc-documentary-gujarat-riots-modi-uk-report.
Balibar, Étienne. "Racism and Nationalism." In *Race, Nation, Class: Ambiguous Identities*, edited by Étienne Balibar and Immanuel Wallerstein, 37–67. New York: Verso, 1991.
Balibar, Étienne, and Immanuel Wallerstein. *Race, Nation, Class: Ambiguous Identities*. New York: Verso, 1991.
Banerjee, Mita. "More than Meets the Eye: Two Kinds of Re-Orientalism in Naseeruddin Shah's What If?" In *Re-Orientalism and South Asian Identity Politics: The Oriental Other Within*, edited by Lisa Lau and Ana Cristina Mendes, 124–143. New York: Routledge, 2011.
Barad, Karen. *Meeting the Universe Halfway: Quantum Physics and the Entanglement of Matter and Meaning*. Durham, NC: Duke University Press, 2008.
———. "What Is the Measure of Nothingness? Infinity, Virtuality, Justice." In *100 Notes—100 Thoughts, dOCUMENTA 13*, 4–17. Berlin: Hatje Cantz Verlag, 2012.
Bascara, Victor. *Model-Minority Imperialism*. Minneapolis: University of Minnesota Press, 2006.
Bateson, Gregory. *Mind and Nature: A Necessary Unity*. New York: E. P. Dutton, 1979.
———. "Problems in Cetacean and Other Mammalian Communication." In *Steps to an Ecology of Mind*, 369–383 New York: Ballantine Books, 1972.
———. *Steps to an Ecology of Mind*. New York: Ballantine Books, 1971.
Beaudoin, Kate. "11 Musicians Who Wish You'd Just Stop with the Cellphones Already." *Mic*, May 15, 2015. https://www.mic.com/articles/118376/11-musicians-who-wish-you-d-just-stop-with-the-cellphones-already.
Beaumont-Thomas, Ben. "Mitski, the US's Best Young Songwriter: 'I'm a Black Hole Where People Dump Their Feelings.'" *Guardian*, February 4, 2022. https://www.theguardian.com/music/2022/feb/04/mitski-us-best-young-songwriter-im-a-black-hole-where-people-dump-feelings.
Beer, Stafford. *Cybernetics and Management*. New York: John Wiley, 1959.
Bella, Timothy. "'Freedom Never Tasted So Good': How Walter Jones Helped Rename French Fries over the Iraq War." *Washington Post*, February 11, 2019. https://www.washingtonpost.com/nation/2019/02/11/freedom-never-tasted-so-good-how-walter-jones-helped-rename-french-fries-over-iraq-war/.
Beller, Jonathan. *The Cinematic Mode of Production: Attention Economy and the Society of the Spectacle*. Dover, NH: Dartmouth College Press, 2006.
———. *The World Computer: Derivative Conditions of Racial Capitalism*. Durham, NC: Duke University Press, 2021.

Benanav, Aaron. "Automation and the Future of Work—1." *New Left Review* 119 (2019): 5–38.
———. "Automation and the Future of Work—2." *New Left Review* 120 (2019): 117–146.
———. "Service Work in the Pandemic Economy." *International Labor and Working-Class History* 99 (2021): 66–74.
———. "A World Without Work?" *Dissent*, Fall 2020. https://www.dissentmagazine.org/article/a-world-without-work/.
Benjamin, Ruha. "Catching Our Breath: Critical Race STS and the Carceral Imagination." *Engaging Science, Technology, and Society* 2 (2016): 145–156.
———. *Race after Technology: Abolitionist Tools for the New Jim Code*. New York: Polity Press, 2019.
Benjamin, Walter. *Illuminations: Essays and Reflections*. Edited by Hannah Arendt. Translated by Harry Zohn. New York: Schocken, 1968.
Bennett, Jane. *Vibrant Matter*. Durham, NC: Duke University Press, 2009.
Benson, Krista L. "Indigenous Reproductive Justice after Adoptive Couple v. Baby Girl." In *Reproductive Justice and Sexual Rights: Transnational Perspectives*, edited by Tanya Saroj Bakhru, 85–104. New York: Routledge, 2019.
Berlant, Lauren. "The Commons: Infrastructures for Troubling Times." *Environment and Planning D: Society and Space* 34, no. 3 (2016): 393–419.
Bérubé, Michael. Introduction to *Frankenstein*, by Mary Shelley, vii–xxi. New York: W. W. Norton, 2021.
Biao, Xiang. *Global "Body Shopping": An Indian Labor System in the Information Technology Industry*. Princeton, NJ: Princeton University Press, 2007.
Biddle, Sam. "Why an 'AI Race' between the U.S. and China Is a Terrible, Terrible Idea." *The Intercept*, July 21, 2019. https://theintercept.com/2019/07/21/ai-race-china-artificial-intelligence/.
Biel, Robert. *The Entropy of Capitalism*. Chicago: Haymarket Books, 2013.
Billboard. "Five Burning Questions: Mitski Debuts in the Billboard 200's Top 5 With 'Laurel Hell.'" *Billboard*, February 15, 2022. https://www.billboard.com/music/chart-beat/mitski-laurel-hell-five-burning-questions-1235032017/.
Birks, Chelsea. "Objectivity, Speculative Realism, and the Cinematic Apparatus." *Cinema Journal* 57, no. 4 (2018): 3–24.
Bisset, Jennifer. "Creator of The Matrix Code Reveals Its Mysterious Origins." *CNET*, October 19, 2017. https://www.cnet.com/culture/entertainment/lego-ninjago-movie-simon-whiteley-matrix-code-creator/.
Black, Fischer. "Noise." *Journal of Finance* XLI, no. 3 (1986): 539.
Bong, Joon-ho, dir. *Parasite*. CJ Entertainment, 2020.
Bow, Leslie. *Racist Love: Asian Abstraction and the Pleasures of Fantasy*. Durham, NC: Duke University Press, 2022.
Brandt, Allan M. "Racism and Research: The Case of the Tuskegee Syphilis Study." *The Hastings Center Report* 8, no. 6 (1978): 21–29.
Breckenridge, Carol A., and Peter Veer, eds. *Orientalism and the Postcolonial Predicament: Perspectives on South Asia*. Philadelphia: University of Pennsylvania Press, 1993.
Breslow, Jacob. *Ambivalent Childhoods: Speculative Futures and the Psychic Life of the Child*. Minneapolis: University of Minnesota Press, 2021.
Brienza, Casey. "Did Manga Conquer America? Implications for the Cultural Policy of 'Cool Japan.'" *International Journal of Cultural Policy* 20, no. 4 (2014): 383–398. https://doi.org/10.1080/10286632.2013.856893.
Brown, Evan Nicole. "How Sci-Fi Films Use Asian Characters to Telegraph the Future While Also Dehumanizing Them." *Hollywood Reporter*, November 16, 2021.

https://www.hollywoodreporter.com/lifestyle/arts/sci-fi-films-asian-characters-representation-movies-appropriation-dehumanization-1235048534/.

Browne, Simone. *Dark Matters: On the Surveillance of Blackness.* Durham, NC: Duke University Press, 2015.

Bruney, Gabrielle. "A 'Black Mirror' Episode Is Coming to Life in China." *Esquire*, March 17, 2018. https://www.esquire.com/news-politics/a19467976/black-mirror-social-credit-china/.

Bruyneel, Kevin. *Settler Memory: The Disavowal of Indigeneity and the Politics of Race in the United States.* Chapel Hill: University of North Carolina Press, 2021.

Bryant, Nathaniel Heggins. "Neutering the Monster, Pruning the Green: The Ecological Evolutions of Nausicaä of the Valley of the Wind." *Resilience: A Journal of the Environmental Humanities* 2, no. 3 (2015): 120–126.

Bui, Long T. *Model Machines: A History of the Asian as Automaton.* Philadelphia: Temple University Press, 2022.

Bulag, Uradyn. "From Yeke-Juu League to Ordos Municipality: Settler Colonialism and Alter/Native Urbanization in Inner Mongolia." *Provincial China* 7, no. 2 (2002): 196–234.

Bullert, B. J. *Public Television: Politics and the Battle over Documentary Film.* New Brunswick, NJ: Rutgers University Press, 1997.

Burke, Verity, and Will Tattersdill. "Introduction: Museums in Science Fiction, Science Fiction in Museums." *Configurations* 30, no. 3 (Summer 2002): 247–256.

Bursztynsky, Jessica. "Oculus Founder Says Best US Minds Need to Work on A.I. Just Like They Did during the Nuclear Arms Race." CNBC, July 19, 2019. https://www.cnbc.com/2019/07/19/palmer-luckey-best-us-minds-need-to-work-on-ai-like-with-nuclear-arms.html.

Bush, George W. "Address to a Joint Session of Congress and the American People." George W. Bush White House Archives, September 20, 2001. https://georgewbush-whitehouse.archives.gov/news/releases/2001/09/20010920-8.html.

"Business of Bodies." ABC, February 15, 2008.

Butler, Octavia E. "Positive Obsession." In *Bloodchild and Other Stories*, 125–136. New York: Seven Stories Press, 2005.

Byler, Darren. *Terror Capitalism.* Durham, NC: Duke University Press, 2022.

Byrd, Jodi. *The Transit of Empire: Indigenous Critiques of Colonialism.* Minneapolis: University of Minnesota Press, 2011.

Caldwell, John Thornton. *Specworld: Folds, Faults, and Fractures in Embedded Creator Industries.* Berkeley: University of California Press, 2023.

Callenbach, Ernest. *Ecotopia.* Berkeley, CA: Banyan Tree Books, 2004.

Canavan, Gerry. "'We Are the Walking Dead': Race, Time, and Survival in Zombie Narrative." *Extrapolation* 51, no. 3 (January 2010): 431–453.

Canossi, Lesly Deschler, and Zoraida Lopez-Diago. *Black Matrilineage, Photography, and Representation: Another Way of Knowing.* Leuven: Leuven University Press, 2022.

Caruth, Cathy. *Unclaimed Experience: Trauma, Narrative, and History.* Baltimore: Johns Hopkins University Press, 1996.

Cazdyn, Eric. *The Already Dead: The New Time of Politics, Culture, and Illness.* Durham, NC: Duke University Press, 2012.

Chadha, Kalyani, and Anandam P. Kavoori. "Exoticized, Marginalized, Demonized: The Muslim 'Other' in Indian Cinema." In *Global Bollywood*, edited by Anandam P. Kavoori and Aswin Punathambekar, 131–145. New York: New York University Press, 2008.

Chalmers, David J. *The Conscious Mind: In Search of a Fundamental Theory.* New York: Oxford University Press, 1997.

Chan, Dawn. "Asia-Futurism." *Artforum* 54, no. 10 (Summer 2016). https://www.artforum.com/print/201606/asia-futurism-60088.

Chang, Alenda Y. *Playing Nature: Ecology in Video Games*. Minneapolis: University of Minnesota Press, 2019.

Chang, Edmond Y. "Queergaming." In *Queer Game Studies*, edited by Bonnie Ruberg and Adrienne Shaw, 15–23. Minneapolis: University of Minnesota Press, 2017.

Chang, Edmond Y., and Larissa Lai. "Circling Asianfuturisms: Edmond Chang and Larissa Lai in Conversation." María Zambrano Fellowship Online Public Lecture Series, University of Huelva, Spain, May 16, 2023. https://eventos.uhu.es/93666/section/41408/larissa-lais-maria-zambrano-fellowship-events-at-the-university-of-huelva.html.

Cheah, Pheng. "Introduction: Situations and Limits of Postcolonial Theory." In *Sitting Postcoloniality: Critical Perspectives from the East Asian Sinosphere*, edited by Pheng Cheah and Caroline Hau, 1–29. Durham, NC: Duke University Press, 2022.

Chen, Mel. *Animacies: Biopolitics, Racial Mattering, and Queer Affect*. Durham, NC: Duke University Press, 2012.

Chen, Tina. "Global Asias." In *Asian American Literature in Transition, 1996–2020*, edited by Betsy Huang and Victor Román Mendoza, 311–330. Cambridge: Cambridge University Press, 2021.

Cheng, Albert Werner. "China 2098: An Unintended Satire." *Medium*, September 6, 2020. https://medium.com/@albertwernercheng_92086/china-2098-an-unintended-satire-c4c6a178a525.

Cheng, Anne Anlin. *Ornamentalism*. New York: Oxford University Press, 2018.

Cheng, John. "Asians and Asian Americans in Early Science Fiction." In *Oxford Research Encyclopedia of Literature*, August 28, 2019. https://oxfordre.com/literature/display/10.1093/acrefore/9780190201098.001.0001/acrefore-9780190201098-e-924.

Chiang, Ted. "ChatGPT Is a Blurry JPEG of the Web." *New Yorker*, February 9, 2023.

———. "Sci-Fi Writer Ted Chiang: 'The Machines We Have Now Are Not Conscious.'" Interview by Murgia Madmunita. *Financial Times*, June 2, 2023.

———. "Story of Your Life." In *Stories of Your Life and Others*, 91–146. New York: Vintage, 2002.

———. "Transcript: Ezra Klein Interviews Ted Chiang." *The Ezra Klein Show*, March 30, 2021. https://www.nytimes.com/2021/03/30/podcasts/ezra-klein-podcast-ted-chiang-transcript.html.

———. "Why Computers Won't Make Themselves Smarter." *New Yorker*, March 30, 2021.

Chin, Frank, and Jeffery Paul Chan. "Racist Love." In *Seeing through Shuck*, ed. Kostelanetz Richard. New York: Ballantine, 1972.

Cho, Grace. *Haunting the Korean Diaspora: Shame, Secrecy and the Forgotten War*. Minneapolis: University of Minnesota Press, 2008.

Choi, Hyaeweol. "'Wise Mother, Good Wife': A Transcultural Discursive Construct in Modern Korea." *Journal of Korean Studies* 14, no. 1 (2009): 1–34.

Cholodenko, Alan. "'First Principles' of Animation." In *Animating Film Theory*, edited by Karen Beckman, 98–110. Durham, NC: Duke University Press, 2014.

Chomsky, Noam. *For Reasons of State*. New York: Penguin Books, 2003.

Chow, Rey. *A Face Drawn in Sand: Humanistic Inquiry and Foucault in the Present*. New York: Columbia University Press, 2021.

Choy, Christine, and Renee Tajima-Peña, dirs. *Who Killed Vincent Chin?* Film News Now Foundation, 1987.

Chua, Beng Huat, and Koichi Iwabuchi, eds. *East Asian Pop Culture: Analysing the Korean Wave*. Hong Kong: Hong Kong University Press, 2008.

Chuang. "The Wandering Earth: A Reflection of the Chinese New Right." *Chuangcn .Org*, 2019. https://chuangcn.org/2019/08/wandering-earth/.

Chuh, Kandice. *Imagine Otherwise: On Asian Americanist Critique*. Durham, NC: Duke University Press, 2003.

Chuki, Sonam, Raju Sarkar, and Ritesh Kurar. "A Review on Traditional Architecture Houses in Buddhist Culture." *American Journal of Civil Engineering and Architecture* 5, no. 3 (2017): 113–123. https://doi.org/10.12691/ajcea-5-3-6.

Chun, Wendy Hui Kyong. *Control and Freedom: Power and Paranoia in the Age of Fiber Optics*. Cambridge, MA: MIT Press, 2006.

——. "Introduction: Race and/as Technology; or, How to Do Things to Race." *Camera Obscura* 24, no. 1 (May 2009): 7–35.

Chung, Lee Isaac, dir. *Minari*. New York: A24 Films, 2021.

Clare, Eli. *Brilliant Imperfection: Grappling with Cure*. Durham, NC: Duke University Press, 2017.

Cloake, Felicity. "Food-Based Fears Are Rarely Rational, but There Is a Long and Vibrant History of Culinary Boycotts—from Adrian Mole to 'Freedom Fries.'" *New Statesman* 149, no. 5512 (March 20, 2020): 49.

Coded Bias, "Coded Bias Activist Toolkit." New York: 7th Empire Media 2020.

Conn, Virginia L. "Photographesomenonic Sinofuturism(s)." *SFRA Review* 50 (2020): 79–85.

Cornum, Lou. "What Are We Really Getting Out of 'Westworld'?" *BuzzFeed News*, June 21, 2018. https://www.buzzfeednews.com/article/loucornum/westworld-season -two-native-american-fantasy.

Coronavirus Resource Center. "Mortality Analyses." Johns Hopkins University, March 16, 2023. https://coronavirus.jhu.edu/data/mortality.

Corrêa, Laura Guimarães. "Intersectionality: A Challenge for Cultural Studies in the 2020s." *International Journal of Cultural Studies* 23, no. 6 (2020): 823–832.

Costa, Stevi, and Edmond Y. Chang. "Fish, Roses, and Sexy Sutures: Disability, Embodied Estrangement, and Radical Care in Larissa Lai's *The Tiger Flu*." In *Project(ing) Human: Representations of Disability in Science Fiction*, edited by Courtney Stanton, 129–148. Wilmington, DE: Vernon Press, 2023.

Couldry, Nick. "Mediatization or Mediation? Alternative Understandings of the Emergent Space of Digital Storytelling." *New Media & Society* 10, no. 3 (2008): 373–391.

Cowan, T. L., and Jas Rault. "Introduction: Metaphors as Meaning and Method in Technoculture." *Catalyst: Feminism, Theory, Technoscience* 8, no. 2 (2022): 1–23.

Coward, Martin. "Hot Spots/Cold Spots: Infrastructural Politics in the Urban Age." *International Political Sociology* 9, no. 1 (2015): 96–99.

Cowen, Deborah. *The Deadly Life of Logistics: Mapping Violence in Global Trade*. Minneapolis: University of Minnesota Press, 2014.

Cox, Daniel, Juhern Navarro-Rivera, and Robert P. Jones. "Race, Religion, and Political Affiliation of Americans' Core Social Networks." PRRI (Public Religion Research Institute), August 3, 2016. https://www.prri.org/research/poll-race-religion-politics -americans-social-networks/.

Cruz, Felipe Fernando. "Napalm Colonization: Native Peoples in Brazil's Aeronautical Frontiers." *Hispanic American Historical Review* 101, no. 3 (2021): 461–489.

Csicsery-Ronay, Istvan. *The Seven Beauties of Science Fiction*. Middletown, CT: Wesleyan University Press, 2008.

Cyberspace Administration of China. "Notice of the Cyberspace Administration of China on the Public Solicitation of Comments on the Measures for the Administration of Generative Artificial Intelligence Services." Translated by Sihao Huang and Justin Curl. Center for Information Technology Policy, Princeton University, April 16, 2023.

Cybriwsky, Roman A. *Roppongi Crossing: The Demise of a Tokyo Nightclub District and the Reshaping of a Global City*. Athens: University of Georgia Press, 2011.

Daliot-Bul, Michael, and Nissim Otmazgin. "The Legacy of Anime in the United States: Anime-Inspired Cartoons." In *The Anime Boom in the United States*, 107–137. Cambridge, MA: Harvard University Asia Center, 2017.

Dargis, Manohla. "Do Androids Dream Of Watching Your Kids?" *New York Times*, March 4, 2022.

Daston, Lorraine, and Katherine Park. *Wonders and the Order of Nature, 1150–1750*. New York: Zone, 2001.

Daum, Jeremy. "China through a Glass, Darkly." *China Law Translate*, December 24, 2017. https://www.chinalawtranslate.com/en/china-social-credit-score/.

Davé, Shilpa S. *Indian Accents: Brown Voice and Racial Performance in American Television and Film*. Urbana: University of Illinois Press, 2013.

Davis, Aeron. *The Mediation of Power*. London: Routledge, 2007.

Day, Iyko. *Alien Capital: Asian Racialization and the Logic of Settler Colonial Capitalism*. Durham, NC: Duke University Press, 2016.

Dery, Mark, Greg Tate, and Tricia Rose. "Black to the Future: Interviews with Samuel R. Delany, Greg Tate, and Tricia Rose." In *Flame Wars: The Discourse of Cyberculture*, edited by Mark Dery, 735–778. Durham, NC: Duke University Press, 1994.

Dick, Philip K. *Do Androids Dream of Electric Sheep?* New York: Del Rey, 1996.

Diem, Ngoc. "Saigon House Maximizes Open Space." *VNExpress International*, May 9, 2023. https://e.vnexpress.net/photo/style/saigon-house-maximizes-open-space-4596775.html.

———. "Saigon House Uses Brick Walls, Plants to Fight Heat." *VNExpress International*, May 23, 2023. https://e.vnexpress.net/photo/style/saigon-house-uses-brick-walls-plants-to-fight-heat-4599030.html.

Dillon, Grace. "Imagining Indigenous Futurisms." In *Walking the Clouds: An Anthology of Indigenous Science Fiction*, edited by Grace Dillon, 1–12. Tucson: University of Arizona Press, 2012.

Disney Enterprises, "Big Hero 6 Press Kit," 2014. https://web.archive.org/web/20181118184455/http://www.wdsmediafile.com/media/bigHero6/writen-material/bigHero6544eac07033e9.pdf.

Disney Wiki, The. "Hiro Hamada," n.d. https://disney.fandom.com/wiki/Hiro_Hamada.

Dorow, Sara K. *Transnational Adoption: A Cultural Economy of Race, Gender, and Kinship*. New York: New York University Press, 2006.

Dubal, Veena. "Digital Piecework." *Dissent*, Fall 2020. https://www.dissentmagazine.org/article/digital-piecework/.

Durkin, Rachael. "Westworld's Player Piano Is the Great Character That Keeps Getting Overlooked." *The Conversation*, April 18, 2018. http://theconversation.com/westworlds-player-piano-is-the-great-character-that-keeps-getting-overlooked-95238.

Dyer-Witheford, Nick. *Cyber-Proletariat: Global Labour in the Digital Vortex*. London, UK: Pluto Press, 2015.

Edelman, Gilad. "Big Tech Targets DC with a Digital Charm Offensive." *Wired*, March 8, 2021. https://www.wired.com/story/big-tech-targets-dc-with-digital-charm-offensive/.

Eelens, Frank, and J. D. Speckmann. "Recruitment of Labor Migrants for the Middle East: The Sri Lankan Case." *International Migration Review* 24, no. 2 (1990): 297–322.

Elfving-Hwang, Joanna, and Jane Chi Hyun Park. "Deracializing Asian Australia? Cosmetic Surgery and the Question of Race in Australian Television." *Continuum* 30, no. 4 (February 2016): 1–11. https://doi.org/10.1080/10304312.2016.1141864.

Eng, David L. *The Feeling of Kinship: Queer Liberalism and the Racialization of Intimacy*. Durham, NC: Duke University Press, 2010.

———. "Transnational Adoption and Queer Diasporas." *Social Text* 21, no. 3 (2003): 1–37, https://doi.org/10.1215/01642472-21-3_76-1.

Engineer, Asghar Ali. "Gujarat Riots in the Light of the History of Communal Violence." *Economic and Political Weekly* 37, no. 50 (December 14, 2002): 5047–5054.

Esaki, Brett J. "Ted Chiang's Asian American Amusement at Alien Arrival." *Religions* 11, no. 2 (February 2020): 56. https://doi.org/10.3390/rel11020056.

Esposito, Elena. "Predicted Uncertainty: Volatility Calculus and the Indeterminacy of the Future." In *Uncertain Futures. Imaginaries, Narratives, and Calculation in the Economy*, edited by Jans Beckert and Richard Bronk, 219–235. New York: Oxford University Press, 2020.

Ess, Charles. "The Online Manifesto: Philosophical Backgrounds, Media Usages, and the Futures of Democracy and Equality." In *The Online Manifesto: Being Human in a Hyperconnected Era*, edited by Luciana Floridi, 89–109. New York: Springer, 2014.

Evans, Karin. *Lost Daughters of China*. New York: J. P. Tarcher/Putnam, 2000.

Exner, Eike. "When Krazy Kat Spoke Japanese." In *Comics and the Origins of Manga: A Revisionist History*, 96–138. New Brunswick, NJ: Rutgers University Press, 2021.

Fan, Christopher T. "Asian/American (Anti-)Bodies: An Introduction." *Post45*, November 23, 2015. https://post45.org/2015/11/asianamerican-anti-bodies-intro/.

———. "The Red Shredding: On Netflix's '3 Body Problem.'" *Los Angeles Review of Books*, March 30, 2024. https://lareviewofbooks.org/article/the-red-shredding-on-netflixs-3-body-problem/.

———. "Science Fictionality and Post-65 Asian American Literature." *American Literary History* 33, no. 1 (2020): 75–102.

Fan, Wennan. Personal communication, February 26, 2024.

Fickle, Tara. *The Race Card: From Gaming Technologies to Model Minorities*. New York: New York University Press, 2019.

Ford, Brody. "IBM to Pause Hiring for Jobs That AI Could Do." *Bloomberg*, May 1, 2023.

Forrester, Katrina, and Moira Weigel. "Bodies on the Line." *Dissent*, Fall 2020. https://www.dissentmagazine.org/article/bodies-on-the-line/.

Foster, John Bellamy. "The New Cold War on China." *Monthly Review* 73, no. 3 (August 2021): 1–20.

Frank, Adam. "Was There a Civilization on Earth Before Humans? A Look at Available Evidence." *The Atlantic*, April 13, 2018. https://www.theatlantic.com/science/archive/2018/04/are-we-earths-only-civilization/557180/.

Franklin, Seb. *Control: Digitality as Cultural Logic*. Cambridge, MA: MIT Press, 2015.

———. *The Digitally Disposed: Racial Capitalism and the Informatics of Value*. Minneapolis: University of Minnesota Press, 2021.

Frederickson, Brooks. Interview by Jaeyeon Yoo, February 7, 2022.
Frelik, Pawel. "Woken Carbon: The Return of the Human in Richard K. Morgan's Takeshi Kovacs Trilogy." In *Beyond Cyberpunk: New Critical Perspectives*, edited by Graham J. Murphy and Sherryl Vint, 173–190. Routledge, 2010.
French, Howard. *China's Second Continent: How a Million Migrants Are Building a New Empire in Africa*. New York City: Knopf, 2014.
Frey, William H. "Even as Metropolitan Areas Diversify, White Americans Still Live in Mostly White Neighborhoods." *Brookings Institution*, March 23, 2020. https://www.brookings.edu/research/even-as-metropolitan-areas-diversify-white-americans-still-live-in-mostly-white-neighborhoods/.
Fujita, Naoya. "The Age of the Japanese Post-human [*Posuto-hyūmanitiizu: Itō Keikaku ikō no SF* [Post-humanities: SF after Project Itoh]. Tokyo: Nan'undō, 2013.
Fussell, Sidney. "This Film Examines the Biases in the Code That Runs Our Lives." *Wired*, November 15, 2020. https://www.wired.com/story/film-examines-biases-code-runs-our-lives/.
Gaillot, Ann-Derrick. "Mitski Wants You to Be Your Own Cowboy." *The Outline*, August 14, 2018. https://theoutline.com/post/5810/mitski-be-the-cowboy-interview.
Galbraith, Patrick W., and Thomas Lamarre. "Otakuology: A Dialogue." *Mechademia* 5, no. 1 (2010): 360–374.
Galison, Peter. "The Ontology of the Enemy: Norbert Wiener and the Cybernetic Vision." *Critical Inquiry* 21, no. 1 (1994): 228–266.
Gallacher, Lesley-Anne. "(Fullmetal) Alchemy: The Monstrosity of Reading Words and Pictures in Shonen Manga." *Cultural Geographies* 18, no. 4 (2011): 457–473.
Garland, Alex, dir. *Ex Machina*. New York: A24 Films, 2015.
Gates, Racquel. *Double Negative: The Black Image and Popular Culture*. Durham, NC: Duke University Press, 2018.
Geal, Robert. *Ecological Film Theory and Psychoanalysis: Surviving the Environmental Apocalypse in Cinema*. New York: Routledge, 2021.
Geoghegan, Bernard Dionysius. "Orientalism and Informatics: Alterity from the Chess-Playing Turk to Amazon's Mechanical Turk." *Ex-Position*, no. 43 (June 2020): 45–90.
Gibson, William. *Neuromancer*. New York: Ace, 1984.
Giroux, Henry A. "Breaking into the Movies: Public Pedagogy and the Politics of Film." *Policy Futures in Education* 9, no. 6 (2011): 686–695.
Giseburt, Annelise. "How a Scientific Approach to Urban Greening Could Cool Japan's Concrete Jungles." *Japan Times*, April 30, 2023. https://www.japantimes.co.jp/news/2023/04/30/national/green-infrastructure-heat-island-effect/.
Gladwin, Thomas N., James J. Kennelly, and Tara-Shelomith Krause. "Shifting Paradigms for Sustainable Development: Implications for Management Theory and Research." *Academy of Management Review* 20, no. 4 (1995): 874–907. https://doi.org/10.5465/amr.1995.9512280024.
Global Green Growth Institute. "Conference on Implementation of Viet Nam Urban Green Growth Development Plan to 2030 and Circular of Urban Green Growth Indicators," October 4, 2018. https://gggi.org/conference-on-implementation-of-viet-nam-urban-green-growth-development-plan-to-2030-and-circular-of-urban-green-growth-indicators/.
Global Times. "Young Artist Paints Epic Sci-Fi Future of China in 2098," May 22, 2022. https://www.globaltimes.cn/page/202205/1266254.shtml.
Goertzel, Ben. "Seeking the Sputnik of AGI." *H+ Magazine*, March 30, 2011.
Goertzel, Ben, and Joel Pitt. "Nine Ways to Bias Open-Source AGI Toward Friendliness." *Journal of Evolution and Technology* 22, no. 1 (2012): 1–26.

Gong, Jing, and Tingting Liu. "Decadence and Relational Freedom among China's Gay Migrants: Subverting Heteronormativity by 'Lying Flat.'" *China Information* 36, no. 2 (2021): 200–220.

Goodley, Dan. "The Dis/Ability Complex." *DiGeSt. Journal of Diversity and Gender Studies* 5, no. 1 (2018): 5–22. https://doi.org/10.11116/digest.5.1.1.

Gossin, Pamela. "Animated Nature: Aesthetics, Ethics, and Empathy in Miyazaki Hayao's Ecophilosophy." *Mechademia: Second Arc* 10 (2015): 209–234. https://doi.org/10.5749/mech.10.2015.0209.

Grossman, Avidan. "The Beginner's Guide to Techwear." *Esquire*, March 5, 2020. https://www.esquire.com/style/mens-fashion/g31213240/best-techwear-brands/.

Grzyb, Liz, and Cat Sparks, eds. *Ecopunk! Speculative Tales of Radical Futures*. Greenwood, Western Australia: Ticonderoga Publications, 2017.

Gu, Hailiang. "The Process and Logic of China's Socialist Market Economy from Mechanism to System." *International Critical Thought* 11, no. 3 (September 2021): 341–356. https://doi.org/10.1080/21598282.2021.1947032.

Guevarra, Anna Romina. *Marketing Dreams, Manufacturing Heroes: The Transnational Labor Brokering of Filipino Workers*. New Brunswick, NJ: Rutgers University Press, 2010.

———. "Mediations of Care: Brokering Labour in the Age of Robotics." *Pacific Affairs* 91, no. 4 (December 2018): 739–758.

Guha, Ramachandra. "Adivasis, Naxalites and Indian Democracy." *Economic and Political Weekly* 42, no. 32 (August 11, 2007): 3305–3312.

Haglund, Elaine. "Japan: Cultural Considerations." *International Journal of Intercultural Relations* 8, no. 1 (1984): 61–76.

Haider, Maheen. "The Racialization of the Muslim Body and Space in Hollywood." *Sociology of Race and Ethnicity* 6, no. 3 (2020): 382–395.

Hall, Don, and Carlos López Estrada, dirs. *Raya and the Last Dragon*. Walt Disney Studios Motion Pictures, 2021. https://www.disneyplus.com/browse/entity-72aceacd-23df-4fc2-89fb-0b9595a764ca.

Hall, Don, and Chris Williams, dirs. *Big Hero 6*. Walt Disney Studios Motion Pictures, 2014. https://www.disneyplus.com/browse/entity-c29f81d8-8c51-4fe7-bb0c-13f099ad3e90.

Halvorson, Britt E., and Joshua O. Reno. *Imagining the Heartland: White Supremacy and the American Midwest*. Berkeley: University of California Press, 2022.

Hand, David, dir. *Snow White and the Seven Dwarfs*. New York: RKO Radio Pictures, 1937. https://www.disneyplus.com/browse/entity-f51f7e6c-2d9a-443c-9831-f3cc22e822b4.

Han-Tani, Melos. *All Our Asias*. V. 1.1. Released February 7, 2018. Windows, macOS, and Linux.

———. Email interview by Edmond Y. Chang, June 15, 2023.

Haraway, Donna. "A Cyborg Manifesto." In *Manifestly Haraway*, 28–37. Minneapolis: University of Minnesota Press, 2016.

———. *Simians, Cyborgs, and Women: The Reinvention of Nature*. New York: Routledge, 1990.

Harris, Felicia. "'Tell Me the Story of Home': Afrofuturism, Eric Killmonger, and Black American Malaise." *Review of Communication* 20, no. 3 (2020): 278–285.

Harris, Tristan, and Aza Raskin. "Bonus—Coded Bias." Center for Humane Technology Podcast: Your Undivided Attention, April 8, 2021. https://www.humanetech.com/podcast/bonus-coded-bias.

Hartman, Saidiya. *Lose Your Mother: A Journey along the Atlantic Slave Route*. New York: Macmillan, 2008.

———. *Scenes of Subjection: Terror, Slavery, and Self-Making in Nineteenth-Century America*. New York: Oxford University Press, 1997.

Harvey, Penny, Caspar B. Jensen, and Atsuro Morita. *Infrastructure and Social Complexity: A Companion*. London: Routledge, 2017.

Harvey, Penny, and Hannah Knox. "The Enchantments of Infrastructure." *Mobilities* 7, no. 2 (2012): 521–536.

Hashimoto, Kazuma. "The Cyberpunk Genre Has Been Orientalist for Decades—but It Doesn't Have to Be." *Polygon*, January 30, 2021. https://www.polygon.com/2021/1/30/22255318/cyberpunk-2077-genre-xenophobia-orientalism.

Hayakawa Shobō, ed. *Itō Keikaku toribyūto* [Project Itoh tribute]. Vol. 2. Tokyo: Hayakawa shobō, 2015.

Hayles, N. Katherine. *How We Became Posthuman: Virtual Bodies in Cybernetics, Literature, and Informatics*. Chicago: University of Chicago Press, 1999.

———. "Virtual Bodies and Flickering Signifiers." *October* 66 (1993): 69–91.

Hays, Toni. "Re-Homing: Dual Orientalisms in Disney's Big Hero 6." In Racial Site: Landedness and Settler Colonial Fantasies of Home. PhD diss., University of California, Irvine, 2025.

He, Wan, Daniel Goodkind, and Paul Kowal. *Asia Aging: Demographic, Economic, and Health Transitions*, U.S. Census Bureau, International Population Reports." Washington, DC: U.S. Government Publishing Office, June 2022. https://www.census.gov/content/dam/Census/library/publications/2022/demo/p95-22-1.pdf.

Hearing to Receive Testimony on Emerging Technologies and Their Impact on National Security, before the Committee on Armed Services, U.S. Senate, February 23, 2021. https://www.armed-services.senate.gov/imo/media/doc/21-05_02-23-2021.pdf.

Hedger, Patrick. "TPA Launches New Ad Featuring Former National Security Advisor." Taxpayers Protection Alliance, March 31, 2022. https://www.protectingtaxpayers.org/technology/tpa-launches-new-ad-featuring-former-national-security-advisor-2/.

Heidegger, Martin. "The Question Concerning Technology." In *Martin Heidegger: Basic Writings*, edited by David Farrell Krell, 287–317. San Francisco: Harper, 1977.

Hessler, Peter. *The Buried: An Archaeology of the Egyptian Revolution*. New York: Penguin Books, 2020.

Hibberd, James. "'Game of Thrones' Creators' Wild Road to Their Biggest Gamble Yet: Netflix's '3 Body Problem.'" *Hollywood Reporter*, January 10, 2024. https://www.hollywoodreporter.com/tv/tv-features/3-body-problem-benioff-weiss-netflix-thrones-interview-1235783117/.

Hicks, Michael J., and Srikant Devaraj. "The Myth and Reality of Manufacturing in America." Muncie, IN: Ball State University Center for Business and Economic Research, 2017.

Hodgson, Kara. "Russia's Colonial Legacy in the Sakha Heartland." The Arctic Institute: Center for Circumpolar Security Studies, November 15, 2022. https://www.thearcticinstitute.org/russias-colonial-legacy-sakha-heartland/.

Holland, Oscar. "Singapore Is Building a 42,000-Home Eco 'Smart' City." CNN, February 1, 2021. https://www.cnn.com/style/article/singapore-tengah-eco-town/index.html.

Holliday, Ruth, and Joanna Elfving-Hwang. "Gender, Globalization and Aesthetic Surgery in South Korea." *Body & Society* 18, no. 2 (2012): 58–81.

Hollinger, Veronica. "Cybernetic Deconstructions: Cyberpunk and Postmodernism." *Mosaic: An Interdisciplinary Critical Journal* 23, no. 2 (1990): 29–44.
Holmberg, Ryan. "Manga Shōnen: Katō Ken'ichi and the Manga Boys." *Mechademia* 8 (2013): 173–193. https://doi.org/10.1353/mec.2013.0010.
Hong, Euny. *The Birth of Korean Cool: How One Nation Is Conquering the World through Pop Culture*. London: Simon & Schuster, 2014.
Hong, Grace Kyungwon. *The Ruptures of American Capital: Women of Color Feminism and the Culture of Immigrant Labor*. Minneapolis: University of Minnesota Press, 2006.
hooks, bell. "Dig Deep: Beyond Lean In." *The Feminist Wire*, October 28, 2013.
Horsley, Jamie. "China's Orwellian Social Credit Score Isn't Real." *Foreign Policy*, November 2018. https://foreignpolicy.com/2018/11/16/chinas-orwellian-social-credit-score-isnt-real/.
Hsu, Allison. "Imagining Asian Futurity: History, Multiplicity, and Racial Solidarity." Color Bloq's WORLDBUILDING Collection. Accessed July 11, 2022. https://www.colorbloq.org/article/imagining-asian-futurity-history-multiplicity-and-racial-solidarity.
Hu, Jane. "Where the Future Is Asian, and the Asians Are Robots." *New Yorker*, March 4, 2022. https://www.newyorker.com/culture/culture-desk/where-the-future-is-asian-and-the-asians-are-robots.
Hu, Zoe. "Mitski Is Much More Than Another Sad Asian American Girl." *Buzzfeed News*, August 24, 2018. https://www.buzzfeednews.com/article/zoehu/mitski-bad-cowboy-asian-american-singer-songwriter.
Huang, Betsy. "Premodern Orientalist Science Fictions." *MELUS* 33, no. 4 (2008): 23–43. https://doi.org/10.1093/melus/33.4.23.
———. *Contesting Genres in Contemporary Asian American Fiction*. New York: Palgrave Macmillan, 2010.
Huang, Betsy, and Victor Román Mendoza, eds. *Asian American Literature in Transition, 1996–2020*, Vol. 4. Cambridge: Cambridge University Press, 2021.
Huang, Michelle N. "The Posthuman Subject in/of Asian American Literature." In *Oxford Research Encyclopedia of Literature*, 1–23. Oxford University Press, 2019. https://doi.org/10.1093/acrefore/9780190201098.013.921.
———. "Inhuman Figures: Robots, Clones, and Aliens." Smithsonian Asian Pacific American Center, accessed July 14, 2024. https://smithsonianapa.org/inhuman-figures/.
Huang, Vivian L. *Surface Relations: Queer Forms of Asian American Inscrutability*. Durham, NC: Duke University Press, 2022.
Hudson, J. Blaine. "Scientific Racism: The Politics of Tests, Race and Genetics." *The Black Scholar* 25, no. 1 (1995): 3–10.
Hudson, Mark J. "Re-Thinking Jōmon and Ainu in Japanese History." *Asia-Pacific Journal* 20, no. 15(2) (2022).
Huesemann, Michael, and Joyce Huesemann. *Techno-Fix: Why Technology Won't Save Us or the Environment*. Gabriola Island, BC: New Society Publishers, 2011.
Hughes, Kayeleigh. "Album Review: Mitski Outdoes Herself on the Stunning Be the Cowboy." *Consequence*, August 14, 2018. https://consequence.net/2018/08/album-review-mitski-outdoes-herself-on-the-stunning-be-the-cowboy/.
Huh, Jinny. "Racial Speculations: (Bio)Technology, Battlestar Galactica, and a Mixed-Race Imagining." In *Techno-Orientalism: Imagining Asia in Speculative Fiction, History, and Media*, edited by David S. Roh, Betsy Huang, and Greta A. Niu, 101–112. New Brunswick, NJ: Rutgers University Press, 2015.
Hung, Ho-Fung. *Clash of Empires: From "Chimerica" to the "New Cold War."* Cambridge: Cambridge University Press, 2022.

———. "Labor Politics under Three Stages of Chinese Capitalism." *South Atlantic Quarterly* 112, no. 1 (2013): 203–212.

Hu Pegues, Juliana. *Space-Time Colonialism: Alaska's Indigenous and Asian Entanglements*. Chapel Hill: University of North Carolina Press, 2021.

Imaginary Worlds. "Episode 193: Asian Futures Without Asians." Accessed August 17, 2023. https://www.imaginaryworldspodcast.org/episodes/asian-futures-without-asians.

Imroz, Parvez, Kartik Murukutla, Khurram Parvez, and Parvaiz Mata. *Alleged Perpetrators: Stories of Impunity in Jammu and Kashmir*. Srinagar: IPTK, 2012.

Itō Keikaku, and Enjō Tō. *Shisha no Teikoku* [Empire of corpses]. Tokyo: Kawade shobō, 2012.

Iwabuchi, Koichi. "Complicit Exoticism: Japan and Its Other." *Continuum* 8, no. 2 (1994): 49–82.

Jacobs, Margaret D. *A Generation Removed: The Fostering and Adoption of Indigenous Children in the Postwar World*. Lincoln: University of Nebraska Press, 2014. https://doi.org/10.2307/j.ctt1d9nmm2.

———. *White Mother to a Dark Race: Settler Colonialism, Maternalism, and the Removal of Indigenous Children in the American West and Australia, 1880–1940*. Lincoln: University of Nebraska Press, 2009.

Jacobson, Heather. *Culture Keeping: White Mothers, International Adoption, and the Negotiation of Family Difference*. Nashville: Vanderbilt University Press, 2008.

Jaffrelot, Christophe. "Communal Riots in Gujarat: The State at Risk?" *Heidelberg Papers in South Asian and Comparative Politics* 17 (2003): 1–21.

Jagoda, Patrick. *Network Aesthetics*. Chicago: University of Chicago Press, 2016.

Jain, Sarah Lochlann. "Living in Prognosis: Toward an Elegiac Politics." *Representations* 98, no. 1 (2007): 77–92.

Jakobson, Mette Mo. "The Relation of Eco-Effectiveness and Eco-Efficiency—an Important Goal in Design for Environment." In *DFX 1999: Proceedings of the 10th Symposium on Design for Manufacturing, Schnaittach/Erlangen*, 101–104. Germany, 1999.

Jamy, Shayan Hassan. "US-China Tech War: Semiconductors at Heart of Competition Driving World towards New Cold War." *South China Morning Post*, August 13, 2023.

Jenkins, Henry. *Convergence Culture: Where Old and New Media Collide*. New York: New York University Press, 2008.

Jeon, Joseph Jonghyun. *Vicious Circuits: Korea's IMF Cinema and the End of the American Century*. 1st ed.. Stanford, CA: Stanford University Press, 2019.

Jeong, Jae-eun, dir. *Take Care of My Cat*. Cinema Service, 2001.

Jia, Qingguo. "The Impact of 9-11 on Sino-US Relations: A Preliminary Assessment." *International Relations of the Asia-Pacific 3*, no. 2 (2003): 159–177.

Jiang, Sisi. "Stray Falls into the Usual Orientalism Pitfalls of the Cyberpunk Genre." *Kotaku*, July 25, 2022. https://kotaku.com/stray-game-annapurna-interactive-cat-cyberpunk-1849328820.

Johnson, Craig. *Another Man's Moccasins: A Longmire Mystery*. New York: Penguin, 2009.

Jones, Nate. "The Cult of A24." *Vulture*, August 24, 2022. https://www.vulture.com/article/a24-movies-cult.html.

Joo, Hee-Jung S. "The Asian (as) Robot: Queer Inhumans in the Works of Margaret Rhee, Greg Pak, and Chang-Rae Lee." *Journal of Asian American Studies* 25, no. 1 (2022): 1–30. https://doi.org/10.1353/jaas.2022.0001.

———. "We Are the World (but Only at the End of the World): Race, Disaster, and the Anthropocene." *Environment and Planning D: Society and Space* 38, no. 1 (February 2020): 72–90. https://doi.org/10.1177/0263775818774046.

Joyce, Kathryn. *The Child Catchers: Rescue, Trafficking, and the New Gospel of Adoption*. New York: PublicAffairs, 2013.

Jung, Girim. "Techno-Orientalism." Political Theology Network, March 29, 2022. https://politicaltheology.com/techno-orientalism/.

Kadogridis, Laeta. *Altered Carbon*. Netflix Streaming Services, 2018.

Kane, David. "A Beginner's Guide to A24, Cinema's Coolest Studio." *Esquire*, May 21, 2021. https://www.esquire.com/uk/culture/a36486289/best-a24-films/.

Kanesaka, Erica. "Racist Attachments: Dakko-Chan, Black Kitsch, and Kawaii Culture." *Positions: Asia Critique* 30, no. 1 (2022): 159–187. https://doi.org/10.1215/10679847-9418007.

Kanesaka Kalnay, Erica. "Yellow Peril, Oriental Plaything: Asian Exclusion and the 1927 U.S.–Japan Doll Exchange." *Journal of Asian American Studies* 23, no. 1 (2020): 93–124. https://doi.org/10.1353/jaas.2020.0003.

Kang, Jaeho. "The Media Spectacle of a Techno-City: COVID-19 and the South Korean Experience of the State of Emergency." *Journal of Asian Studies* 79, no. 3 (August 2020): 589–598. https://doi.org/10.1017/S0021911820002302.

Kang, Je-kyu, dir. *Shiri*. Samsung Entertainment, 1999.

Karuka, Manu. *Empire's Tracks: Indigenous Nations, Chinese Workers, and the Transcontinental Railroad*. Berkeley: University of California Press, 2019.

Keegan, Rebecca. "MOVIES; It's a Whole New World for Animated Filmmakers; Disney's 'Big Hero 6' Married Computer Programs with Architectural Creativity to Create Its San Fransokyo." *Los Angeles Times*, October 26, 2014. https://www.proquest.com/newspapers/movies-whole-new-world-animated-filmmakers/docview/1616304976/se-2.

Keenan, Jesse M. "A Climate Intelligence Arms Race in Financial Markets: Public Policy Grapples with Private 'Black Box' Models." *Science* 365, no. 6459 (September 20, 2019): 1240.

Keevak, Michael. *Becoming Yellow: A Short History of Racial Thinking*. Princeton, NJ: Princeton University Press, 2011.

Kelly, James P., and John Kessel, eds. *Rewired: The Post-Cyberpunk Anthology*. San Francisco: Tachyon Publications, 2007.

Kendall, Lauren. *Under Construction: The Gendering of Modernity, Class and Consumption in the Republic of Korea*. Honolulu: University of Hawai'i Press, 2001.

Khan, Amir. "Technology Fetishism in *The Wandering Earth*." *Inter-Asia Cultural Studies* 21, no. 1 (2020): 20–37.

Khubchandani, Kareem. *Ishtyle: Accenting Gay Indian Nightlife*. Ann Arbor: University of Michigan Press, 2020.

Kim, Christine. "Figuring North Korean Lives and Human Rights." In *The Subject(s) of Human Rights: Crises, Violations, and Asian/American Critique*, edited by Cathy J. Schlund-Vials, Guy Beauregard, and Hsiu-Chuan Lee, 217–232. Philadelphia: Temple University Press, 2020.

Kim, Elaine H., and Chungmoo Choi, eds. *Dangerous Women: Gender and Korean Nationalism*. New York: Routledge, 1998.

Kim, Eleana J. "Wedding Citizenship and Culture: Korean Adoptees and the Global Family of Korea." *Social Text* 21, no. 1 (March 2003): 57–81. https://doi.org/10.1215/01642472-21-1_74-57.

Kim, Kristin Yoonsoo. "Mitski Is Not Here to Save Indie Rock (but She Might Save You)." *Complex*, June 17, 2016. https://www.complex.com/music/2016/06/mitski-puberty-2-profile.

Kim, Leo. "How *After Yang* Subverts Sci-Fi's Fetishistic 'Hollow Asian' Trope." *Polygon*, March 10, 2022. https://www.polygon.com/22971003/after-yang-kogonada-interview-asian-robot-trope.

———. "On Techno-Orientalism." *Real Life*. Accessed July 11, 2022. https://reallifemag.com/on-techno-orientalism/.

Kim, Scarlet, and Graham Webster. "The Data Arms Race Is No Excuse for Abandoning Privacy." *Foreign Policy*, August 14, 2018. https://foreignpolicy.com/2018/08/14/the-data-arms-race-is-no-excuse-for-abandoning-privacy/.

Kim, Soyoung. *Korean Cinema in Global Contexts: Post-Colonial Phantom, Blockbuster, and Trans-Cinema*. Amsterdam: Amsterdam University Press, 2022.

Kim, Suk-Young. "Disastrously Creative: K-Pop, Virtual Nation, and the Rebirth of Culture Technology." *TDR: The Drama Review* 64, no. 1 (2020): 22–35.

———. "Postornamentality: Ajumma Fabulosity and the Art of Wearing a Visor with Ferocity." *Prism* 19, no. 1 (March 1, 2022): 203–214. https://doi.org/10.1215/25783491-9645992.

Kim, Taeyon. "Neo-Confucian Body Techniques: Women's Bodies in Korea's Consumer Society." *Body and Society* 9, no. 2 (2003): 97–113.

Kim, Youna, ed. *The Korean Wave: Korean Media Go Global*. London: Routledge, 2013.

Kindler, Benjamin. "Maoist Miniatures: The Proletarian Everyday, Visual Remediation, and the Politics of Revolutionary Form." *Modern China* 48, no. 5 (2022): 911–947.

Kirk, Robert. "Sentience and Behaviour." *Mind* 83, no. 329 (1974): 43–60.

Klein, Christina. "'Copywood' No Longer: The South Korean Film Industry Shows It Can Hold Its Own by Combining Local Themes with Hollywood Style." *YaleGlobal Online*, October 11, 2004. https://archive-yaleglobal.yale.edu/content/copywood-no-longer.

Knox, Hannah. "Affective Infrastructures and the Political Imagination." *Public Culture* 29, no. 2 (2017): 363–384.

Kogonada, dir. *After Yang*. New York: A24 Films, 2022.

Kolko, Beth, Lisa Nakamura, and Gilbert Rodman. Introduction to *Race in Cyberspace*, edited by Beth Kolko, Lisa Nakamura, and Gilbert Rodman, 1–14. New York: Routledge, 2000.

Koreeda, Hirokazu, dir. *Air Doll*. Asmik Ace Entertainment, 2009.

Körperwelten. "Body Worlds FAQs," n.d. https://bodyworlds.com/about/faq/.

Kostka, Genia. "China's Social Credit Systems and Public Opinion: Explaining High Levels of Approval." *New Media & Society* 21, no. 7 (2019): 1565–1593.

Kostka, Genia, Léa Steinacker, and Miriam Meckel. "Between Security and Convenience: Facial Recognition Technology in the Eyes of Citizens in China, Germany, the United Kingdom, and the United States." *Public Understanding of Science* 30, no. 6 (2021): 671–690.

Kotliar, Dan M. "Data Orientalism: On the Algorithmic Construction of the Non-Western Other." *Theory and Society* 49, no. 5 (October 1, 2020): 919–939. https://doi.org/10.1007/s11186-020-09404-2.

Kozen, Cathleen. "Traces of the Transpacific U.S. Empire: A Japanese Latin American Critique." *Amerasia Journal* 42, no. 3 (2016): 108–28.

Kroker, Arthur, and Michael A. Weinstein. *Data Trash: The Theory of Virtual Class*. New York: St. Martin's Press, 1994.

Kumar, Sanjeev. "Metonymies of Fear: Islamophobia and the Making of Muslim Identity in Hindi Cinema." *Society and Culture in South Asia* 2, no. 2 (2016): 233–255.

Kwak, Sung-sun. "Foreign Patients Receiving Plastic Surgery Hit Record High in 2018." *Korea Biomedical Review*, October 16, 2019. https://www.koreabiomed.com/news/articleView.html?idxno=6611.

Kwan, Daniel, and Daniel Scheinert, dirs. *Everything Everywhere All at Once*. New York: A24 Films, 2022.

Lai, Larissa. *The Tiger Flu*. Vancouver: Arsenal Pulp Press, 2018.

Lai, Paul. "Stinky Bodies: Mythological Futures and the Olfactory Sense in Larissa Lai's 'Salt Fish Girl.'" *MELUS* 33, no. 4 (2008): 167–187.

LAist. "How the Dark Future of Blade Runner's 2019 Los Angeles Looks in the Light of Actual Today," November 18, 2019. https://laist.com/news/entertainment/blade-runners-2019-los-angeles-then-now-future-past.

LaMarre, Thomas. "Animation and Animism." In *Animals, Animality, and Literature*, edited by Brian Massumi, Bruce Boehrer, and Molly Hand, 284–300. New York: Cambridge University Press, 2018.

———. *The Anime Ecology: A Genealogy of Television, Animation and Game Media*. Minneapolis: University of Minnesota Press, 2018.

———. *The Anime Machine: A Media Theory of Animation*. Minneapolis: University of Minnesota Press, 2009.

Landers, Ashley L. "Abuse after Abuse: The Recurrent Maltreatment of American Indian Children in Foster Care and Adoption." *Child Abuse & Neglect* 111 (January 2021): 104805. https://doi.org/10.1016/j.chiabu.2020.104805.

Landon, Brooks. "Extrapolation and Speculation." In *The Oxford Handbook of Science Fiction*, edited by Rob Latham, 23–34. New York: Oxford University Press, 2014.

Larkin, Brian. *Signal and Noise: Media, Infrastructure and Urban Culture in Nigeria*. Durham, NC: Duke University Press, 2008.

———. "The Politics and Poetics of Infrastructure." *Annual Review of Anthropology* 42 (2013): 327–343.

Last.fm. "Mitski," December 9, 2023. https://www.last.fm/music/Mitski/+wiki.

Latham, Rob. *The Oxford Handbook of Science Fiction*. Oxford: Oxford University Press, 2014.

Latour, Bruno. *Pandora's Hope: Essays on the Reality of Science Studies*. Cambridge, MA: Harvard University Press, 1999.

———. *Science in Action: How to Follow Scientists and Engineers through Society*. Cambridge, MA: Harvard University Press, 1987.

Lau, Lisa. "Re-Orientalism: The Perpetration and Development of Orientalism by Orientals." *Modern Asian Studies* 43, no. 2 (2009): 571–590.

Lau, Lisa, and Ana Cristina Mendes. "Introducing Re-Orientalism: A New Manifestation of Orientalism." In *Re-Orientalism and South Asian Identity Politics: The Oriental Other Within*, edited by Lisa Lau and Ana Cristina Mendes, 1–14. New York: Routledge, 2011.

Lauro, Sarah Juliet, and Karen Embry. "A Zombie Manifesto: The Nonhuman Condition in the Era of Advanced Capitalism." In *Zombie Theory: A Reader*, edited by Sarah Juliet Lauro and Karen Embry, 395–412. Minneapolis: University of Minnesota Press, 2017.

Lavender III, Isiah, ed. *Dis-Orienting Planets: Racial Representations of Asia in Science Fiction*. Jackson: University Press of Mississippi, 2021.

Le, Quynh Nhu. *Unsettled Solidarities: Asian and Indigenous Cross-Representations in the Americas*. Philadelphia: Temple University Press, 2019.

Le Guin, Ursula K. *The Word for World Is Forest*. 2nd ed. New York: Tor Books, 2010.
Lee, Ching Kwan. "Ching Kwan Lee: The Specter of Global China." *Made in China Journal* 2, no. 4 (October–December 2017). https://madeinchinajournal.com/2017/12/24/ching-kwan-lee-the-specter-of-global-china/#:~:text=In%20her%20new%20book%20The%20Specter%20of%20Global,fixates%20on%20China%20as%20a%20new%20colonial%20power.
———. *The Specter of Global China*. Chicago: University of Chicago Press, 2018.
Lee, Dawn, and Betsy Huang. "The Challenges and Joys of Adoption, Part 1." *From Here*, podcast, February 23, 2021. https://www.buzzsprout.com/653773/8018519-the-challenges-and-joys-of-adoption-part-1.
———. "The Challenges and Joys of Adoption, Part 2." *From Here*, podcast, March 1, 2021. https://www.buzzsprout.com/653773/8042291-the-challenges-and-joys-of-adoption-part-2.
Lee, James Kyung-Jin. "An Adopter and the Ends of Adoption." *Adoption & Culture* 6, no. 2 (2018): 282–291.
Lee, Rachel C. *The Americas of Asian American Literature*. Princeton, NJ: Princeton University Press, 1999.
Lee, Robert G. *Orientals: Asian Americans in Popular Culture*. Philadelphia: Temple University Press, 2000.
Lee, Sharon Heijin. "Beauty between Empires: Global Feminism, Plastic Surgery, and the Trouble with Self-Esteem." *Frontiers* 37, no. 1 (2016): 1–31.
———. "Gender, Beauty, and Plastic Surgery: Towards a Transpacific Korean/American Studies." In *A Companion to Korean American Studies*, edited by Rachael Joo and Shelley Sang-Hee Lee, 475–502. Leiden: Brill, 2018.
Lee, Summer Kim. "Staying In: Mitski, Ocean Vuong, and Asian American Asociality." *Social Text* 37, no. 1 (March 2019): 27–49.
Leonard, Dickens. "Towards a Caste-Less Community." *Economic and Political Weekly* 54, no. 21 (2019): 47.
Leskosky, Richard J. "Review of Big Hero 6, Directed by Don Hall and Chris Williams." *Science Fiction Film and Television* 10, no. 1 (2017): 137–141.
Leys, Ruth. "The Turn to Affect: A Critique." *Critical Inquiry* 37, no. 3 (2011): 434–472.
Lindquist, Johan, Biao Xiang, and Brenda S. A. Yeoh. "Opening the Black Box of Migration: Brokers, the Organization of Transnational Mobility and the Changing Political Economy in Asia." *Pacific Affairs* 85, no. 1 (2012): 7–19.
Lister, Martin, Jon Dovey, Seth Giddings, Iain Grant, and Kieran Kelly. *New Media: A Critical Introduction*. New York: Routledge, 2008.
Liu, Andrew. "Lab-Leak Theory and the 'Asiatic' Form: What Is Missing Is a Motive." *n+1*, no. 42 (Spring 2022). https://www.nplusonemag.com/issue-42/politics/lab-leak-theory-and-the-asiatic-form/.
Liu, Cixin. *Wandering Earth*. Translated by Ken Liu. New York: Tor, 2021.
Liu, Warren. "Queer Excavations: Technology, Temporality, Race." In *Techno-Orientalism: Imagining Asia in Speculative Fiction, History, and Media*, edited by David S. Roh, Betsy Huang, and Greta A. Niu, 64–75. New Brunswick, NJ: Rutgers University Press, 2015.
Liu, Xiao, and Shuang Shen. "Introduction: Infrastuctures as An Inter-Asia Method." *Interventions* 25, no. 3 (2023): 297–305.
Lodi-Ribeiro, Gerson. *Solarpunk: Ecological and Fantastical Stories in a Sustainable World*. Translated by Fábio Fernandez. Albuquerque, NM: World Weaver Press, 2018
López, Iván Chaar. "Latina/o/e Technoscience? Labor, Race, and Gender in Cybernetics and Computing." *Social Studies of Science* 52, no. 6 (2022): 829–852.

Lopez, Lori Kido. *Race and Digital Media: An Introduction*. London: Polity Press, 2023.
Louie, Andrea. *How Chinese Are You? Adopted Chinese Youth and Their Families Negotiate Identity and Culture*. New York: New York University Press, 2015.
Lowe, Lisa. *Immigrant Acts: On Asian American Cultural Politics*. Durham, NC: Duke University Press, 1996.
Lu, Sidney Xu. *The Making of Japanese Settler Colonialism: Malthusianism and Trans-Pacific Migration, 1868–1961*. Cambridge: Cambridge University Press, 2019.
Lye, Colleen. *America's Asia: Racial Form and American Literature, 1893–1945*. Princeton, NJ: Princeton University Press, 2004.
Lyman, Stanford M. "The 'Yellow Peril' Mystique: Origins and Vicissitudes of a Racist Discourse." *International Journal of Politics, Culture, and Society* 13, no. 4 (2000): 683–747.
Maçães, Bruno. "China's Black Box Superiority." *Politico*, November 12, 2018. https://www.politico.eu/blogs/the-coming-wars/2018/11/china-black-box-superiority-cybersecurity-artificial-intelligence-ai/
MacKenzie, Donald. *Inventing Accuracy: A Historical Sociology of Nuclear Missile Guidance*. Cambridge, MA: MIT Press, 1990.
MacWilliams, Mark. *Japanese Visual Culture: Explorations in the World of Manga and Anime*. London: Routledge, 2008.
Magnan-Park, Aaron Han Joon. "*Shiri* (1999) and the Reunifying Korean Romantic Fantasy of Namnambungnyŏ." *Quarterly Review of Film and Video* 37, no. 4 (2019): 363–383.
Maizland, Lindsay. "India's Muslims: An Increasingly Marginalized Population." Council on Foreign Relations, last modified March 18, 2024. https://www.cfr.org/backgrounder/india-muslims-marginalized-population-bjp-modi.
Mamdani, Mahmood. "Good Muslim, Bad Muslim: A Political Perspective on Culture and Terrorism." *American Anthropologist* 104, no. 3 (2002): 766–775.
Manalansan, Martin F., IV. "Queer." In *Keywords for Asian American Studies*, edited by Cathy J. Schlund-Vials, Linda Trinh Võ, and K. Scott Wong, 197–202. New York: New York University Press, 2015.
Mannathukkaren, Nissim. "The Banality of Evil." *The Hindu*, March 22, 2014. https://www.thehindu.com/features/magazine/the-banality-of-evil/article5818580.ece.
Marshall, Kingsley. "Music as a Source of Narrative Information in HBO's Westworld." In *Reading Westworld*, edited by Alex Goody and Antonia Mackay, 97–118., Cham, Switzerland: Palgrave, 2019.
Martin, Fran. *Dreams of Flight: The Lives of Chinese Women Students in the West*. Durham, NC: Duke University Press, 2021.
Martoccio, Angie. "Mitski Cuts Sharp and Sure With 'Working for the Knife.'" *Rolling Stone*, October 5, 2021. https://www.rollingstone.com/music/music-features/mitski-working-for-the-knife-new-music-1236649/.
Marzzarella, William. "Culture, Globalization, Mediation." *Annual Review of Anthropology* 33 (2004): 345–367.
Massumi, Brian. "The Autonomy of Affect." *Cultural Critique* 31 (1995): 83–109.
Matsakis, Louise. "How the West Got China's Social Credit System Wrong." *Wired*, July 29, 2019. https://www.wired.com/story/china-social-credit-score-system/.
McCormack, Michael Brandon. "'Your God Is a Racist, Sexist, Homophobic, and a Misogynist.... Our God Is Change': Ishmael Reed, Octavia Butler and Afrofuturist Critiques of (Black) American Religion." *Black Theology* 14, no. 1 (2016): 6–27.
McDonough, William, and Michael Braungart. *Cradle to Cradle: Remaking the Way We Make Things*. New York: North Point Press, 2002.

McGee, Ebony Omotola. *Black, Brown, Bruised: How Racialized STEM Education Stifles Innovation*. Cambridge, MA: Harvard Education Press, 2021.

McGrath, Jason. *Postsocialist Modernity: Chinese Cinema, Literature, and Criticism in the Market Age*. Stanford, CA: Stanford University Press, 2010.

McGray, Douglas. "Japan's Gross National Cool." *Foreign Policy* 130 (2002): 44–54.

McKee, Kimberly D. *Adoption Fantasies: The Fetishization of Asian Adoptees from Girlhood to Womanhood*. Columbus: The Ohio State University Press, 2023.

———. *Disrupting Kinship: Transnational Politics of Korean Adoption in the United States*. Urbana: University of Illinois Press, 2019.

McKinley, Alexander. "Merchants, Maidens, and Mohammedans: A History of Muslim Stereotypes in Sinhala Literature of Sri Lanka." *Journal of Asian Studies* 81, no. 3 (2022): 1–18.

McLuhan, Marshall. *Understanding Media: The Extensions of Man*. Cambridge, MA: MIT Press, 1994.

McNeill, Dougal. "Future City: Tokyo After Cyberpunk." *Review of Asian and Pacific Studies* 44 (2019): 89–108.

Melamed, Jodi. "Racial Capitalism." *Critical Ethnic Studies* 1, no. 1 (2015): 76–85.

———. *Represent and Destroy: Rationalizing Violence in the New Racial Capitalism*. Minneapolis: University of Minnesota Press, 2011.

Merish, Lori. "Cuteness and Commodity Aesthetics: Tom Thumb and Shirley Temple." In *Freakery: Cultural Spectacles of the Extraordinary Body*, edited by Rosemarie Garland Thomson, 185–203. New York: New York University Press, 1996.

Metz, Cade. "Away from Silicon Valley, the Military Is the Ideal Customer." *New York Times*, February 26, 2021. https://www.nytimes.com/2021/02/26/technology/anduril-military-palmer-luckey.html.

Midford, Paul. "Japan's Response to Terror: Dispatching the SDF to the Arabian Sea." *Asian Survey* 42, no. 3 (2003): 329–351. https://doi.org/10.1525/as.2003.43.2.329.

Miller, Chris. *Chip War: The Fight for the World's Most Critical Technology*. New York: Scribner, 2022.

Miller, Robert J. "The Doctrine of Discovery: The International Law of Colonialism." *Indigenous Peoples' Journal of Law, Culture and Resistance* 5, no. 1 (2019): 35–42.

Minch-de Leon, Mark. "Race and the Limitations of 'the Human.'" In *After the Human: Culture, Theory and Criticism in the 21st Century*. Edited by Sherryl Vint. Cambridge: Cambridge University Press, 2020.

Mindell, David. *Between Human and Machine: Feedback, Control, and Computing before Cybernetics*. Baltimore: Johns Hopkins University Press, 2002.

Ministry of Education, Sports, Culture, Science, and Technology. "White Papers: Japanese Government Policies in Education, Science, Sports and Culture 2000," 2000. https://warp.da.ndl.go.jp/info:ndljp/pid/11402417/www.mext.go.jp/b_menu/hakusho/html/hpae200001/index.html.

Mitchell, David T., and Sharon L. Snyder. *Narrative Prosthesis, Disability and the Dependencies of Discourse*. Ann Arbor: University of Michigan Press, 2000.

Mitchell, Timothy. "Infrastructures Work on Time." *E-Flux*, 2020. https://www.e-flux.com/architecture/new-silk-roads/312596/infrastructures-work-on-time/.

Mitchell, W.J.T., and Mark B. N. Hansen, eds. Introduction to *Critical Terms for Media Studies*, vii–xxii. Chicago: University of Chicago Press, 2010.

Miura, Sean. "San Fransokyo." *Discover Nikkei*, April 14, 2015. https://www.discovernikkei.org/en/journal/2015/4/14/san-fransokyo.

Miyawaki, Mitski. "When I'm on Stage and Look to You but You're Gazing into a Screen, It Makes Me Feel as Though Those of Us on Stage Are Being Taken from and

Consumed as Content, Instead of Getting to Share a Moment with You." Twitter, February 24, 2022.

Monroe, Shayla. "Tradition, Purpose, and Technology: An Archaeological Take on the Role of Technological Progress in Black Panther." In *Afrofuturism in Black Panther: Gender, Identity, and the Re-Making of Blackness*, edited by Renee T. White and Karen A. Ritzenhoff, 245–266. Lanham, MD: Lexington Books, 2021.

Moodley, Jacqueline, and Lauren Graham. "The Importance of Intersectionality in Disability and Gender Studies." *Agenda* (Durban, South Africa) 29, no. 2 (2015): 24–33. https://doi.org/10.1080/10130950.2015.1041802.

Moon, Kat. "*Shang-Chi and the Legend of the Ten Rings* Made Me Feel Seen Like No Other Hollywood Blockbuster Has." *Time*, September 3, 2021. https://time.com/6095108/shang-chi-asian-representation/.

Moon, Seungsook. "Begetting the Nation: The Androcentric Discourse of National History and Tradition in South Korea." In *Dangerous Women: Gender and Korean Nationalism*, edited by Chungmoo Choi and Elaine Kim, 33–66. New York: Routledge, 1998.

———. *Militarized Modernity and Gendered Citizenship in South Korea*. Durham, NC: Duke University Press, 2005.

Moreland, Quinn. "Album Review: Be the Cowboy." *Pitchfork*, August 17, 2018. https://pitchfork.com/reviews/albums/mitski-be-the-cowboy/.

Morgan, Richard K. *Altered Carbon*. London: Gollancz, 2002. Reprint, New York: Del Rey, 2006.

Mori Building Company. "Countermeasure to Urban Heat Island Phenomenon," 2023. Accessed January 17, 2023.

Morley, David, and Kevin Robins. *Spaces of Identity: Global Media, Electronic Landscapes and Cultural Boundaries*. London: Routledge, 1995.

Morozov, Evgeny. "Critique of Techno-Feudal Reason." *New Left Review* 133, no. 4 (2022): 89–126.

Mosco, Vincent, and Janet Wasko, eds. *The Political Economy of Information*. Madison: University of Wisconsin Press, 1988.

Moseley, Roger. *Keys to Play: Music as a Ludic Medium from Apollo to Nintendo*. Oakland: University of California Press, 2016.

Mozur, Paul. "Google's AlphaGo Defeats Chinese Go Master in Win for A.I." *New York Times*, May 23, 2017. https://www.nytimes.com/2017/05/23/business/google-deepmind-alphago-go-champion-defeat.html.

Mozur, Paul, and John Markoff. "Is China Outsmarting America in A.I.?" *New York Times*, May 27, 2017. https://www.nytimes.com/2017/05/27/technology/china-us-ai-artificial-intelligence.html.

Mueller, Benjamin, and Eleanor Lutz. "U.S. Has Far Higher Covid Death Rate Than Other Wealthy Countries." *New York Times*, February 1, 2022.

Müller-Wille, Staffan, and Hans-Jörg Rheinberger. *A Cultural History of Heredity*. Chicago: University of Chicago Press, 2012.

Muñoz, José Esteban. *Cruising Utopia: The Then and There of Queer Futurity*. New York: New York University Press, 2009.

———. *Disidentifications: Queers of Color and the Performance of Politics*. Minneapolis: University of Minnesota Press, 1999.

Musk, Elon (@elonmusk). "At Risk of Stating the Obvious, Unless Something Changes to Cause the Birth Rate to Exceed the Death Rate, Japan Will Eventually Cease to Exist. This Would Be a Great Loss for the World." Twitter, May 7, 2022, 5:02 P.M. https://twitter.com/elonmusk/status/1523045544536723456?.

———. "Broad Subject Interview with @DavidFaber," May 16, 2023.

———. "Buying Twitter Is an Accelerant to Creating X, the Everything App." Twitter, October 4, 2022, 6:39 P.M. https://twitter.com/elonmusk/status/1577428272056389633?s=20&t=ywTmQoTb5iku3Lie6YHO_A.

———. "Most People Still Think China Has a One-Child Policy. China Had Its Lowest Birthdate Ever Last Year, despite Having a Three-Child Policy! At Current Birth Rates, China Will Lose 40% of People Every Generation! Population Collapse." Twitter, June 6, 2022, 9:11 A.M. https://twitter.com/elonmusk/status/1533798671984119808.

Musker, John, and Ron Clements, dirs. *The Little Mermaid*. Buena Vista Pictures Distribution, 1989. https://www.disneyplus.com/browse/entity-f7643452-fe64-4b05-8f09-c8bea9b2dd60.

Myers, Kit. "Complicating Birth-Culture Pedagogy at Asian Heritage Camps for Adoptees." *Adoption & Culture* 7, no. 1 (2019): 67–94.

Nagahama, Hiroshi, dir. *Mushishi*. Showgate, 2006.

Nagamatsu, Sequoia. *How High We Go in the Dark*. New York: William Morrow, 2022.

Nakamura, Lisa. *Cybertypes: Race, Ethnicity, and Identity on the Internet*. New York: Routledge, 2002.

———. *Digitizing Race: Visual Cultures of the Internet*. Minneapolis: University of Minnesota Press, 2007.

———. "Race in/for Cyberspace: Identity Tourism and Racial Passing on the Internet." In *CyberReader*, edited by Victor J. Vitanza 442–453. New York: Allyn Bacon, 1999.

Napier, Susan J. *Anime from Akira to Princess Mononoke: Experiencing Contemporary Japanese Animation*. New York: Palgrave, 2000.

Nash, Roderick Frazier. *The Rights of Nature: A History of Environmental Ethics*. Madison: University of Wisconsin Press, 1989.

Naughton, John. "Cold War 2.0 Will Be a Race for Semiconductors, Not Arms." *Guardian*, February 18, 2023.

Nelson, Kim Park. "'Loss Is More than Sadness': Reading Dissent in Transracial Adoption Melodrama in 'The Language of Blood' and 'First Person Plural.'" *Adoption & Culture* 1, no. 1 (2007): 101–128.

Neves, Joshua. *Underglobalization*. Durham, NC: Duke University Press, 2020.

Ngai, Sianne. "The Cuteness of the Avant-Garde." *Critical Inquiry* 31, no. 4 (2005): 811–847.

———. *Our Aesthetic Categories: Zany, Cute, Interesting*. Cambridge, MA: Harvard University Press, 2012.

———. *Ugly Feelings*. Cambridge, MA: Harvard University Press, 2007.

Nguyen, Clinton. "China Might Use Data to Create a Score for Each Citizen Based on How Trustworthy They Are." *Business Insider*, October 26, 2016. https://www.businessinsider.com/china-social-credit-score-like-black-mirror-2016-10.

Nguyen, Thu, and Siddartha Bhatla. "Green Buildings in Vietnam: How Sustainable Are They?" *Vietnam Briefing*, March 28, 2022. https://www.vietnam-briefing.com/news/green-buildings-in-vietnam-how-sustainable-are-they.html.

Nguyen, Viet Thanh. "On Art and Politics." Interviewed by Karen Tei Yamashita. Bay Area Book Festival, May 7, 2018. https://www.youtube.com/watch?v=m4Rf_DIBmO8&t=3698s.

Nieuwenhuis, Paul, A. Touboulic, and Lee Matthews. "Is Sustainable Supply Chain Management Sustainable?" In *Sustainable Development Goals and Sustainable*

Supply Chains in the Post-Global Economy, edited by Natalia Yakovleva, Regina Frei, and Sudhir Rama Murthy, 13–30. New York: Springer, 2019.

Nishime, LeiLani. *Undercover Asian: Multiracial Asian Americans in Visual Culture*. Champaign: University of Illinois Press, 2013.

———. "Whitewashing Yellow Futures in *Ex Machina*, *Cloud Atlas*, and *Advantageous*: Gender, Labor, and Technology in Sci-Fi Film." *Journal of Asian American Studies* 20, no. 1 (February 2017): 29–49.

Niu, Greta Aiyu. "Techno-Orientalism, Nanotechnology, Posthumans, and Post-Posthumans in Neal Stephenson's and Linda Nagata's Science Fiction." *MELUS* 33, no. 4 (2008): 73–96.

Nye, David. *American Technological Sublime*. Cambridge: MIT Press, 1996.

Odell, Colin, and Michelle Le Blanc. *Studio Ghibli: The Films of Hayao Miyazaki and Isao Takahata*. Harpenden, UK: Kamera Books, 2010.

Okawada, Akira. *'Sekai naisen' to wazuka na kibō* ["Global civil war" and a glimmer of hope]. Tokyo: Shoen shinsha, 2013.

Ollstein, Alice Miranda. "Comity Crumbles on Congress' Covid Committee." *Politico*, August 6. 2023. https://www.politico.com/news/2023/08/06/congress-covid-wars-test-bipartisan-bond-00109926.

Ong, Aihwa. *Fungible Life Experiment in the Asian City of Life*. Durham, NC: Duke University Press, 2016.

Ooi, Yen. "Chinese Science Fiction: A Genre of Adversity." *SFRA* 50, no. 2–3 (2020): 141–148.

Ortiga, Yasmin. *Emigration, Employability and Higher Education in the Philippines*. New York: Routledge, 2019.

Ouellette, Laurie. *Viewers Like You: How Public TV Failed the People*. New York: Columbia University Press, 2002.

Palmiste, Claire. "From the Indian Adoption Project to the Indian Child Welfare Act: The Resistance of Native American Communities." *Indigenous Policy Journal* XXII, no. 1 (2011): 10.

Pang, Laikwan. "China's Post-Socialist Governmentality and the Garlic Chives Meme: Economic Sovereignty and Biopolitical Subjects." *Theory, Culture & Society* 39, no. 1 (2022): 81–100.

Paperson, La. "A Ghetto Land Pedagogy: An Antidote for Settler Environmentalism." *Environmental Education Research* 20, no. 1 (2014): 115–130.

Parikka, Jussi. *Operational Images: From the Visual to the Invisual*. Minneapolis: University of Minnesota Press, 2023.

Park, Jane Chi Hyun. *Yellow Future: Oriental Style in Hollywood Cinema*. Minneapolis: University of Minnesota Press, 2010.

Park, Lorraine, and Katharine Park. *Wonder and the Order of Nature, 1150–1750*. New York: Zone, 2001.

Park, Sang Un. "'Beauty Will Save You': The Myth and Ritual of Dieting in Korean Society." *Korea Journal* 47, no. 2 (2007): 41–71.

Parray, M. Imran. "Choking the 'Periphery': Pride and Prejudice in India's Globalizing Internet Imaginary." *Internet Histories* 5, no. 3–4 (2021): 323–340.

Parthasarathi, Vibodh. "Between Strategic Intent and Considered Silence: Regulatory Contours of the TV Business." In *The Indian Media Economy: Industrial Dynamics and Cultural Adaptation*, edited by Adrian Athique, Vibodh Parthasarathi, and S. V. Srinivas, 1:144–166. Oxford: Oxford University Press, 2018.

Pasquale, Frank. *Black Box Society: The Secret Algorithms that Control Money and Information*. Cambridge, MA: Harvard University Press, 2015.

Patterson, Christopher B. "Asian Americans and Digital Games." In *Oxford Research Encyclopedia of Literature*, July 30, 2018. https://doi.org/10.1093/acrefore/9780190201098.013.859.

———. *Open World Empire: Race, Erotics, and the Global Rise of Video Games*. New York: New York University Press, 2020.

Pauleit, Winfried. "Video Surveillance and Postmodern Subjects: The Effects of the Photographesomenon, an Image Form in the Futur Anteriéur." In *Ctrl [Space]: Rhetorics of Surveillance from Bentham to Big Brother*, edited by Thomas Y. Levin, Irsula Frohne, and Peter Weibel, 464–479. Cambridge, MA: MIT Press, 2002.

Penley, Constance, Andrew Ross, and Donna Haraway. "Cyborgs at Large: Interview with Donna Haraway." *Social Text*, no. 25/26 (1990): 8–23.

Pershing, Linda, and Margaret R. Yocom. "The Yellow Ribboning of the USA: Contested Meanings in the Construction of a Political Symbol." *Western Folklore* 55, no. 1 (1996): 41. https://doi.org/10.2307/1500148.

Peters, Michael A., Roderigo Britiz, and Ergin Bulut. "Cybernetic Capitalism, Informationalism and Cognitive Labour." *Geopolitics, History, and International Relations* 1, no. 2 (2009): 11–40.

Petersen, Casper Skovgaard. "Interview: Kim Stanley Robinson." *Farsight*, August 10, 2022. https://farsight.cifs.dk/interview-kim-stanley-robinson/.

Petrick, Elizabeth R. "Building the Black Box: Cyberneticians and Complex Systems." *Science, Technology, & Human Values* 45, no. 4 (2019): 575–595.

Pfohl, Stephen. "New Global Technologies of Power: Cybernetic Capitalism and Social Inequality." In *The Blackwell Companion to Social Inequalities*, edited by Mary Romero and Eric Margolis, 546–592. Hoboken, NJ: John Wiley and Sons, 2005.

Pinch, Trevor J. *Confronting Nature: The Sociology of Solar-Neutrino Detection*. Dordrecht, Netherlands: D. Reidel, 1986.

———. "Opening Black Boxes: Science, Technology and Society." *Social Studies of Science* 22, no. 3 (1992): 487–510.

Ponce, Martin Joseph. *Beyond the Nation: Diasporic Filipino Literature and Queer Reading*. New York: New York University Press, 2012.

Ponciano, Jonathan. "Everything in the $1.2 Trillion Infrastructure Bill: New Roads, Electric School Buses and More." *Forbes*, November 15, 2021. https://www.forbes.com/sites/jonathanponciano/2021/11/15/everything-in-the-12-trillion-infrastructure-bill-biden-just-signed-new-roads-electric-school-buses-and-more/?sh=66723020161f.

Popovich, Nadja, and Brad Plumer. "How the New Climate Bill Would Reduce Emissions." *New York Times*, August 2, 2022. https://www.nytimes.com/interactive/2022/08/02/climate/manchin-deal-emissions-cuts.html.

Posadas, Baryon Tensor. "Beyond Techno-Orientalism: Virtual Worlds and Identity Tourism in Japanese Cyberpunk." In *Dis-Orienting Planets: Racial Representations of Asia in Science Fiction*, edited by Isiah Lavender III, 144–159. Jackson: University Press of Mississippi, 2017.

Price, Margaret. "The Bodymind Problem and the Possibilities of Pain." *Hypatia* 30, no. 1 (2015): 268–284.

Project Itoh. *Harmony*. San Francisco: VIZ Media, 2010.

PRRI (Public Religion Research Institute). "PRRI Survey: Friendship Networks of White Americans Continue to Be 90% White." PRRI, May 24, 2022. https://www.prri.org/press-release/prri-survey-friendship-networks-of-white-americans-continue-to-be-90-white/.

Puar, Jasbir K. "Critical Disability Studies and the Question of Palestine: Toward Decolonizing Disability." In *Crip Genealogies*, edited by Mel Y. Chen, Alison Kafer, Eunjung Kim, and Julie Avril Minich, 117–134. Durham, NC: Duke University Press, 2023.

———. *Terrorist Assemblages: Homonationalism in Queer Times*. Durham, NC: Duke University Press, 2007.

———. *The Right to Maim: Debility, Capacity, Disability*. Durham, NC: Duke University Press, 2017.

Quan, H.L.T. *Growth Against Democracy: Savage Developmentalism in the Modern World*. Lanham, MD: Lexington Books, 2012.

Quint, The. "Bollywood's Role in India's Heated Political Environment." *The Quint's Films & Politics Roundtable*, January 2, 2020. https://www.youtube.com/watch?v=N8M4BgaTm38.

Raimon, Eva Allegra. *The "Tragic Mulatta" Revisited: Race and Nationalism in Nineteenth-Century Antislavery Fiction*. New Brunswick, NJ: Rutgers University Press, 2004.

Raleigh, Elizabeth. *Selling Transracial Adoption: Families, Markets, and the Color Line*. Philadelphia: Temple University Press, 2017.

Ramirez, Catherine S. "Afrofuturism/Chicanafuturism." *Aztlán: A Journal of Chicano Studies* 33, no. 1 (2008): 185–194.

Rao, Sonia. "How the Indie Studio behind 'Moonlight,' 'Lady Bird' and 'Hereditary' Flourished While Breaking Hollywood Rules." *Washington Post*, August 5, 2019. https://www.washingtonpost.com/lifestyle/style/how-the-indie-studio-behind-moonlight-lady-bird-and-hereditary-flourished-while-breaking-hollywood-rules/2019/08/01/47094878-a4dc-11e9-bd56-eac6bb02d01d_story.html.

Ren, Mike Yi. *Hazy Days*. Released April 1, 2020. Windows, macOS, and Linux.. https://mikeyren.itch.io/hazydays.

———. "Interview: Mike Ren Yi,." By Christopher B. Patterson. *First Person Scholar*, May 27, 2020. http://www.firstpersonscholar.com/interview-mike-ren-yi/.

Renfro, Kim. "Here's Why Modern Songs Play on the Saloon's Piano in 'Westworld.'" *Business Insider*, October 11, 2016, https://www.businessinsider.com/westworld-piano-songs-2016-10.

Richards, Joanne. "An Institutional History of the Liberation Tigers of Tamil Eelam (LTTE)." Graduate Institute of International and Development Studies. Geneva: Geneva Graduate Institute, 2014.

Riendeau, Danielle. "A Game That Uses PS1 Visuals to Tell a Story about Asian American." *Vice*, February 9, 2018. https://www.vice.com/en/article/pampqg/all-our-asias-ps1-sean-han-tani.

Rippa, Alessandro, Galen Murton, and Matthaus Rest. "Building Highland Asia in the Twenty-First Century." *Verge* 6, no. 2 (2020): 83–111.

Rivera, Takeo. *Model Minority Masochism: Performing the Cultural Politics of Asian American Masculinity*. New York City: Oxford University Press, 2022.

Robertson, Jennifer. "Furusato Japan: The Culture and Politics of Nostalgia." *International Journal of Politics, Culture, and Society* 1, no. 4 (1988): 494–518.

Robinson, Kim Stanley. *The Ministry for the Future*. New York: Orbit, 2020.

Roh, David, Betsy Huang, and Greta Niu. "Desiring Machines, Repellant Subjects: A Conclusion." In *Techno-Orientalism: Imagining Asia in Speculative Fiction, History, and Media*, edited by David Roh, Betsy Huang, and Greta A. Niu, 221–226. New Brunswick, NJ: Rutgers University Press, 2015.

———. "Technologizing Orientalism: An Introduction." In *Techno-Orientalism: Imagining Asia in Speculative Fiction, History, and Media*, edited by David Roh, Betsy Huang, and Greta A. Niu, 1–20. New Brunswick, NJ: Rutgers University Press, 2015.

———. *Techno-Orientalism: Imagining Asia in Speculative Fiction, History and Media*. New Brunswick, NJ: Rutgers University Press, 2015. https://doi.org/10.36019/9780813570655.

Rolston, Bill, and Amaia Alvarez Berastegi. "Taking Murals Seriously: Basque Murals and Mobilisation." *International Journal of Politics, Culture, and Society* 29, no. 1 (2016): 33–56.

Roose, Kevin, and Casey Newton. "Bluesky Has the Juice, A.I. Jobs Apocalypse and Hard Questions." *New York Times*, May 5, 2023. https://www.nytimes.com/2023/05/05/podcasts/hard-fork-bluesky-ai-jobs.html.

Roosth, Sophia, and Astrid Schrader. "Feminist STS Out of Science: An Introduction." *Differences* 25, no. 3 (2012): 1–8.

Roper, Caitlin. "Big Hero 6 Proves It." *Wired*, October 21, 2014. https://www.wired.com/2014/10/big-hero-6/.

Roquet, Paul. "Ambient Literature and the Aesthetics of Calm: Mood Regulation in Contemporary Japanese Fiction." *Journal of Japanese Studies* 35, no. 1 (2009): 111–187.

Rosenblueth, Arturo, Norbert Wiener, and Julian Bigelow. "Behavior, Purpose, and Teleology." *Philosophy of Science* 10, no. 1 (1943): 18–24.

Rossignol, Derrick. "Mitski Announces an Intimate New Album with The Epic Single 'Geyser.'" *Uproxx*, May 14, 2018. https://uproxx.com/music/mitski-be-the-cowboy-new-album-geyser-song/.

Ruberg, Bonnie. "Straight Paths Through Queer Walking Simulators: Wandering on Rails and Speedrunning in *Gone Home*." *Games and Culture* 15, no. 6 (March 2019): 1–21.

Rupprecht, Christoph, Deborah Cleland, Norie Tamura, Rajat Chaudhuri, and Sarena Ulibarri. *Multispecies Cities: Solarpunk Urban Futures*. Albuquerque, NM: World Weaver Press, 2021.

Ryōtarō, Makihara, dir. *Empire of Corpses*. Wit Studio, 2015.

Sachar Committee Report, The. "Social, Economic and Educational Status of the Muslim Community of India: A Report." *Contemporary Education Dialogue* 4, no. 2 (January 2007).

Sacks, Samm, and Graham Webster. "The Trump Administration's Approach to Huawei Risks Repeating China's Mistake." *Slate*, May 21, 2019. https://slate.com/technology/2019/05/u-s-china-huawei-executive-order-foreign-adversary-national-security.html.

Saich, Tony. "The Fourteenth Party Congress: A Programme for Authoritarian Rule." *China Quarterly* 132 (December 1992): 1136–1160. https://doi.org/10.1017/S0305741000045574.

Said, Edward W. *Orientalism*. New York: Vintage Books, 1978.

Saito, Kohei. *Marx in the Anthropocene*. Cambridge: Cambridge University Press, 2022.

Sajed, Alina. "Between Algeria and the World: Anticolonial Connectivity, Aporias of National Liberation and Postcolonial Blues." *Postcolonial Studies* 26, no. 1 (2023): 1–19.

Sakai, Naoki. *The End of Pax Americana: The Loss of Empire and Hikikomori Nationalism*. Durham, NC: Duke University Press, 2022.

Samuels, Ellen. "Six Ways of Looking at Crip Time." *Disability Studies Quarterly* 37, no. 3 (August 31, 2017). https://dsq-sds.org/index.php/dsq/article/view/5824

Sato, Kumiko. "How Information Technology Has (Not) Changed Feminism and Japanism: Cyberpunk in Japanese Contexts." *Comparative Literature Studies* 41, no. 3 (August 2004): 335–355.

Schaffer, Simon. "Enlightened Automata." In *The Sciences in Enlightened Europe*, edited by William Clark, Jan Golinski, and Simon Schaffer, 126–165. Chicago: University of Chicago Press, 1999.

Schalk, Sami. *Bodyminds Reimagined: (Dis)Ability, Race, and Gender in Black Women's Speculative Fiction*. Durham, NC: Duke University Press, 2018.

Schmeink, Lars. "Embodiment in Altered Carbon." In *Sex, Death and Resurrection in Altered Carbon: Essays on the Netflix Series*, edited by Aldona Kobus and Łukasz Muniowski, 67–80. Jefferson, NC: McFarland, 2020.

Schodt, Frederik L. *Dreamland Japan: Writings on Modern Manga*. Berkeley, CA: Stone Bridge Press, 2011.

Schonhardt, Sara, and E&E News. "China Invests $546 Billion in Clean Energy, Far Surpassing the U.S." *Scientific American*. https://www.scientificamerican.com/article/china-invests-546-billion-in-clean-energy-far-surpassing-the-u-s/.

Schwenkel, Christina. "Spectacular Infrastructure and Its Breakdown in Socialist Vietnam." *American Ethnologist* 42, no. 3 (2015): 520–534.

Scott, Ridley, dir. *Blade Runner: The Final Cut*. Warner Home Video, 2007.

Searle, John. "Minds, Brains and Programs." *Behavioral and Brain Sciences* 3, no. 3 (1980): 417–457.

Sedgwick, Eve Kosofsky. *Tendencies*. Durham, NC: Duke University, 1993.

See, Sarita Echavez. *The Filipino Primitive: Accumulation and Resistance in the American Museum*. New York: New York University Press, 2017.

Selman, Peter. "Intercountry Adoption after the Haiti Earthquake: Rescue or Robbery?" *Adoption & Fostering* 35, no. 4 (December 2011): 41–49. https://doi.org/10.1177/030857591103500405.

———. "International Adoption from China and India 1992–2018." In *Social Welfare in India and China: A Comparative Perspective*, edited by Jianguo Gao, 393–415. Singapore: Palgrave Macmillan, 2020. https://doi.org/10.1007/978-981-15-5648-7_22.

Shamoon, Deborah. "The Yōkai in the Database: Supernatural Creatures and Folklore in Manga and Anime." *Marvels & Tales* 27, no. 2 (2013): 276–289. https://doi.org/10.13110/marvelstales.27.2.0276.

Shanfeld, Ethan, and Meredith Woerner. "How Netflix's 'Coded Bias' Breaks Down the Frightening Race and Gender Biased Algorithms That Run the World," n.d. https://variety.com/video/coded-bias-shalini-kantayya-documentary/.

Shannon, Claude. "A Mathematical Theory of Communication." *The Bell System Technical Journal* 27, no. 3 (October 1948): 379–423.

———. "A Mathematical Theory of Communication." *The Bell System Technical Journal* 27, no. 4 (October 1948): 623–656.

Shannon, Claude, and Warren Weaver. *The Mathematical Theory of Communication*. Chicago: University of Chicago Press, 1949.

Shaviro, Steven. *Cinematic Body*. Minneapolis: University of Minnesota Press, 1993.

———. "Unpredicting the Future." *Alienocene*, stratus 1, 2018. https://alienocene.files.wordpress.com/2018/04/unpredicting-to-print.pdf.

Sherman, Justin. "Reframing the U.S.-China AI 'Arms Race.'" *New America*, March 2019. http://newamerica.org/cybersecurity-initiative/reports/essay-reframing-the-us-china-ai-arms-race/.

Shin, Chi-Yun, and Julian Stringer. "Storming the Big Screen: The Shiri Syndrome." In *Seoul Searching: Culture and Identity in Contemporary Korean Cinema*, edited by Frances Gateward, 55–72. Albany: SUNY Press, 2007.

Shinji, Miyadai. *Owaranaki Nichijō o Ikirō* [Live the endless everyday]. Tokyo: Chikuma shobô, 1995.

Shiu, Anthony Sze-Fai. "What Yellowface Hides: Video Games, Whiteness, and the American." *Journal of Popular Culture* 39, no. 1 (February 2006): 109–125.

Shouse, Eric. "Feeling, Emotion, Affect." *M/C Journal* 8, no. 6 (2005). https://doi.org/10.5204/mcj.2443.

Shūkō, Murase, dir. *Genocidal Organ*. Geno Studio, 2017.

Singapore Green Plan. "A City of Possibilities," accessed December 30, 2022. https://www.greenplan.gov.sg.

Singh, Vandana. "The Occasional Writer: An Interview with Science Fiction Author Ted Chiang." *The Margins*, October 3, 2012. https://aaww.org/the-occasional-writer-an-interview-with-science-fiction-author-ted-chiang/.

Singh, Vijaita. "Only Six Religion Options Make It to Next Census Form." *The Hindu*, May 26, 2023.

Skloot, Rebecca. *The Immortal Life of Henrietta Lacks*. New York: Random House, 2011.

Smedley, Audrey. *Race in North America: Origin and Evolution of a Worldview*. Boulder, CO: Westview Press, 2007.

Smith, Rhianedd. "'I Just Don't Think About It': Engaging Students with Critical Heritage Discourse through Science Fiction." *Configurations* 30, no. 3 (2002): 349–355.

Smithson, Robert. "A Sedimentation of the Mind: Earth Projects." *Artforum* 7, no. 1 (1968): 44–50.

Sohn, Stephen Hong. "Introduction: Alien/Asian: Imagining the Racialized Future." *MELUS* 33, no. 4 (2008): 5–22.

———. *Inscrutable Belongings: Queer Asian North American Fiction*. Stanford, CA: Stanford University Press, 2018.

Sollors, Werner. *Neither Black Nor White Yet Both: Thematic Explorations of Interracial Literature*. Oxford: Oxford University Press, 1997.

Song, Min Hyoung. "Becoming Planetary." *American Literary History* 23, no. 3 (2011): 555–573.

———. *The Children of 1965: On Writing, and Not Writing, as an Asian American*. Durham, NC: Duke University Press, 2013.

Song, Mingwei. "After 1989: The New Wave of Chinese Science Fiction." *China Perspectives* 1, (2015), 7–14.

Soromenho-Marques, Viriato. "'Walden': A Tale on the 'Art of Living.'" *Configurações* 25 (2020): 25–35.

SP Group. "Empowering My Tengah," May 27, 2023. https://www.mytengah.sg.

Spice, Anne. "Fighting Invasive Infrastructures: Indigenous Relations against Pipelines." *Environment and Society* 9 (2018): 40–56.

Spillius, Alex. "Obama Administration Embroiled in Illegal Removal of Haitian Children." *Telegraph* (London), February 28, 2010. https://www.telegraph.co.uk/news/worldnews/northamerica/usa/7338739/Obama-administration-embroiled-in-illegal-removal-of-Haitian-children.html.

Spivak, Gayatri Chakravorty. "The Author in Conversation." In *Imaginary Maps*, edited by Mahasweta Devi, translated by Gayatri Chakravorty Spivak, ix–xxii. New York: Routledge, 1997.

Srinivasan, Meera. "No Action from Govt. on Tamils' Concerns, Says TNA." *The Hindu*, January 5, 2023.

Stacey, Kiran, and Caitlin Gilbert. "Big Tech Increases Funding to US Foreign Policy Think-Tanks." *Financial Times*, February 1, 2022. https://www.ft.com/content/4e4ca1d2-2d80-4662-86d0-067a10aad50b.

Stanton, Courtney, ed. "Estrangement, and Radical Care in Larissa Lai's *The Tiger Flu*." In *Project(ing) Human: Representations of Disability in Science Fiction*, 129–147. Wilmington, DE: Vernon Press, 2023.

Star, Susan Leigh. "The Ethnography of Infrastructure." *American Behavioral Scientist* 43, no. 3 (1999): 377–391.

Star, Susan Leigh, and Karen Ruhleder. "Steps toward an Ecology of Infrastructure." *Information Systems Research* 7, no. 1 (1996): 111–134.

Stearns, Elizabeth, Claudia Buchmann, and Kara Bonneau. "Interracial Friendships in the Transition to College: Do Birds of a Feather Flock Together Once They Leave the Nest?" *Sociology of Education* 82, no. 2 (April 2009): 173–195. https://doi.org/10.1177/003804070908200204.

Steeves, Paulette. *The Indigenous Paleolithic of the Western Hemisphere*. Lincoln: University of Nebraska Press, 2021.

Steinacker, Léa, Miriam Meckel, Genia Kostka, and Damian Borth. "Facial Recognition: A Cross-National Survey on Public Acceptance, Privacy, and Discrimination." In *Proceedings of the 37th International Conference on Machine Learning*, 2020. https://arxiv.org/abs/2008.07275.

Ström, Timothy Erik. "Capital and Cybernetics." *New Left Review* 135 (2022): 23–41.

Strong, Turner and Pauline. "To Forget Their Tongue, Their Name, and Their Whole Relation: Captivity, Extra-Tribal Adoption, and the Indian Child Welfare Act." In *Relative Values: Reconfiguring Kinship Studies*, edited by Sarah B. Franklin and Susan McKinnon, 468–494. Durham, NC: Duke University Press, 2001.

Sullivan, Shannon. *The Physiology of Sexist and Racist Oppression*. Oxford: Oxford University Press, 2015.

Summers, Tim. "China's 'New Silk Roads': Sub-National Regions and Networks of Global Political Economy." *Third World Quarterly* 37, no. 9 (2016): 1628–1643.

Sun, Mengtian. "Alien Encounters in Liu Cixin's *The Three-Body Trilogy* and Arthur C. Clarke's *Childhood's End*." *Frontiers of Literary Studies in China* 12, no. 4 (2018): 610–644.

Suparak, Astria. "Asian Futures, Without Asians: An Illustrated Talk by Astria Suparak." The Wattis Institute, June 28, 2021. https://wattis.org/browse-the-library/watch-listen/past-events/asian-futures-without-asians-an-illustrated-talk-by-astria-suparak.

———. "Seedy Space Ports and Colony Planets: Asian Canonical Hats in Cinematic Dystopias," *Seen* 2 (Spring 2021). https://blackstarfest.org/seen/read/issue-002/seedy-space-ports-and-colony-planets.

Svensson, Patrik. "From Optical Fiber to Conceptual Cyberinfrastructure." *Digital Humanities Quarterly* 5, no. 1 (2011). http://www.digitalhumanities.org/dhq/vol/5/1/000090/000090.html.

Swanson, Ana. "The CHIPS Act Is about More Than Chips: Here's What's in It." *New York Times*, February 28, 2023. https://www.nytimes.com/2023/02/28/business/economy/chips-act-childcare.html.

Tanaka, Shouhei. "The Great Arrangement: Planetary Petrofiction and Novel Futures." *Modern Fiction Studies* 66, no. 1 (2020): 190–215.

Taneja, Hemant, and Fareed Zakaria. "AI and the New Digital Cold War." *Harvard Business Review*, September 6, 2023.

Tchen, John Kuo Wei, and Dylan Yeats, eds. *Yellow Peril! An Archive of Anti-Asian Fear*. London: Verso, 2014.

Tech Transparency Project. "Funding the Fight Against Antitrust: How Facebook's Antiregulatory Attack Dog Spends Its Millions," May 17, 2022. https://www.techtransparencyproject.org/articles/funding-fight-against-antitrust-how-facebooks-antiregulatory-attack-dog-spends-its-millions.

Teo, Yugin. "Recognition, Collaboration and Community: Science Fiction Representations of Robot Carers in *Robot & Frank*, *Big Hero 6* and *Humans*." *Medical Humanities* 47, no. 1 (2021): 95–102. https://doi.org/10.1136/medhum-2019-011744.

Terranova, Tiziana. "Attention, Economy, and the Brain." *Culture Machine* 13 (2012): 1–19.

———. "Free Labor: Producing Culture for the Digital Economy." *Social Text* 18, no. 2 (2000): 33–58.

Thakur, Tanul. "Review: *Family Man 2* Is a Worthy Sequel, Even If It Lacks Intimate Dilemmas of First Season." *The Wire*, June 5, 2021. https://thewire.in/culture/review-family-man-season-2.

Thieret, Adrian. "Society and Utopia in Liu Cixin." *China Perspectives*, no. 2015 (2015): 33–39.

Thompson, Nicholas, and Ian Bremmer. "The AI Cold War that Threatens Us All." *Wired*, October 23, 2018. https://www.wired.com/story/ai-cold-war-china-could-doom-us-all/.

Thomson, Rosemarie Garland. *Extraordinary Bodies: Figuring Physical Disability in American Culture and Literature*. New York: Columbia University Press, 2017.

Thoreau, Henry David. "Walden." In *The Portable Thoreau*, edited by Carl Bode, 258–572. New York: The Viking Press, 1964.

Tillet, Salamishah. "'Black Panther' Brings Hope, Hype and Pride." *New York Times*, February 9, 2018. https://www.nytimes.com/2018/02/09/movies/black-panther-african-american-fans.html.

Tiqqun. *The Cybernetic Hypothesis*. Translated by Robert Hurley. South Pasadena, CA Semiotext(e), 2020.

Tran, Julie Ha. "Thinking about Bodies, Souls, and Race in Gibson's Bridge Trilogy." In *Techno-Orientalism: Imagining Asia in Speculative Fiction, History, and Media*, edited by David S. Roh, Betsy Huang, and Greta A. Niu, 139–150. New Brunswick, NJ: Rutgers University Press, 2015.

Trieu, Monica Mong. *Fighting Invisibility: Asian Americans in the Midwest*. New Brunswick, NY: Rutgers University Press, 2023.

Turing, Alan. "Computing Machinery and Intelligence." *Mind*, n.s. 59, no. 236 (October 1950): 433–460.

Turner, Fred. *From Counterculture to Cyberculture: Stewart Brand, the Whole Earth Network, and the Rise of Digital Utopianism*. Chicago: University of Chicago Press, 2006.

Tuvel, Rebecca. "In Defense of Transracialism." *Hypatia* 32, no. 2 (2017): 263–278. 20/20.

Tykwer, Tom, Lana Wachowski, and Lilly Wachowski, dirs. *Cloud Atlas*. Warner Bros. Pictures, 2012.

Ueno, Toshiya. "Japanimation and Techno-Orientalism." In *The Uncanny: Experiments in Cyborg Culture*, edited by Ed Bruce Grenville, 223–236. Vancouver: Arsenal Pulp Press, 2002.

———. "Techno-Orientalism and Media-Tribalism: On Japanese Animation and Rave Culture." *Third Text* 47 (1999): 95–106.

United Nations Office for Project Services (UNOPS). "Infrastucture," 2022. https://www.unops.org/expertise/infrastructure.

U.S. Census Bureau. "Census Demographic Profile." Explore Census Data, 2020. https://data.census.gov/table?g=040XX00US26_310XX00US11460&d=DEC+Demographic+Profile&tid=DECENNIALDP2020.DP1.

Van Noort, Carolijn. "On the Use of Pride, Hope and Fear in China's International Artificial Intelligence Narratives on CGTN." *AI & Society* 39 (2022): 295–307. https://doi.org/10.1007/s00146-022-01393-3.

Villeneuve, Denis, dir. *Blade Runner 2049*. Warner Brothers Home Entertainment, 2017.

Vinge, Vernor. "Technological Singularity." *Whole Earth Review* 81 (1993): 88–95.

Vint, Sherryl. *Biopolitical Futures in Twenty-First-Century Speculative Fiction*. Cambridge University Press, 2021.

———. "Long Live the New Flesh: Race and the Posthuman in Westworld." In *Reading Westworld*, edited by Alex Goody and Antonia Mackay, 141–160. Cham, Switzerland: Palgrave, 2019.

Vukovich, Daniel. *China and Orientalism: Western Knowledge Production and the PRC*. New York: Routledge, 2012.

Wagner, Kurt. "Mark Zuckerberg Says Breaking Up Facebook Would Pave the Way for China's Tech Companies to Dominate." *Vox*, July 18, 2018. https://www.vox.com/2018/7/18/17584482/mark-zuckerberg-china-antitrust-breakup-artificial-intelligence.

Walt Disney Animation Studios. *Baymax!* Walt Disney Animation Studios, 2022. https://www.disneyplus.com/browse/entity-bbbb983c-20a5-4b8f-bcb6-653613b80f19.

———. "*Big Hero 6*: The Series." Walt Disney Animation Studios, 2017–2020. https://www.disneyplus.com/browse/entity-79f140c4-c27f-42ab-8c8b-7a277e17ac76.

———. "Disney's Hyperion Renderer," July 2015. https://www.disneyanimation.com/technology/hyperion/.

Walter, W. Grey. *The Living Brain*. New York: W. W. Norton, 1953.

Wang, Ban. *Illuminations from the Past: Trauma, Memory, and History in Modern China*. Stanford, CA: Stanford University Press, 2004.

Wang, Leslie K. *Outsourced Children: Orphanage Care and Adoption in Globalizing China*. Stanford University Press, 2016.

Wang, Xin. "Asian Futurism and the Non-Other." *E-Flux Journal*, no. 81 (April 2017): 1–10.

Warner, Kristen J. "In the Time of Plastic Representation." *Film Quarterly* 71, no. 2 (2017): 32–37.

Washington, Haydn, Bron Taylor, Helen Kopnina, Paul Cryer, and John J. Piccolo. "Why Ecocentrism Is the Key Pathway to Sustainability." *Ecological Citizen* 1 (2017): 35–41.

Washington Post. "Transcript of Mark Zuckerberg's Senate Hearing," April 10, 2018. https://www.washingtonpost.com/news/the-switch/wp/2018/04/10/transcript-of-mark-zuckerbergs-senate-hearing/.

Watanabe, Scott. "Don wanted to figure out a logical explanation for how a mash-up city like this could exist. I came up with the idea that, after the 1906 earthquake in San Francisco, Japanese immigrants rebuilt the place using techniques that allow movement and flexibility in a seismic event." Tumblr, February 17, 2015. https://disneyanimation.tumblr.com/post/111288640767/don-wanted-to-figure-out-a-logical-explanation.

———. "Work: Big Hero 6," n.d. https://www.scottwatanabeart.com/big-hero-6.

Webb, Amy. *The Big Nine: How the Tech Titans and Their Thinking Machines Could Warp Humanity*. New York: Public Affairs, 2018.

Weheliye, Alexander. *Habeas Viscus: Racializing Assemblages, Biopolitics, and Black Feminist Theories of the Human*. Durham, NC: Duke University Press, 2014.

Wei, Lingling, Yoko Kubota, and Dan Strumpf. "China Locks Information on the Country Inside a Black Box." *Wall Street Journal*, April 30, 2023. https://www.wsj.com/articles/china-locks-information-on-the-country-inside-a-black-box-9c039928.

Weinstein, Alexander. *Children of the New World: Stories*. 1st ed. New York: Picador, 2016.

White, Lynn, Jr. "The Historical Roots of Our Ecologic Crisis." *Science* 155, no. 3767 (1967): 1203–1207. https://doi.org/10.1126/science.155.3767.1203.

Wiener, Norbert. *Cybernetics: Or, Control and Communication in the Animal and the Machine*. 2nd ed. Cambridge, MA: MIT Press, 1961.

Wikipedia. "Mitski," accessed April 25, 2023. https://en.wikipedia.org/wiki/Mitski.

Williams, Raymond. *Marxism and Literature*. Oxford: Oxford University Press, 1977.

Williams, Stereo. "What 'Black Panther' Means to Black Boys and Girls." *Daily Beast*, February 18, 2018. https://www.thedailybeast.com/what-black-panther-means-to-black-boys-and-girls.

Willmore, Alison. "Orientalism Is Alive and Well in American Cinema." *BuzzFeed News*, April 4, 2018. https://www.buzzfeednews.com/article/alisonwillmore/isle-of-dogs-jared-leto-orientalism.

Wills, Jenny, ed. *Adoption and Multiculturalism: Europe, the Americas, and the Pacific*. Ann Arbor: University of Michigan Press, 2020. https://doi.org/10.3998/mpub.10032835.

Wilson, Ara. "The Infrastructure of Intimacy." *Signs: Journal of Women in Culture and Society* 41, no. 2 (2016): 247–280.

Wing Luke Museum of the Asian Pacific American Experience. *Asian American Arcade: The Art of Video Games*, 2012.

Winner, Langdon. "Upon Opening Up the Black Box and Finding It Empty: Social Constructivism and the Philosophy of Technology." *Science, Technology, & Human Values* 18, no. 3 (1993): 362–378.

Winnubst, Shannon. "The Many Lives of Fungibility: Anti-Blackness in Neoliberal Times." *Journal of Gender Studies* 29, no. 1 (January 2020): 102–112.

Wissinger, Elizabeth. "The Sociology of Self-Tracking and Embodied Technologies: How Does Technology Engage Gendered, Raced, and Datafied Bodies?" In *The Oxford Handbook of Digital Media Sociology*, edited by Deana A. Rohlinger and Sarah Sobieraj, 316–336. New York: Oxford University Press, 2020.

Wong, Danielle. "Dismembered Asian/American Android Parts in Ex Machina as 'Inorganic' Critique." *Transformations* 29 (2017): 34–51.

Worlds Beyond Here: The Expanding Universe of APA Science Fiction. Presented at the Wing Luke Museum of the Asian Pacific American Experience, Seattle, October 12, 2018.

Wu, Tim. "Don't Fall for Facebook's 'China Argument.'" *New York Times*, December 10, 2018. https://www.nytimes.com/2018/12/10/opinion/facebook-china-tech-competition.html.

Wyver, Richey. "'More Beautiful than Something We Could Create Ourselves': Exploring Swedish International Transracial Adoption Desire." PhD diss., University of Auckland, 2020.

Xaxa, Virginius. "Tribes as Indigenous People of India." *Economic and Political Weekly*, December 18, 1999, 3589–3595.

Xiang, Biao. *Global Body Shopping: An Indian Labor System in the Information Technology Industry*. Princeton, NJ: Princeton University Press, 2008.

Xiang, Sunny. *Tonal Intelligence: The Aesthetics of Asian Inscrutability during the Long Cold War*. New York: Columbia University Press, 2020.
Yamashita, Karen Tei. *Brazil Maru*. Minneapolis: Coffee House Press, 1992.
———. *Through the Arc of the Rainforest*. Minneapolis: Coffee House Press, 1990.
Yang, Andrew. *The War on Normal People: The Truth About America's Disappearing Jobs and Why Universal Basic Income Is Our Future*. New York: Hachette Books, 2018.
———. "Yes, Robots Are Stealing Your Job." *New York Times*, November 14, 2019. https://www.nytimes.com/2019/11/14/opinion/andrew-yang-jobs.html.
Yang, George. "Orientalism, 'Cyberpunk 2077,' and Yellow Peril in Science Fiction." *Wired*, December 8, 2020. https://www.wired.com/story/orientalism-cyberpunk-2077-yellow-peril-science-fiction/.
Yang, So-hyang, and Kim Kyŏng-hŭi. "Ilbon aenimeisyŏn <kangch'ŏl ŭi yŏn'gŭm sulssa> sŭt'orit'elling yŏn'gu: yŏngung k'aerikt'ŏ rŭl chungsim ŭro" [Research on the storytelling of the Japanese animation "Fullmetal Alchemist": Focusing on the hero characters]. *Kŭllobŏl munhwa k'ont'ench'ŭ hakhoe haksul taehoe*, no. 2 (2019): 117–120.
Yano, Christine R. *Pink Globalization: Hello Kitty's Trek across the Pacific*. Durham, NC: Duke University Press, 2013.
Ye, Shana. "'Paris' and 'Scar': Queer Social Reproduction, Homonormative Division of Labour and HIV/AIDS Economy in Postsocialist China." *Gender, Place and Culture* 28, no. 12 (2021): 1778–1798.
Yellow Horse, Angie J., Karen J. Leong, and Karen Kuo. "Introduction: Viral Racisms: Asian Americans and Pacific Islanders Respond to COVID-19." In "Viral Racisms," special issue, *Journal of Asian American Studies* 23, no. 3 (October 2020): 313–318.
Yoneyama, Lisa. *Cold War Ruins: Transpacific Critique of American Justice and Japanese War Crimes*. Durham, NC: Duke University Press, 2016.
Youn, J. K., dir. *Ode to My Father*. CJ Entertainment, 2014.
Yu, Charles, and Leslie Bow. "An Interview with Charles Yu." *Contemporary Literature* 58, no. 1 (2017): 1–17.
Yu, Timothy. "Oriental Cities, Postmodern Futures: 'Naked Lunch,' 'Blade Runner,' and 'Neuromancer.'" *MELUS* 33, no. 4 (2008): 45–71.
Yuen, Laura. "Why 'Shang-Chi' Matters to My Asian American Boys in the Era of #StopAsianHate." *Minnesota Star Tribune*, September 10, 2021. https://www.startribune.com/laura-yuen-when-we-needed-him-most-a-hero-brings-humanity-to-asian-american-narratives-laura-yuen-wh/600095774/.
Zhang, Leslie. "The Corruption of Techwear Fashion Design, Cyberpunk Media and Orientalism." *Grailed*. Accessed July 21, 2022. https://www.grailed.com/drycleanonly/techwear-cyberpunk-orientalism.
Zhang, Muqiang. "What Western Media Got Wrong about China's Blockbuster 'The Wandering Earth.'" *Vice*, April 2, 2019.
Zhao, Junfu. "The Political Economy of the U.S.-China Technology War." *Monthly Review* 73, no. 23 (2021): 112–126.
Zhou, Yangyang, Hanping Hu, Jin Diao, and Yitao Chen. "The Entropy of Stochastic Processes Based on Practical Considerations." *AIP Advances* 10 (2020): 045321-1–045321-5.
Zhu, Ping. "From Patricide to Patrilineality: Adapting *The Wandering Earth* for the Big Screen." *Arts* 9, no. 3 (September 2020): 1–12.

Zia, Helen. "The Vincent Chin Legacy Guide: Asian Americans and Civil Rights." 2nd ed. Vincent Chin Institute: 2022–2023. https://www.vincentchin.org/legacy-guide/english.

Zizek, Slavoj. "Discipline between the Two Freedoms: Madness and Habit in German Idealism." In *Mythology, Madness, and Laughter: Subjectivity in German Idealism*, edited by Markus Gabriel and Slavoj Zizek, 95–121. London: A&C Black, 2009.

Zoladz, Lindsay. "Mitski Is More than TikTok." *New York Times*, March 11, 2022. https://www.nytimes.com/interactive/2022/03/11/magazine/mitski.html.

Zuboff, Shoshana. *The Age of Surveillance Capitalism: The Fight for a Human Future at the New Frontier of Power*. New York: Public Affairs, 2019.

Zurcher, Jessica D., Pamela Jo Brubaker, Sarah M. Webb, and Tom Robinson. "Parental Roles in 'The Circle of Life': Representations of Parents and Parenting in Disney Animated Films from 1937 to 2017." *Mass Communication and Society* 23 (2019): 128–150. https://doi.org/10.1080/15205436.2019.1616763.

Zylinska, Joanna, and Sarah Kember. *Life after New Media: Mediation as a Vital Process*. Cambridge, MA: MIT Press, 2012.

Notes on Contributors

JUSTIN MICHAEL BATTIN is senior lecturer of communication at RMIT University in Ho Chi Minh City, Vietnam. His research intersects place-oriented strands of phenomenology with the everyday uses of mobile media technologies and mobile social media. He is the author of *Mobile Media Technologies and Poiēsis* and coeditor of *We Need to Talk about Heidegger: Essays Situating Heidegger in Contemporary Media Studies* and *Reading Black Mirror: Insights into Technology and the Post-Media Condition.*

EDMOND Y. CHANG is associate professor of English at Ohio University. He earned his PhD in English at the University of Washington. Recent publications include "Imagining Asian American (Environmental) Games" in *AMSJ*, "Why are the Digital Humanities So Straight?" in *Alternative Historiographies of the Digital Humanities*, and "Queergaming" in *Queer Game Studies*. He is an editor for *Analog Game Studies* as well as the website *Gamers with Glasses*.

CHRISTOPHER T. FAN is associate professor of English at the University of California, Irvine. He is the author of *Asian American Fiction After 1965: Transnational Fantasies of Economic Mobility*, and has written extensively on science fiction and techno-Orientalism.

ANNA ROMINA GUEVARRA is associate professor and founding director of the Global Asian Studies Program at the University of Illinois Chicago. Her interdisciplinary scholarship, teaching, and community-engaged work focus on immigrant and transnational labor, the geopolitics of carework, the Philippine diaspora, and critical race/ethnic studies. In addition to publishing numerous articles in interdisciplinary journals, she is the author of *Marketing Dreams, Manufacturing Heroes: The Transnational Labor Brokering of Filipino Workers.*

BETSY HUANG is professor of English at Clark University. She is the author of *Contesting Genres in Contemporary Asian American Fiction* and coeditor of three essay collections: *Techno-Orientalism: Imagining Asia in Speculative Fiction, History, and Media*, *Diversity and Inclusion in Higher Education and Societal Contexts*, and *Asian American Literature in Transition, 1996–2020*. Her work has appeared in *The Cambridge Companion to Asian American Literature*, *The Cambridge Companion to American Horror*, *Race and Utopian Desire in American Literature and Society*, *Journal of Asian American Studies*, and *MELUS.*, among others.

WON JEON is a PhD candidate in the History of Consciousness department at the University of California, Santa Cruz. Her research draws on the history of science, psychoanalysis, and political economy to examine the contemporary relevance of cybernetics, particularly on the development of informatics in a capitalist context.

AGNIESZKA KIEJZIEWICZ holds a PhD in the arts and humanities (with an emphasis on film and media) from the Jagiellonian University in Poland; and she is the author of *Japanese Cyberpunk: From Avant-garde Transgressions to the Popular Cinema*, *Japanese Avant-garde and Experimental Film*, and *Completed in Apparent Incompletion: The Sculpture Art of Wojciech Sęczawa*. She is the author of over forty peer-reviewed articles concerning Asian film, media, culture, and art. For the past few years, she has been working at the University of Gdansk, Poland, teaching Asian cinema, new media technologies, academic writing, and film analysis. In her research activities, Agnieszka focuses on the role of new technologies in film art, Asian cinema, Nordic cinema, audiovisual experiments, Asian popular culture, and games narratives. She currently works as a lecturer in the Game Design program at RMIT University in Vietnam.

ADHY KIM is assistant professor in the Department of Literatures in English at Cornell University. His scholarly work navigates the intersection of Asian and Asian American literary studies and focuses on transimperial relationships among Japan, South Korea, and the United States. Their manuscript-in-progress studies how speculative literature engages with natural history to rethink post-1945 U.S.-Northeast Asian geopolitical relations.

CLARE S. KIM is assistant professor of history and global Asian studies at the University of Illinois Chicago. Her research examines the history of twentieth- and twenty-first-century mathematical sciences and its relation to the formation and representation of social differences. She is currently writing a book on the interplay of calculation practices and twentieth-century U.S.-Asian relations on the racialized dynamics of knowledge exchange.

JUNG SOO LEE is a PhD candidate in the English department at the University of California, Irvine. She has served as editor for *The Journal of Japanese and Korean Cinema* discussing the Japanese and Korean cultural exchange during the late 1990s to early 2000s. Her dissertation project is titled "Take Care: Making and Re-making Disability in Contemporary Science Fiction," where she discusses the intersections of dis/ability and race, gender, and sexuality through works of science fiction written by women and writers of color.

LORI KIDO LOPEZ is professor of communication arts and associate dean of social sciences in the College of Letters and Science at the University of Wisconsin–Madison. She has published many books on the subject of race and media, including *Asian American Media Activism: Fighting for Cultural Citizenship*, *Race and Media: Critical Approaches*, and *Micro Media Industries: Hmong American Media Innovation in the Diaspora*. Her research examines the way that minority groups use media in the fight for social justice, particularly focusing on Asian American communities and media cultures.

KIMBERLY D. MCKEE is an associate professor in the School of Interdisciplinary Studies at Grand Valley State University. She is the author of *Adoption Fantasies: The Fetishization of Asian Adoptees from Girlhood to Womanhood*, *Disrupting Kinship: Transnational Politics of Korean Adoption in the United States*, and coeditor of *Degrees of Difference: Reflections of Women of Color on Graduate School*. McKee serves as a cochair of the executive committee for the Alliance of the Study of Adoption and Culture. She was a 2023–2024 U.S. Fulbright Scholar to South Korea (Sogang University).

GRETA AIYU NIU earned her PhD in English from Duke University. She has taught courses in literature; film, media and digital media studies; gender, film and postcolonial theory; and Asian American studies at State University of New York (SUNY) Brockport, the University of Rochester, SUNY College at Geneseo , and St. John Fisher University, among other places. She is an independent scholar based in Rochester, NY.

JANE CHI HYUN PARK is a senior lecturer in the Gender and Cultural Studies Department at the University of Sydney. Her research explores changing notions of race and gender in film and popular culture in the United States, Australia, and South Korea, focusing on representations of East Asian and Asian diasporic identities and cultures. Jane has published work in a wide range of journals including *Cultural Studies*, *World Literature Today*, *Inter-Asia Cultural Studies*, and *Educational Philosophy and Theory* as well as anthologies on film, media, and popular culture. Her monograph *Yellow Future: Oriental Style in Hollywood Cinema* explores the growing popularity of East Asian bodies and styles in

Hollywood films from the 1980s to the early 2000s. Her current research projects examine the development of antiracist thought and practice in Australia and the role of compassion in critical pedagogy. Jane is also a cultural consultant who has worked with organizations including Space Doctors, Proctor & Gamble, CISCO Technologies, Goalpost Pictures, Sydney Film Festival, BrisAsia, and OzAsia.

M. IMRAN PARRAY is assistant professor at the School of Liberal Arts and Humanities, Woxsen University, Hyderabad, India. He previously taught at the University of Delhi and Jamia Millia Islamia, New Delhi. His publications have appeared in *Internet Histories*, *Media History*, *Society and Culture in South Asia*, *Economic and Political Weekly*, and other journals. A 2023 Journalismfund Europe grantee and 2013 ICRC awardee, his investigative journalism has featured in *The Scroll*, *Mongabay*, *Dialogue Earth*, and other publications.

BARYON TENSOR POSADAS is an associate professor in the Division of Humanities at the Hong Kong University of Science and Technology. He is the author of *Double Visions, Double Fictions: The Doppelganger in Japanese Film and Literature* and the translator of Aramaki Yoshio's science fictional novel *The Sacred Era: A Novel*. He is currently completing a second book project provisionally titled *Science Fiction, Empire, Japan*.

DAVID S. ROH is professor and chair of English at the University of Utah. He is the author of *Minor Transpacific: Triangulating American, Japanese, and Korean Fictions*, *Illegal Literature: Toward a Disruptive Creativity*, and coeditor of *Techno-Orientalism: Imagining Asia in Science Fiction, History, and Media*.

THOMAS XAVIER SARMIENTO (they/he) is an award-winning associate professor of English and affiliated faculty member of gender, women, and sexuality studies at Kansas State University. Tom specializes in diasporic Filipinx American literature and media cultures, queer-feminist theories, and cultural representations of the U.S. Midwest. Their research appears in the journals *Alon: Journal for Filipinx American and Diasporic Studies*, *Amerasia Journal*, *American Studies*, *Asian American Literature: Discourses & Pedagogies*, *MELUS.: Multi-Ethnic Literature of the United States*, and *Women, Gender, and Families of Color*, and in the edited collections *Asian American Feminisms* and *Women of Color Politics*, *Curricular Innovations: LGBTQ Literatures and the New English Studies*, *The Oxford Encyclopedia of Asian American Literature and Culture*, and *Q & A: Voices from Queer Asian North America*.

GERALD SIM is professor in the School of Communication and Multimedia Studies at Florida Atlantic University. He is the author of *The Subject of Film and Race: Retheorizing Politics, Ideology, and Cinema*, *Postcolonial Hangups in*

Southeast Asian Cinema: Poetics of Space, Sound, and Stability, and *Screening Big Data: Films that Shape Our Algorithmic Literacy*.

LELAND TABARES is assistant professor of race, ethnicity, and migration studies at Colorado College. His book manuscript examines how Asian Americans' access to increasingly diverse professions in the twenty-first century enculturates new meanings of race, generationality, and solidarity. His research is published in *Profession*, *Journal of Asian American Studies*, *Verge: Studies in Global Asias*, *Arizona Quarterly*, *ASAP/J*, *Lateral: Journal of the Cultural Studies Association*, *Hyphen*, and *The Recipes Project*. He currently serves as cochair for the Circle for Asian American Literary Studies.

RACHEL TAY is a PhD student in Duke University's Graduate Program in Literature. Spanning the intersections of aesthetics, computational culture, and the philosophies of mind and technology, her current research considers the mediating role of distraction in our contemporary paradigms of thought. Her writing can be found in *Post45* and *Reading in Translation*, and is forthcoming in a special issue of *Communications +1* on media aesthetics.

IAN LIUJIA TIAN is assistant professor of global equity studies in the Department of Women's Studies at Mount Saint Vincent University, Canada. Their ethnographical research explores gendered and racialized labor in China, with a particular focus on the relations between everyday social reproduction and infrastructure.

CHARLES M. TUNG is professor of English at Seattle University, where he teaches courses on twentieth- and twenty-first-century literature, temporal scale, and representations of racial anachronism. He is the author of *Modernism and Time Machines*. His current project, *Big Clocks and Ethnofuturist Timescales*, focuses on contemporary extinction fictions, the representation of expanded temporal scales, and racial futures.

SHANA YE is assistant professor of women and gender studies at the University of Toronto Scarborough and the Women and Gender Studies Institute at the University of Toronto. Her research lies at the intersection of transnational feminism, queer studies, post/socialist studies, affect studies, feminist technology study, and speculative fiction. Her first monograph, *Queer Chimerica: A Speculative Auto/Ethnography of the Cool Child*, examines how queerness has been "produced" through the interdependence of China and the United States since the late Cold War, drawing connections between the discourse of queer fluidity and capital's demands for labor flexibility. Shana's second book project, tentatively titled *Silicon Yellow: Sojourner Colonialism and the Aesthetics of Techno-Chimerica*, explores the racialized construction of Chineseness in relation to techno-gentrification, material agency, heteropatriarchy, and racial-ethnic extractive capitalism. Shana is also working on a coedited volume on the queer archive and several side projects on topics including COVID-19 affect and memory, generative AIs and the digital afterlife, and queer quantum theory.

JAEYEON YOO is a PhD student in literature at Duke University. Her research interests include Asian America and Asian diaspora studies, the intersections of postcolonial and translation theory, and contemporary literature and popular music. Her writing can be found in *Asymptote*, *Words Without Borders*, and other public-facing journals. She is also a contributing writer for *Electric Literature* and an editor-at-large for *Barricade: A Journal of Antifascism and Translation*.

Index

Page numbers in italics indicate illustrations.

Abel, Jonathan, 12
Adivasis, 92, 96, 97n9; Othering of, 94
adoption: in *After Yang*, 57, 59–60; transracial, 57, 73
adoption, transnational, 61n5, 73; as assimilation, 51, 61n4; as child trafficking, 53, 62n9; multiculturalism and, 52; rescue narratives, 62n9; as soft power, 51
adoption studies, 14
affect, 121, 134n4; infrastructure and, 124–125, 126–134
Afrofuturism, 13, 267–268; Asianfuturism and, 269; sovereignty and, 276; technology and, 270–271, 282
After Yang (film), 2, 13–14, 52–53, 56–60, 206, 214; adoption in, 57, 59–60; AI in, 28–30; anti-Asian racism in, 57, 60; Asianness in, 53, 57, 59; Asian obsolescence in, 25; cloning in, 54–55, 58, 60; colorblindness and, 56; healing anime style, 215; postracial family in, 56–57; racial difference in, 59–60; racialization in, 32–34, 58, 59; reproductive futurity in, 54–55; yellow peril in, 33–34
agency: Asian, 24; black boxes and, 111–112; stereotypes and, 175–176
Ahmad, Irfan, 85
Ahmed, Hilal, 90
Ahmed, Sara, 309

Akimoto, Daisuke, 214, 218n36
Alaimo, Stacey, 69
Allen, Kathryn, 42
Allison, Graham, 139, 144, 147
All Our Asias (video game), 283, 291–292, *292*, 294
AlphaGo, 107, 147
Alsultany, Evelyn, 277
Altered Carbon (novel and television series), 6, 14, 67–69, 77, 77–79, *79*; transracial transfer in, 77–80
American Edge Project (AEP), 140, 142
animation: racialization of, 230; animetic spectatorship, 230
anime: healing anime, 214
Appadurai, Arjun, 84, 85
Arendt, Hannah, 95
artificial intelligence (AI), 5, 19n11; in *After Yang*, 28–30; black boxes and, 105–106; ChatGPT, 34–35; Chinese, 8, 15, 144; Chinese Room Argument, 106; documentaries on, 141–150; in *Ex Machina*, 28–31; fear of, 8, 137, 141; generative, 34–36; human trafficking and, 10; informatics and, 106–107; racialization and, 29; singularity, 30; as threat, 15; Turing Test, 106; U.S.–China "arms race," 140
Ashby, W. Ross, 105

367

Asian America, 13, 59, 291, 305; as subjectless discourse, 198–199
Asian American Arcade (exhibition), 283, *289*, 289–290
Asian American literature: mainstreaming of, 318; science fiction and, 285
Asian American studies: techno-Orientalism and, 3–4, 11–12, 112. *See also* techno-Orientalism studies
Asian Cool, 173
Asianfuturism, 18, 279, 281; Afrofuturism and, 269; as COVID-19 pandemic response, 281; definitions of, 281–283, 287, 295; as horizon, 295; kinship and healing, 269; optimism of, 276–277, 278; queerness of, 286, 295; science fiction and, 283–285; technology, ambivalence toward, 274, 278; utopianism of, 275–276; video games and, 289–294
Asianness, 53, 236; in *After Yang*, 53, 57, 59; alien, 239; Bering Strait migration and, 243; commodification of, 199; labor and, 11, 24; obsolescence and, 25; as queer, 300; technology and, 31, 300; techno-Orientalism and, 300
Asians: as forever foreigners, 56; as machines, 29–30, 40, 105, 112, 176, 236, 268; as robots, 70, 176, 268, 304
Asian studies: techno-Orientalism and, 11–12, 112
Atanasoski, Neda, 148
Atwood, Jeff, 107
authenticity: *hallyu* and, 174–175; Mitski and, 194–195; originality and, 174; race and, 59; singer-songwriters and, 193
automation: capitalism and, 25, 26–28; disposability and, 56; labor and, 148–149; obsolescence and, 24–26; underemployment and, 25, 34–35
automatons, 103, 115n8; Japanese as, 105
Azuma, Eiichiro, 240
Azuma Hiroki, 230

Bae Doona, 182–183, 185
Bahng, Aimee, 239, 288
Bakunin, Mikhail, 95
Banerjee, Mita, 85
Bano, Bilkis, 91
Barad, Karen, 42
Barrat, James, 141
Bascara, Victor, 76
Bateson, Gregory, 46, 50n20
Battin, Justin, 16
Baymax! (television series), 299, 310–311
Bear Woman (Ungnyeo) myth, 178
beauty standards, Korean, 12, 179
Benanav, Aaron, 25, 27
Benjamin, Ruha, 114
Berlant, Lauren, 125, 127
Bérubé, Michael, 31, 37n39
Bharatiya Janata Party (BJP), 96
Biden, Joe, 122
Biel, Robert, 45
Big Hero 6 (film), 6, 18, 58, 298–301, *303*, *305*; cuteness in, 303–304; empathy in, 310; family and kinship in, 306–310; Japanese/Japanese American aesthetics of, *301*, 306; plot, 298–299; as queer, 298, 299–300, 304; as queer assemblage, 310; queering of techno-Orientalism, 298, 299–300, 305–306, 307, 308–311; racial identity in, 298; utopian aesthetics of, 302
Big Tech: lobbying for, 140–142; nationalism and, 141–142; political mobilization of, 140; regulatory resistance of, 138–140, 150
biopolitics, 112, 223; technopolitics and, 10
Black, Fischer, 41, 50n11
black boxes, 14–15; AI and, 105–106; as analytic, 108–111; Asian "inscrutability" and, 101; cybernetics and, 104–105, 109; as descriptive technologies, 113; as epistemic space, 102–103; figurations of, 100–101; history of, 102–108; Mechanical Turk, 103, *104*; migration research and, 110; racialization of, 15, 101–111, 113–114; racialized logics of, 108–111; white boxes, 104
Black Mirror (television series), 144, 145
blackness: negative images of, 277–278; technology and, 31
Black Panther (film), 216, 267; technology in, 271, 273–274
Black Panther: Latinx and Indigenous cultures in, 271–272; technology in, 273–274; *Wakanda Forever* (film), 267
Blackpink, 6
Blade Runner (film), 183, 268, 277, 302; aesthetics of, 305
Blair, Tony, 90
bodies, Asian: disposability and expendability of, 57, 69; fungibility of, 14, 69–70, 72, 74–75, 80; interchangeability of, 190;

labor and, 11, 74–75; as machines, 29–30, 40, 105, 112, 176; obsolescence of, 30–34; racialization of, 24, 28, 30; in *The Tiger Flu*, 288; women's, 16, 174; yellowface and, 78. *See also* labor, Asian
bodies, black: fungibility of, 75–76
bodies, dis/abled, 249, 262n3; in *Fullmetal Alchemist*, 250, *254*, 256, 260, 261; trauma and, 255. *See also* dis/ability
bodies, multiracial, 190–191, 192, 199n6
BODIES: *The Exhibition*, 14, *71*; skinning and, 71–72
Bollywood film: Islamophobia in, 88–89; Orientalism in, 85
Bong Joon-Ho, 16
Bow, Leslie, 10, 29
Breckenridge, Carol, 84
Bremmer, Ian, 139
Brienza, Casey, 250
Broussard, Meredith, 143
Brubaker, Pamela Jo, 307
BTS, 6, 174
Bui, Long, 105, 176, 311
Buolamwini, Joy, 143
Bush, George W., 90, 253
Butler, Octavia, 270, 295
Byrd, Jodi, 236

Callenbach, Ernest, 205–206, 215
Canavan, Gerry, 231
capitalism: American Dream and, 33; automation and, 25, 26–28; China and, 156–157; COVID-19 and, 4; cybernetic capitalism, 40, 47–48, 49n4, 50n11; erotics of, 11; infrastructure and, 122; innovation and, 28; labor and, 107, 116n28; obsolescence and, 47; Orientalism and, 320–321; racial capitalism, 149, 192; racial production and, 198; surveillance and, 148; technology and, 138; techno-Orientalism and, 40–41; zombies and, 17
Caruth, Cathy, 163
Cazdyn, Eric, 225, 226
Chan, Dawn, 282
Chan, Jeffrey Paul, 10
Chang, Edmond Y., 10, 18
ChatGPT, 34–35
Chen, Mel, 261
Chen, Tina, 322–323
Cheng, Anne Anlin, 9–10, 16; ornamentalism, 191–192, 198, 312n12

Chiang, Ted, 35, 138, 290; colorblindness in, 285–286; "Story of Your Life," 285–286
Chicanx futurism, 17
Chin, Frank, 10
Chin, Vincent, 13, 58–59, 63n20
China, 6; AI development in, 8; global capitalism and, 156–157; Made in China initiative, 8; move to global stage, 155–158, 160–161, 164–168; Social Credit System, 137, *146*; social upheaval in, 130–131, 136n36; as surveillance state, 137–138, 144–146, 148–149, 158; techno-Orientalism of, 157; techno-Orientalist discourse on, 144; as threat to West, 83, 154, 268; transnational adoption and, 51
China 2098 (Fan Wennan series), 15, 121–124, *123*, *130*, *131*, *132*; affective politics of, 126–134; cinematic technique of, 129; as dystopia, 132; humans, absence of, 132; infrastructure in, 123–124; as infrastructure sublime, 124; optimism in, 127–129; as propaganda, 121; socialist slogans in, 129–133; storyline, 123–124, 126–128, 135n12; temporality of, 129–131
Chineseness, 321
Cholodenko, Alan, 228
Chow, Rey, 193
Chuh, Kandace, 196–197
Chun, Wendy Hui Kyong, 80, 193, 238
Clare, Eli, 253–255
climate change, 1–2, 5; *China 2098* and, 126; sustainability and, 208–212; urban planning and, 208–212
Clinton, Hilary Rodham, 5
cloning: in *After Yang*, 54–55, 58, 60; in *Cloud Atlas*, 183, 185, 186; in "Saying Goodbye to Yang," 54, 55, 60; in *Shiri*, 175, 186; in *The Tiger Flu*, 287; white supremacy and, 53–55, 58, 60
Cloud Atlas (film), 16, 174, 183–186; cloning in, 183, 185, 186; cosmetic surgery in, 175; imitation in, 184; race in, 183; romantic plot of, 183; transracial performance in, 183; unhappy ending of, 174; yellowface in, 183, 184
Cobra Kai (television series), 324n8
Coded Bias (documentary film), 15, 138, 142–144; facial recognition technology in, 142, *146*; techno-Orientalist tropes of, 143–144

colonialism: Bering Strait migration and, 243–244; futurism and, 17; genocide and, 241; individualism and, 30; sojourner colonialism, 154; zombies and, 17
Columbus (film), 7
Confucian ethics, 163
Conn, Virginia L., 123
Cornum, Lou, 70
Corrêa, Laura Guimarães, 97n7
cosmetic surgery: in *Cloud Atlas*, 175; deracialization and, 178; imitation and, 178; Korean, 176, 178–179, 185; in *Shiri*, 175, 180
COVID-19 pandemic, 1–2, 316; American Orientalism and, 8–9; anti-Asian animus and, 13, 232, 281; anti-Asian violence and, 232; Asianfuturism as response to, 281; capitalism and, 4; economic stagnation and, 27; *Fancang* quarantine facilities, *165*, 165; globalization and, 4; nationalism and, 8; population decline and, 23; U.S. exceptionalism and, 8
Cox, Renee, 270
Crackdown (web series), 14, 87–88, 90
Crazy Rich Asians (film), 7
critical race theory: STS and, 112–114
Csicsery-Ronay, Istvan, 219
cybernetic capitalism, 40, 47–48, 49n4, 50n11
cybernetics: black boxes and, 104–105, 109; cybernetic capitalism, 40, 47–48, 49n4, 50n11
cyberpunk: aesthetic, 323; as dystopic, 212, 213, 215; Japan and, 231, 304; Japanese, 152, 157–158, 304; Lone Ranger trope, 42; postcyberpunk, 205, 217n29; transhumanism and, 68; trauma and, 218n30; wetware, 31
Cyberpunk 2077 (video game), 6
cyberspace: race in, 76, 78
cyborgs: limits of, 224; as predecessors to zombies, 220, 227–231

Daum, Jeremy, 137–138
Davé, Shilpa, 312n12
Day, Iyko, 74–76
Dear Alice (animated short), 218n35
Deng Xiaoping, 51, 61n1
deracialization: of Asian women, 192; authenticity and, 194; cosmetic surgery and, 178; of Mitski, 194, 196, 197–198. *See also* racialization

Dery, Mark, 270, 282, 295
Deus Ex: Human Revolution (video game), 11
Devi, Mahasweta, 93–94
dis/ability: gender and, 249; race and, 249. *See also* bodies, dis/abled
disability justice, 255
dispossession: of Asian labor, 25; Indigenous, 241, 245; of Japanese Americans, 236; of land, 161; techno-environmentalism and, 245
Do You Trust This Computer? (documentary film), 141
Dubal, Veena, 27
Duterte, Rodrigo, 122

ecofuturism, 205–206
ecology: techno-Orientalism and, 16
Ecotopia, 205
Elfving-Hwang, Joanna, 178, 179
Embry, Karen, 220, 224–225, 226
Empire of Corpses (Project Itoh), 220–229; animated adaptation of, 225, 227, *228, 229*; reanimation in, 222; zombies in, 221–223
Eng, David, 73
EngKey, 107, *108*
Enjō Tō, 222–223
Ess, Charles, 102
Evans, Karin, 73
Everything Everywhere All at Once (film), 2, 7, 321, 324n8; language in, 44; noise in, 42, 44, 46, 47; nothingness in, 13–14, 39–40, 42, 44–48; queerness in, 44; stochastic processes in, 42–48; success of, 318–320
exceptionalism, Chinese, 131–132, 155–156
exceptionalism, U.S, 5; COVID-19 and, 8; Silicon Valley and, 26; tech deregulation and, 139
Ex Machina (film), 3, 9–10, 13; AI in, 28–31; Asian obsolescence in, 25; Japan as unseen antagonist, 31; racialization in, 29–31; Western imperialism and, 31; whiteness in, 31

facial recognition technology, 144–146, *146*; in AI documentaries, 142
Family Man, The (television series), 14, 84, 86–96, *87, 89, 93*; Indian Muslims in, 89–90; Indigenous and Adivasi populations in, 92; re-Orientalization in, 91, 93; Tamils in, 92; techno-Orientalism of, 91; violence in, 94–95
Fan, Christopher T., 26, 69, 248

Fanchen, 70
Fan Wennan, 15, 121. See also *China 2098* (Fan Wennan series)
Farewell, The (film), 7, 48n1
Farocki, Harun, 200n9
femininity, Asian: consumption of, 190, 199n6
feminist criticism, Asian American, 16
Fickle, Tara, 10–11, 290, 294
Foster, John Bellamy, 28
Foucault, Michel, 223
Frankenstein (Shelley), 31, 34, 221
Franklin, Seb, 75
Frelik, Pawel, 77, 79
Fresh off the Boat (television series), 7
Fullmetal Alchemist (manga series), 17, 249–261, *257*; alchemy/alkahestry in, 249, 250–251, 253, 256, 258, 259–260, 262n2, 263n28; automail in, 250, 262n14; dis/abled bodies in, 250, *254*, 256, 260, 261; historical context of, 250, 255; Orientalist tendencies of, 260; prosthetics in, 252, 259; scholarship on, 251; tamed techno-Orientalism in, 259
fungibility: of Asian bodies, 14, 69–70, 72, 74–75; of black bodies, 75–76; neoliberal, 76; racialization and, 75–76
furusato, 207, 210, 211–212, 214
futurism: Asian American, 279; Chicanx, 17; colonialism and, 17; ecofuturism, 205–206; Indigenous, 13, 17, 267–268, 271–272, 278; Latinx, 267–268, 271–272, 278; Sinofuturism, 159. *See also* Afrofuturism; Asianfuturism
futurity: nostalgia and, 5; reproductive, 51–55; of science fiction, 217n29; techno-Orientalist, 54; yellow peril and, 69

Galison, Peter, 104
game studies, 10, 287. *See also* video games
Gardens by the Bay, 207–208
Garland-Thomson, Rosemarie, 252
Gates, Racquel, 277–278
Geal, Robert, 212, 218n30
Gebru, Timnit, 143
Geller, Arnie, 71
gender: beauty standards, 12, 179; femininity, Asian, 9; in *Liulang Diqiu*, 163–164; masculinity, Asian American, 291–292; masculinity, white, 53; Orientalism and, 61n5; race and, 249; robotics and, 9; in South Korea, 177–179; technology and, 154

genocide, 241, 271, 272, 324n6
Geoghgan, Bernand Dionysius, 103
geopolitics: post-9/11, 251–252, 259
Ghaywan, Neeraj, 89
Global Asias, 322
globalization: COVID-19 and, 4; deglobalization, 4; of *hallyu*, 185; imitation and, 176; neoliberalism and, 231
Goertzel, Ben, 147
Gossin, Pamela, 214
Gourley, Sean, 141
Guevarra, Anna Romina, 14–15
Guha, Ramachandra, 94
Gwo, Frant, 153

Habermas, Jürgen, 142
Half of It, The (film), 48n1
Hall, Don, 300
hallyu (Korean Wave), 7, 16, 173–179, 316; copying and, 174, 176–177; cultural authenticity and, 174–175; culture hybridity of, 176; globalization of, 185; imitation and, 175–176
Hao Jingfan, 122
Haraway, Donna, 41, 113; on cyborgs, 220, 224, 231
Harmony (Project Itoh), 223; animated adaptation of, 225–227, *226*, *227*; epilogue, 232; medical surveillance in, 223, 224; philosophical zombies in, 224; plot, 223–224; posthuman consciousness in, 223–224; zombies in, 224
Hartman, Saidiya, 75
Hazy Days (video game), 283, *292*, 292–294
Heidegger, Martin, 207
heteronormativity, 18, 282; human repopulation and, 159; in Liu, 162, 164; queerness and, 309; science fiction and, 285; in *Wandering Earth* films, 154, 164
Hindu Orientalism, 88
history of science, 101; whiteness and, 106
Holliday, Ruth, 179
Hong, Euny, 174
hooks, bell, 26
How High We Go in the Dark (Nagamatsu), 17, 242–245; ecomodernism of, 244–245; indigeneity in, 245; Japanese American artifacts in, 242–243; Japanese settlers in, 243–244; pandemic in, 242; settler colonialism in, 244, 248n59

How to Live Safely in a Science Fictional Universe (Yu), 320
Hu, Jane, 52
Huang, Betsy, 40, 57, 68–69, 70, 83, 85, 301; on Asian workers, 255–256; on constraints placed on Asian American writers, 285, 286; on shifts in techno-Orientalist logics, 159
Huang, Dinglong, 101
Huang, Michelle, 74, 236
Huawei, 8, 12
Huesemann, Michael and Joyce, 209

identity: codability of, 73; race and, 78; technology and, 40
identity politics, 190; Mitski and, 193–194
imaginable ageography, 322–323
imitation: in *Cloud Atlas*, 184; cosmetic surgery and, 178; creative imitation, 174; globalization and, 176; *hallyu* and, 175–176
imperialism: erotics of, 11; Japanese, 237; Russian, 247n47
imperialism, U.S.: techno-Orientalism and, 235–236
Imroz, Parvez, 95
India: Gujarat riots, 90–91; indigeneity in, 84, 96n5; partition, 85–86, 275, 276; peripheral and marginal communities, 84, 96, 97n7, 97n9; techno-Orientalism and, 83–86; violence against minorities, 94–95
indigeneity: Asian, 243, 247n47; in *How High We Go*, 245; in India, 84, 96n5; *mestizaje* and, 247n25
Indigenous futurism, 13, 17, 267–268, 271–272, 278; sovereignty and, 276, 282
industrial overcapacity, 27
informatics, 43, 47; AI and, 106–107; of domination, 41; of value, 75
infrastructure, 15; affect and, 124–125, 126–134; capitalism and, 122; in *China 2098*, 123–124; hypervisibility of, 125; as mediation, 125–126; politics of, 121–122, 125; representation of, 124, 126; of settler colonialism, 125
infrastructure, Chinese: Belt and Road Initiative, 122, 156; in *China 2098*, 123–124
In the Age of AI (documentary film), 15, 138, 144–149, *145*, *149*; China as surveillance state in, 148–149; China as threat in, 148–150; facial recognition technology in, 142
Islamophobia, 276; in Bollywood film, 88–89; Orientalism and, 88; technology and, 88
Ito Satoshi, 17
Iwabuchi, Koichi, 96

Japan: "cool Japan," 250; economic miracle, 6; as economic threat to U.S., 220, 231; Japan Panic, 6; modernity and, 144; modernization and, 250; population decline in, 23; postwar soft power, 17; sustainability and, 218n32; techno-Orientalist discourse on, 144; as threat to West, 83, 231, 268
Japanese incarceration, 236, 245
Japaneseness: cuteness and, 304; indigenization of, 243
Japanese studies: science fiction and, 219–220
Jeon, Won, 13–14
Jia, Qingguo, 260
Joji, 6
Joo, Hee-Jung S., 60, 311
Joy, Lisa, 70
Joyrich, Lynne, 193
Joy Ride (film), 7

Kanesaka, Erica, 304
Kantayya, Shalini, 142–143
Kanter, Jonathan, 138
Kathmandu Connection (web series), 14, 87–88
Kellogg, Keith, 140–141
Kelly, James, 217n29
Kessel, John, 217n29
Khan, Amir, 159, 162
Khan, Lina, 138
Khubchandani, Kareem, 312n12
Kiejziwicz, Agnieszka, 16
Kim, Adhy, 17
Kim, Christine, 177
Kim, Clare S., 14–15
Kim, Jodi, 177
Kim, Shiyoon, 300
Kim, Suk-Young, 11–12
Kim Dae Jung, 175
Kim Yun-jin, 180, 182, 185
Kirk, Robert, 224
Kogonada, 2, 14, 16, 54, 59. See also *After Yang* (film)
Kolko, Beth, 78

Korean New Cinema, 174
Korean studies, 16
Korean Wave. See *hallyu* (Korean Wave)
K-pop, 4; neoliberalism and, 12; visual iconography of, 11–12
Krishna, Arvind, 34–35
Kroker, Arthur, 145

labor: abstract labor, 74–75; automation and, 148–149; capitalism and, 107, 116n28; care labor, 52; colonialism and, 107, 116n28; racialized, 76; reproductive, 52, 163
labor, Asian, 235; Asianness and, 11, 24; dehumanization of, 68–69; disposability of, 25, 75; dispossession of, 25; white anxiety and, 56; yellow peril and, 255–256. See also bodies, Asian
Lai, Larissa, 291, 295
LaMarre, Thomas, 221, 229, 230
LaMyers, Tonya, 143
Laputa: Castle in the Sky (film), 213–214, 218n36
Latinx futurism, 267–268, 271–272, 278
Latour, Bruno, 109–110
Lau, Lisa, 84
Lauro, Sarah Juliet, 220, 224–225, 226
Le, Quynh Nhu, 239
Lee, Bruce, 182, 322, 324nn7–8
Lee, Jung Soo, 17
Lee, Kai-Fu, 144, 146–147
Lee, S. Heijin, 12, 178, 179
Lee, Summer Kim, 193
Lee Sedol, 107, 147
Le Guin, Ursula K., 16, 205, 218n36
Lenin, Vladimir, 4
Leonard, Dickens, 92
Leskosky, Richard, 306
Leys, Ruth, 134n4
Liang Qichao, 122
Liberation Tigers of Tamil Eelam (LTTE), 92
libertarianism: of Silicon Valley, 138–139
Lindquist, Johan, 110
Liu, Andrew, 9
Liu Cixin, 8, 15, 122, 159; "dark forest" concept, 324n6; heteronormativity in, 162, 164; utopianism of, 158. See also *Liulang Diqiu* (Liu novella); *Wandering Earth, The* (films)
Liulang Diqiu (Liu novella), 15, 153–155, 158; gender and sexuality in, 154, 163–164; history in, 160–161, 162; population control in, 160; queer world of, 162–164; relationality in, 161–162; reproductive labor in, 163; suspicion of technology in, 159–160
Longmire mysteries, 315–316, 317
López, Iván Chaar, 113
Lopez, Lori Kido, 17
Love, Robot (film), 3
Luckey, Palmer, 140
Lunch atop a Skyscraper (photograph), 165–166, *167*
Lye, Colleen, 236

machine learning, 106–107, 116n24
Magnan-Park, Aaron Han Joon, 181
Mamdani, Mahmood, 90
manga: cool Japan in, 250; sociopolitical contexts of, 250; as soft power, 250, 262n8
Mannathukkaren, Nissim, 95
Marcos, Ferdinand "Bongbong," Jr., 5
Marshall, Kingsley, 70, 81n13
martial arts, 47; commodification of, 173, 322, 324n8
Marvel Cinematic Universe (MCU), 17, 267; villains of color, 278; whiteness of, 277, 279
masculinity, Asian American, 291–292
masculinity, white: techno-Orientalism and, 53
Massumi, Brian, 134n4
Matrix, The (film), 42, 73, 74, 183, 184, 268, 322; as postcyberpunk, 217n29
McGray, Douglas, 250
McKee, Kimberley D., 14, 307
McLuhan, Marshall, 102, 269–270
media: operationalization of, 191, 200n9
Melamed, Jodi, 32, 148
Melzian, Harley, 105
Mendes, Ana Cristina, 84
Merton, Robert C., 50n11
mestizaje, 242, 247n25
migration: Asian, 110–111; Japanese, 240; Korean, 176; transhemispheric, 243–244
militarism, U.S., 261; post-9/11, 251–253, 259
Miller, Chris, 8
Minari (film), 48n1, 177
Ministry for the Future, The (Robinson), 1

Mitchell, David, 250
Mitski, 189–199; authenticity and, 194–195; *Be the Cowboy*, 192–193, 194, 195; commodification of, 190, 192–194, 196, 198; deracialization of, 194, 196, 197–198; identity politics and, 193–194; *Laurel Hell*, 195, 201n28; online sampling of, 196–197, 201n30; racialization of, 189–194, 196, 197–198; sound production of, 194–196, 197; "Strawberry Blonde," 196–197; Twitter bot, 198
Miura, Sean, 305–306
Miyazaki, Hayao, 16, 206, 213–214
model minority discourse, 268, 315; automation threat narratives, 25–26; masochism and, 11; model machine minorities, 176, 311, 315; science fiction and, 28; techno-Orientalism and, 11, 24
modernity: Asian American, 237–238; Japanese, 144; Orientalism and, 4–5; technology and, 269; techno-Orientalism and, 9, 40, 237
Modi, Narendra, 5, 91
Monae, Janelle, 270
Monroe, Shayla, 271
Moon, Seungsook, 178
Moravec, Hans, 68
Morley, David, 6, 83, 113, 220
Morozov, Evgeny, 25, 27
Mozur, Paul, 147
Ms. Marvel (television series), 17, 273–276, 278; family reunification in, 276; generational trauma in, 275; Islamophobia and, 276; partition in, 275, 276; technology in, 273–275
multiculturalism, U.S., 32–33; transnational adoption and, 52
Muñoz, José Esteban, 295, 300; on disidentification, 301–302
Musk, Elon, 23–24, 35; purchase of Twitter, 34
Muslims: racialization of, 88–89; surveillance of, 110
Muslims, Indian: in *The Family Man*, 89–90; Othering of, 14; Orientalizing and techno-Orientalizing of, 84–85, 86–96; Sachar Committee Report, 88, 97n20; stereotyped as terrorists, 86–89, 93

Nagamatsu, Sequoia, 237
Nakamura, Lisa, 78
nationalism: Big Tech and, 141–142; COVID-19 and, 8; cultural, 177; cybernationalism, 158; Orientalism and, 85
nationalism, Chinese, 15, 121, 153; technonationalism, 164–165
nature: civilization opposed to, 209–210; in solarpunk, 213–214
Nausicaä of the Valley of the Wind (film), 206, 213
neocolonialism, Chinese, 156
neoliberalism, 5; globalization and, 231; K-pop and, 12; multiculturalism and, 32; Silicon Valley and, 138–139; Sino–U.S., 158; skinning and, 14; technology and, 32; zombies and, 231
New Cold War, 2, 6, 25, 28, 128
Ngai, Sianne, 230–231, 256, 304
Nguyen, Viet Thanh, 246n20
Nishime, LeiLani, 78, 185, 190–191, 192
Nishizawa, Ryue, 210
Nishizawa Garden & House, 210–211
Niu, Greta A., 40, 57, 69, 76, 83, 85, 301; on shifts in techno-Orientalist logics, 159
Noble, Safiya, 143
noise, 41–42, 43, 49n10; finance and, 46–47
Nolan, Jonathan, 70, 81n13
North Korea: yellow peril discourse and, 176
Nye, David, 124

obsolescence: Asian, 24–26, 28–34; automation and, 24–26; capitalism and, 47; whiteness and, 148
Okja (film), 206, 214–215
Olympics, Beijing 2008, 165, *166*
O'Neil, Cathy, 143
Orientalism (Said), 3, 83. *See also* Said, Edward
Orientalism, 268, 322; American, 8–9; anti-Asian animus of, 235; capitalism and, 320–321; and cybernetic technology, 41; discourse of, 83; European power and, 138; gender and, 61n5; Hindu Orientalism, 88; Islamophobia and, 88; modernity and, 4–5; nationalism and, 85; necro-Orientalism, 231; ornamentalism and, 191–192; re-Orientalism, 91, 93. *See also* techno-Orientalism
Oriental style, 173; in Hollywood film, 184
ornamentalism, 198, 312n12; Orientalism and, 191–192; techno-ornamentalism, 191–192, 199

Page, Larry, 30
Parasite (film), 48n1, 177
Parikka, Jussi, 200n9
Paris Climate Accords, 5
Park, Jane Chi Hyun, 16
Park, Sang Un, 178
Parray, M. Imran, 14
Parthasarathi, Vibodh, 89
Pasqualeh, Frank, 110
Patterson, Christopher, 10, 11, 290
Pegues, Juliana Hu, 235
Pence, Mike, 137
Petrick, Elizabeth R., 105
Pinch, Trevor, 109–110
population: decline: in Asia, 23–24; population control, 160, 163; repopulation, 159
Posadas, Baryon Tensor, 16–17
posthumanism, 10, 112, 220; in *Harmony*, 223–224; post-posthumanism, 76; post-racialism and, 67–68; zombies and, 224–225, 231
postracialism: in *After Yang*, 56–57; posthumanism and, 67–68
Potter, Ryan, 298, 312n2
presence, 6
Project Itoh (Itō Keikaku), 16–17, 220; animated adaptations of, 221, 225–227; legacy of, 220; philosophical zombies of, 221; zombies in, 220–224, 232. See also *Empire of Corpses* (Project Itoh); *Harmony* (Project Itoh)
Psy, 6
Puar, Jasbir, 252, 253, 255, 310

queerness: of Asianfuturism, 286, 295; as assemblage, 309–310; of *Big Hero 6*, 298–311; of *Everything Everywhere All at Once*, 44; futurity of, 300–301; horizon of, 295, 300; in *Liulang Diqiu*, 162–164; technology and, 154; of *The Tiger Flu*, 287; of video games, 292; whiteness and, 307

race: adoption and, 51–52, 61n5; Asian hyperethnicity, 40; authenticity and, 59; in *Cloud Atlas*, 183; codability of, 76; dis/ability and, 249; identity and, 78, 298; interchangeability and, 190; reproduction and, 53; skinning and, 69; technology and, 40, 111–112, 114, 192; transferability of, 68–69. See also racialization; racism
racial capitalism, 149, 192, 317

racialization: in *After Yang*, 32–34, 58, 59; AI and, 29; of animation, 230; anti-Asian, 318; of Asian bodies, 24, 28, 30; of Asian labor, 24; of black boxes, 15, 101–111, 113–114; doll-ness and, 9; in *Ex Machina*, 29–31; fungibility and, 75–76; in gaming, 10; of machine systems, 29; of Mitski, 189–192, 193–194, 196, 197–198; of Muslims, 88–89; neoliberal, 76; roboticism and, 9; skinning and, 72–73; speculative fiction and, 4; of technology, 59; techno-Orientalism as mode of, 317. See also bodies, Asian; deracialization
racism, 4; anti-Asian, 53, 55–56, 318; cultural, 4; scientific racism, 270; structural, 4; technology and, 270; techno-Orientalism as, 315; Trump and, 19n10. See also racialization; stereotypes, Asian; yellow peril
Racist Love (Bow), 10
racist love, 10
Ramirez, Catherine, 271
Red Star (Bogdanov), 133
regulation: of Big Tech, 138–140
Ren Yi, Mike, 292–294
re-Orientalism: of *The Family Man*, 91, 93
reproduction: race and, 53
Rivera, Takeo, 10, 11
Robins, Kevin, 6, 83, 113, 220
Robinson, Kim Stanley, 1, 18n3
Robinson, Tom, 307
robots: Asians as, 26, 70, 176, 268, 304; Asian-presenting, 28–30; diasporic Asians as, 26
Rodman, Gilbert, 78
Roh, David S., 40, 57, 83, 85, 301; on shifts in techno-Orientalist logics, 159
Roosth, Sophia, 112–113
Roppongi Hills, 207, 208, 218n32
Ruberg, Bo, 292

Said, Edward, 83, 138, 316, 322–323. See also *Orientalism* (Said)
Sangh, Rashtriya Swayamsevak, 96
Sarmiento, Thomas, 18, 58
"Saying Goodbye to Yang" (Weinstein), 14, 37n40, 52–56; adoption in, 59–60; anti-Asian racism in, 55–56, 57, 60; Asianness in, 53; cloning in, 53–55, 60; racial codes in, 53–54; reproductive futurity in, 54–55; yellow peril rhetoric in, 56

Schaffer, Simon, 103
Schalk, Sami, 249, 262n3, 285
Schmeink, Lars, 78–79
Schmidt, Eric, 139, 144, 147
Scholes, Myron, 50n11
Schrader, Astrid, 112–113
science and technology studies (STS): black box metaphor in, 109, 110–111; critical race theory and, 112–114; feminist, 113, 114; techno-Orientalism in, 101
science fiction (SF): Asianfuturism and, 283–285; Chinese, 122–123, 153, 157–158; Chinese New Wave, 157–158; endless expansion fantasy, 26; heteronormativity of, 285; model minority discourse and, 28; as mode of discourse, 219; people of color in, 285; postcyborgian, 221; post-Project Itoh, 220–221, 231; race in, 286; Western colonization narratives, 158–159; whitewashing in, 79
scientific racism, 270
Searle, John, 74; Chinese Room Argument, 106
Sedgwick, Eve Kosofsky, 286
semiconductors, 7–8, 19nn15–16
settler colonialism, 161; Asian migrants and, 236–237; empire and, 255; in *How High We Go*, 244, 248n59; infrastructure of, 125; Japanese, 237, 240–242; militarism and, 255; settler Orientalism and, 244; spacetime dimensions of, 236–237; U.S., 237
settler Orientalism, 235, 244; in *How High We Go*, 245; in *Through the Arc of the Rainforest*, 245
Shamoon, Deborah, 306
Shang-Chi and the Legend of the Ten Rings (film), 7, 17, 269, 272–273, 278; familial trauma in, 275; family reunification in, 276; technology, ambivalence toward, 274; technology in, 273–275
Shannon, Claude, 49n10, 50n19
Shaviro, Steven, 230
Shin, Chi-Yun, 180
Shinji, Miyadai, 231
Shiri (film), 16, 174, 175, 179–182; cloning in, 175, 186; cosmetic surgery in, 175, 180; critical and commercial reception, 180; heterosexual romance in, 181; Korean unification and, 182; unhappy ending of, 174

Shōgun (television series), 319
Shouse, Eric, 134n4
Silicon Valley (television series), 5
Silicon Valley, 5; individualistic ideology of, 26; libertarianism of, 138–139; neoliberalism of, 138–139; technological innovation in, 26
Sim, Gerald, 15
Singh, Vandana, 286
Sinofuturism, 15
Sinophobia, 269
skinning, 14; *BODIES: The Exhibition* and, 71–72; race and, 69; racialization and, 72–73; techno-Orientalist, 73–74; in *Westworld*, 72
Snyder, Sharon, 250
social justice, 4
Soft Science (film), 2
Sohn, Stephen, 69
sojourners, 15, 154–155, 158, 160; history of, 156; overseas laborers as, 165–166; sojourner colonialism, 154
solarpunk: aesthetics of, 206, 213; belonging in, 213; characteristics of, 205–206; cinematic, 205, 218n32; climate change and, 212; eco-centrism of, 212; fan-driven development of, 206, 215; genealogy of, 16, 205–206; hopefulness of, 214; land development in, 215; narrative solutions in, 212–213; nature in, 213–214; sustainability and, 206–207; urban planning and, 205, 206–207, 214
Song, Mingwei, 157
Song, Min Hyoung, 238, 285, 318, 319, 322
South Korea: cosmetic surgery in, 178, 185; neo-Confucian gender ideals, 177–178
Special Ops (web series), 14, 87
speculative fiction: as "designed experience," 287; racialization and, 4; Sinophone, 157; techno-Orientalism and, 2–3, 84. *See also* science fiction (SF)
speculative narratives: game design and, 18
Spivak, Gayatri Chakravorty, 94
Squid Games (film), 48n1
Steeves, Paulette, 243, 244
stereotypes, Asian, 175–176; forever foreigners, 56; as machines, 29–30, 40, 105, 112, 176, 236, 268; robots, 70, 176, 268, 304. *See also* model minority discourse; Muslims, Indian

stochastic processes, 42–43; in *Everything Everywhere All at Once*, 42–48
"Story of Your Life" (Chiang), 285–286
Stringer, Julian, 180
Ström, Timothy Erik, 49n4
Studio Ghibli, 206, 212
subjectivity: Asian American, 194, 198, 199, 237–238; Asian/American, 311; Japanese American, 243; stereotyping and, 175–176
Suparak, Astria, 3
surveillance: capitalism and, 148; in China, 137–138, 144–146, 148–149, 158; in *Harmony*, 223, 224; of Muslims, 110
sustainability: climate change and, 208–212; degradability and, 215; economics of, 209; Japan and, 218n32; in private housing, 210–211; solarpunk and, 206–207; urban planning and, 207–209

Tabares, Leland, 13, 58
Tamils, 92, 94
Tani, Han, 291
Taxpayers Protection Alliance (YPA), 140–142, *141*
Tay, Rachel, 16
techlash, 138, 142, 148
techno-environmentalism, 236–238; Asian American, 238; colonial reproduction of, 238; dispossession and, 245; land theft and, 245; as speculative natural history, 242; temporality of, 243; utopian aspirations of, 238, 245
technology: Afrofuturism and, 270–271, 282; Asianness and, 31; Blackness and, 31; capitalism and, 138; care labor and, 52; conceptualizations of, 269–270; identity and, 40; Islamophobia and, 88; Japaneseness and, 113–114; modernity and, 269; in *Ms. Marvel*, 273–275; neoliberalism and, 32; people of color and, 269; queerness and, 154; race and, 40, 111–112, 114, 192; racialization of, 59; racism and, 270
techno-optimism, 40, 48
techno-Orientalism: anti-Asian animus of, 235; Asian American studies and, 3–4, 11–12, 112; Asianness and, 300; capitalism and, 40–41; definitions of, 83; dehumanization and, 236, 268; discourse of, 2–4,

13, 83; disidentification and, 301–302; ecology and, 16; futurity, 54; India and, 83–86; as land occupation policy, 235–236; literature and, 12; model minority discourse and, 11, 24; as mode of critique, 315; as mode of racialization, 317; as mode of revelation, 3; modernity and, 9, 40, 237; Orientalism of, 7; queering of, 298, 299–300, 305–306, 307, 308–311; as racism, 315; as recoding of Japan, 219; Sinopostcoloniality and, 154; skinning and, 73–74; as style of thought, 316; tamed, 250, 252, 253, 255, 257, 259, 261; temporality of, 243; threat of, 260–261; uncertainty, as response to, 320; U.S. imperialism and, 235–236; visual language of, 3; white supremacy and, 14, 235. *See also* Orientalism
techno-Orientalism studies, 2–4, 16; as critique of racism, 317; expanding field of, 12–13; future of, 320–321, 323; interdisciplinarity of, 3; success of, 315–320
temporality: of *China 2098*, 129–131; crisis and, 5; of disability, 225; new chronic, 225; of techno-Orientalism, 243
Terranova, Tiziana, 201n33
Thakur, Tanul, 93
Thiel, Peter, 30–31, 140
Thompson, Nicholas, 139
Thoreau, Henry David, 210
Three-Body Problem, The (Liu), 8, 158, 167, 319, 321
Through the Arc of the Rainforest (Yamashita), 17, 238–242, 245, 246n20; alien Asianness in, 239; genocide in, 241; Indigenous dispossession in, 241; Japanese settler colonialism and, 240–242; settler Orientalism in, 245
Thu Duc City, 207, 208, 211
Tian, Ian Liujia, 15
Tiger Flu, The (Lai), 18, 283, 286, 287–289; Asian bodies in, 288; cloning in, 287; queerness of, 287
TikTok, 12; Mitski samples on, 196–197
Tiqqun, 49n4
Tongyi Qianwen, 35
Tragic Mulattas, 174–175
transhumanism: cyberpunk and, 68; in *Westworld*, 67

Trump, Donald: isolationism of, 148; 2016 election, 5
Tufekci, Zeynep, 143
Tung, Charles M., 14
Turing, Alan: Turing Test, 106
Turning Red (film), 7
Twitter (X), 23, 24, 34; Mitski bot, 196

Ueno, Toshiya, 85
urban planning: climate change and, 208–212; eco-efficiency and eco-effectiveness, 209–210, 211–212; solarpunk and, 205, 206–207, 214; sustainability and, 207–209

van der Veer, Peter, 84
video games: Asianfuturism and, 289–294; Asianfuturist, 291; designed experiences of, 289, 290; game design, 18; ludo-Orientalisms of, 290; queerness of, 292
Vinge, Vernor, 30
Vint, Sherryl, 70
violence: anti-Asian, 232; anti-Black, 270; anti-Muslim, 91; in *The Family Man*, 94–95; against Indian minorities, 94–95; state, 253, 255, 259
"Vitruvian Man" (da Vinci), 67, *68*, 71, 72
Vora, Kalindi, 148
Vukovich, Daniel, 321

Walter, W. Grey, 105
Wand, Xin, 283
Wandering Earth, The (film), 15; collective action, emphasis on, 159; differences from novella, 155; heteronormativity of, 154, 164; Liu's endorsement of, 157–158; population control in, 160; Sinofuturism of, 159; technofetishism of, 162; Third World solidarity in, 161; uncertainty in, 157; Western skepticism about, 153–154, 168n1; yellow peril discourse and, 168n1. See also *Liulang Diqiu* (Liu novella); *Wandering Earth II, The* (film)
Wandering Earth II, The (film), 15, *164*, 164, *166*; collective action, emphasis on, 159; differences from novella, 155; heteronormativity of, 154, 164; Liu's endorsement of, 157–158; population control in, 160; techno-Orientalist ambiguities of, 165–166; Third World solidarity in, 161. See also *Liulang Diqiu* (Liu novella); *Wandering Earth, The* (film)
Warner, Kristen, 277
War on Terror, 253
Watanabe, Scott, 305, 313n36
Weaver, Warren, 50n19
Webb, Amy, 143
Webb, Sarah M., 307
web series, Indian, 86–96
WeChat, 34
Weheliye, Alexander, 72
Weinstein, Michael, 145
Wennan, Fan, 15
Westworld (television series), 14, 67–72; Chinese, absence of, 73; player piano as robot, 70–71; skinning in, 72; Vitruvian Man in, 67, *68*, 71, 72
White, Lynn, Jr., 209
whiteness, 106, 282; in *Ex Machina*, 31; history of science and, 106; humanness and, 103; of MCU, 277, 279; naturalizing of, 283; obsolescence and, 148; queerness and, 307
white supremacy, 272; cloning and, 53–55, 58, 60; reproductive futurity and, 52–53; techno-Orientalism and, 14, 235
whitewashing, 79; of Asian women, 185; in science fiction, 79; yellowface and, 78, 185
Wiener, Norbert, 104, 105, 138
Williams, Raymond, 134n4
Winner, Langdon, 109
Winnubst, Shannon, 75–76
Wissinger, Elizabeth, 80
Wong, Danielle, 30
Word for the World is Forest, The (Le Guin), 213, 218n37
Worlds Beyond Here: The Expanding Universe of APA Science Fiction (exhibition), 283–285, *284*
Wu, Tim, 138, 139

X (formerly Twitter), 23, 24, 34
Xiang, Biao, 110, 111
Xi Jingping, 5; Made in China initiative, 8

Yamashita, Karen Tei, 237
Yang, Andrew, 24–25, 26–27, 28; on failure of technology, 35
Yano, Christine R., 250, 304

Ye, Shana, 15
yellowface, 78; in *Cloud Atlas*, 183, 184; whitewashing and, 185
yellow peril, 18, 24, 268, 315; in *After Yang*, 33–34; Asian labor and, 255–256; futurity and, 69; Japanese settler colonialism and, 240; North Korea and, 176; in "Saying Goodbye to Yang," 56; *The Wandering Earth* and, 168n1
Yeoh, Brenda S. A., 110
Yoo, Jaeyon, 16

Zhang, Muqiang M., 168n1
Zhao, Junfu, 27
Zhou, Xander, 3
Zizek, Slavoj, 230
zombies: animation as, 227–228; capitalism and, 17; colonialism and, 17; neoliberalism and, 231; philosophical zombies, 221, 224; as posthuman, 224–225, 231; transition to cyborgs, 220, 227–231
Zuboff, Shoshana, 148
Zurcher, Jessica D., 307

Available titles in the Asian American Studies Today series:

Chien-Juh Gu, *The Resilient Self: Gender, Immigration, and Taiwanese Americans*
Stephanie Hinnershitz, *Race, Religion, and Civil Rights: Asian Students on the West Coast, 1900–1968*
Jennifer Ann Ho, *Racial Ambiguity in Asian American Culture*
Helene K. Lee, *Between Foreign and Family: Return Migration and Identity Construction among Korean Americans and Korean Chinese*
Melody Yunzi Li, *Transpacific Cartographies: Narrating the Contemporary Chinese Diaspora in the United States*
Huping Ling, *Asian American History*
Huping Ling, *Chinese Americans in the Heartland: Migration, Work, and Community*
Haiming Liu, *From Canton Restaurant to Panda Express: A History of Chinese Food in the United States*
Jun Okada, *Making Asian American Film and Video: History, Institutions, Movements*
Kim Park Nelson, *Invisible Asians: Korean American Adoptees, Asian American Experiences and Racial Exceptionalism*
Zelideth María Rivas and Debbie Lee-DiStefano, eds., *Imagining Asia in the Americas*
David S. Roh, Betsy Huang, and Greta A. Niu, eds., *Techno-Orientalism: Imagining Asia in Speculative Fiction, History, and Media*
David S. Roh, Betsy Huang, Greta A. Niu, and Christopher T. Fan, eds., *Techno-Orientalism 2.0: New Intersections and Interventions*
Corinne Mitsuye Sugino, *Making the Human: Race, Allegory, and Asian Americans*
Leslie Kim Wang, *Chasing the American Dream in China: Chinese Americans in the Ancestral Homeland*
Jane H. Yamashiro, *Redefining Japaneseness: Japanese Americans in the Ancestral Homeland*